C0-AYI-643

Climates and Societies –
A Climatological Perspective

The GeoJournal Library

Volume 36

The titles published in this series are listed at the end of this volume.

Climates and Societies – A Climatological Perspective

A Contribution on Global Change and Related Problems Prepared by the Commission on Climatology of the International Geographical Union

edited by

MASATOSHI YOSHINO

Institute of Geography, Aichi University, Toyohashi, Japan

MANFRED DOMRÖS

Department of Geography, Mainz University, Germany

ANNICK DOUGUÉDROIT

Institute of Geography d'Aix University, France

JANUSZ PASZYŃSKI

Institute of Geography and Spatial Organisation, Polish Academy of Science, Warsaw, Poland

LAWRENCE C. NKEMDIRIM

Department of Geography, University of Calgary, Canada

KLUWER ACADEMIC PUBLISHERS

DORDRECHT / BOSTON / LONDON

Library of Congress Cataloging-in-Publication Data

Climates and societies : a climatological perspective : a contribution on global change and related problems / prepared by the Commission on Climatology of the International Geographical Union ; edited by Masatoshi Yoshino ... [et al.].
p. cm. -- (GeoJournal library ; v. 36)
ISBN 0-7923-4324-7 (hb : alk. paper)
1. Climatic changes--Social aspects. 2. Human beings--Effect of climate on. I. Yoshino, Masatoshi, 1928- . II. International Geographical Union. Commission on Climatology. III. Series.
QC981.8.C5C5117 1997
304.2'5--dc21 96-48830

ISBN 0-7923-4324-7

Published by Kluwer Academic Publishers,
P.O. Box 17, 3300 AA Dordrecht, The Netherlands

Sold and distributed in the U.S.A. and Canada
by Kluwer Academic Publishers,
101 Philip Drive, Norwell, MA 02061, U.S.A.

In all other countries, sold and distributed
by Kluwer Academic Publishers Group,
P.O. Box 322, 3300 AH Dordrecht, The Netherlands

Printed on acid-free paper

Printed in the Netherlands

TABLE OF CONTENTS

PART 3: REGIONAL SCALE CLIMATES

PART 4: LOCAL SCALE CLIMATES

PREFACE

The impact of climate on human activities and the effect of humans on climate are two of the most important areas of inquiry in climatology. These interactions conducted through physical, chemical and biological processes were described as early as Roman and Greek times. Marcus Vitruvius (75–25 B.C.), a famous Roman engineer and architect, made the following observation about the climatic conditions necessary for founding a city:

Land ideal for the health is slightly elevated and there should be neither fog nor frost. The direction of the slope and the distance to the swamps, lakes, and beaches must also be considered. The prevailing wind directions, observed by a wind tower at the center of the city, like Horologium at Athens, should be taken into consideration in city planning. The main and narrow streets should be placed in the middle angle of the two prevailing wind directions. Then the location of the Pantheons and squares should be decided.

The influence of humans on climate was a major subject for discussion in the 19th century, inspired in part, by the rapid industrial growth and expanding deforestation of the time. D.L. Howard wrote brilliant pieces on the climate of London in the 1830s, while G.P. Marsh discussed the effects of forests on precipitation in the U.S.A. in the second half of the 19th century. On both sides of the Atlantic, human intervention was the critical factor in environmental change. In London, it was population growth. In the United States, population growth and land use change were driving factors.

Climatology has gone through several stages of development since the late 19th century. Climatography, whose golden era tailed off towards the end of the 19th century, described climate at any point. Micro- or local-scale climatology, with its emphasis on human environments, developed strongly in the 1930s and 1940s. Meso-scale climatology gained prominence in the 1950s.

Since the beginning of the later half of the 19th century, the science has made great progress especially in two areas, large-scale meteorology, developed as a part of atmospheric physics; and micro- and local-scale environ-

M. Yoshino et al. (eds.), Climates and Societies – A Climatological Perspective, vii–xi.

mental sciences, developed within ecology and geography. The term 'topoclimatology' is often used in contemporary climatological literature to describe some climates at the micro and local scales. R. Geiger proposed it in the first half of this century for a climate affected strongly by topography. Topoclimate and topoclimatology were discussed in considerable detail by C.W. Thornthwaite in 1953.

In the 1970s, supported by developments in computers and remote sensing, climatological studies advanced in many fronts. In the introductory chapter to this book, Yoshino describes the very significant role played by the World Climate Programme (WCP) in promoting climatological research through the 1980s. Research from then on, focused significantly on global change and climate change.

Even if confidence in the estimates is low, climate change of the magnitude predicted by GCMs will cause serious social and economic disruptions that go beyond what is experienced with climate variations today. The severity of impact will vary globally.

Early discussions on climate change were confined, in the most part, to scientific groups and agencies involved in environmental programmes. The World Meteorological Organisation (WMO) through the World Climate Research Programme (WCRP), and the International Council of Scientific Unions (ICSU) through the International Geosphere-Biosphere Programme (IGBP) provided early leadership in this area. Those discussions later spread into social and economic domains. Politics quickly followed. They now occur at the highest national and international levels. The Intergovernmental Panel on Climate Change (IPCC) has recently updated its evaluation of climate change research.

The increasing sensitivity to threats to global environments due to climate change led to a meeting in March 1989 of 24 heads of state in The Hague, Netherlands. There, they discussed how the earth's atmosphere can be protected. Following the second World Climate Conference held in Geneva, Switzerland in November 1990, the Assembly of Ministers of governments upgraded the strategy developed by the IPCC for safeguarding the atmosphere into an Action Plan. That plan became the reference document for the United Nations Conference on Environment and Development held in Rio de Janeiro, Brazil, in 1992. Practical decisions, however, depend on national policies, which in many cases, do not entirely fit the general recommendations agreed to at the international level. As well, the North and the South differ, often strongly, in their views on strategies needed to stem the long-term effects of climate change.

Global change has five main characteristics: (1) complexity, because interactions among the participating variables are tangled; (2) non-reversibility, because life and the earth change in one direction; (3) limited in character, because the earth is a closed system; (4) non-certainty, because many of its features, causes and effects are still unknown; and, (5) urgent in nature, because the speed of environmental change is faster than corresponding changes in living conditions.

Global warming will impact agriculture and industrial production. Understanding the dimension of the impacts requires improved knowledge of weather-crop and weather-industrial production relationships. It demands more research on the impact of climate change on land use, on product mixes, on changes in agricultural calendars, on production systems, on market, on demands, etc. How will humans themselves adapt to climate change? What economic, social, and technological adjustments are needed to accommodate the new climatic reality? Behavioral modification and changes in sanitation methods may be part of adaptation programmes. Adaptation itself is a developing science which, at this point, lacks quantitative and systematic rigor. But even at a qualitative level, the science still serves a useful purpose.

The International Geosphere-Biosphere Programme began its work with climate firmly recognized as one of the basic components of human-environment relationships. The role played by land use practice and land cover change is singled out as the most important component in the relationships. This component named "Human Dimensions of the Global Environmental Change Programme (HDP)" developed a global model which connects the biogeochemical cycles and physical climate systems to external forces driven by volcanoes and the sun. These forces, systems and cycles influence the physical and chemical dynamics of the atmosphere. Through various loops and feedbacks, the interactions developed among them impact earth processes and cause a change in climate.

Terrestrial ecosystems respond, often negatively, to human use of Earth's land resources. Examples of such response are found in the relationship between land use change and climate at the local level. In this book, the processes involved are highlighted in those chapters which deal with local climates. We encourage more work in this area and suggest the following methodological and topical approaches: (1) division of the target area into local climate zone drawn to coincide with patterns of land use/cover; (2) statistical analysis of the role played by land use/cover and micro-topography in local climate differentiations; (3) quantitative descriptions of land use/cover conditions; (4) numerical simulations of climate at the local scale; (5) incorporation of remote sensing methodology in local studies; and, (6) development of both mapping techniques and advanced visual tools. We also recommend that more work be done on quantitative expression of the geometry and fabric of local areas (building density, street canyons, parks, etc.). Structural differences in human habitats and urban function types (industrial, university, transportation centre, commercial, etc.) should be recognized as well.

The development of human activity based climatology must not ignore the role played by other physical landscape variables in the evolution of local climates; orography, vegetation, and soils, to name a few. We recommend, therefore, that climatic impacts on socio-economic and political systems and on human well-being be studied across disciplinary lines. Human bioclimatology, for example, has its roots in ecology rather than climatology since it deals primarily with the influence of climate on the human physiology.

Local air pollution modifies topoclimates in urban and industrial areas. Air pollutants attenuate solar radiation, increase atmospheric turbidity, and provide an abundance of nuclei which may seed clouds. The result may be an increase in both the amount and frequency of precipitation. J. Paszynski and others have done useful work in this area. Their work on energy balance, land surface interactions and topography, begun in the seventies, has contributed enormously to our understanding of the processes involved.

Regional scale climatic impacts are also addressed in the book with examples drawn from the Caribbean, Asia, and the Polar region. They complement earlier studies by Domroes, Suppiah and Yoshino in Sri Lanka in the nineteen seventies. More recently, Manfred Domroes has written on deforestation in the tropics and Annick Douguedroit on forests in the Mediterranean area. Both studies were done in the 1990s.

In 1978, the International Geographical Union (IGU) established the "Working Group on Tropical Climatology and Human Settlements" chaired by Masatoshi Yoshino. Two other IGU Working Groups, one on "Climatic Change" (Chairman: Stanley Gregory) and the other on "Methods of Topoclimatological Mapping and Investigation" (Chairman: Janus Paszynski) were subsequently formed. These Groups merged into one "Commission on Climatology" in 1988 under Yoshino's chairmanship. Manfred Domroes succeeded Yoshino in 1992. Meanwhile, the International Commission on Climate of the International Association on Meteorology and Physics (IAMAP) continues to work actively on climate related issues. Symposia and meetings are held biannually. In recent years, there has been a remarkable growth in collaboration between the geographer dominated IGU commission and its meteorologist controlled IAMAP counterpart to the benefit of the science.

In view of the pressing need to understand human-biosphere impact, the IGU Commission on Climatology focused its activities in three main areas: (1) tropical climatology; (2) climatic fluctuations; and (3) small scale and local climates. These sub areas are closely related and together provide a framework for research on the impacts of humans on climate and the impact of climate change on life.

At business meetings of the Commission, we decided to write a book about "Climates and Societies". After several discussions on structure, topics, and authors, a final form was agreed to in 1990. The result is this authoritative book which deals with complex interactions between climate and people.

We thank all authors who contributed chapters for this book. We express our deepest gratitude to our colleagues who participated in our meetings and symposia, shared their research and science with us, and contributed to published proceedings of our many conferences. Without their support, this book could not have been written.

Dr. Hye-Sook Park Ono, Mie University typed and word processed the original manuscripts; Mr. S. Osaki of the University of Tsukuba drew the diagrams; and Mrs. Debbie Little of the University of Calgary prepared the electronic version of the manuscripts. Dr. Wolf Tietze, the former editor of GeoJournal, and Ms. Petra D. van Steenbergen, Kluwer Academic Publishers, provided helpful advice to us at all stages of this book. To this support and very supportive group, we express our indebtedness.

February, 1996

PART 1

INTRODUCTION

1.1. HUMAN ACTIVITIES AND ENVIRONMENTAL CHANGE: A CLIMATOLOGIST'S VIEW

MASATOSHI YOSHINO

1. Introduction

The impacts of human activities on our environments and its corollary, the impacts of the environment on human activities, have been major topics in geography since ancient times. Because climate is arguably the most important element of the natural environment, it has occupied a central position in both discussions.

Following the First World Climate Conference in 1979, the United Nations Environment Programme (UNEP) adopted "the impacts of climate change and variability" as one of its areas of responsibility. This responsibility is administered under UNEP's World Climate Impacts Studies Programme (WCIP), one of the four components of the World Climate Programme (WCP). In 1985, an important international conference on greenhouse gases was held in Villach, Austria. Since these initiatives, the impacts question has been addressed in increasing details at numerous meetings and conferences. In Autumn 1990, the Second World Climate Conference was held in Geneva, Switzerland. It reviewed the work done within the World Climate Programme including the WCIP over the preceding 10 years. In 1992, WCP component programmes were reorganised as follows (WMO 1992): The World Climate Data and Monitoring Programme (WCDMP), the World Climate Applications and Services Programme (WCASP), the World Climate Impact Assessment and Response Strategies Programme (WCIRP), and the World Climate Research Programme (WCRP). Within this structure, work on the interactions between human activities and climatic environments is vested in the WCIRP, and remains the responsibility of UNEP.

Paralleling these developments, the International Geographical Union (IGU) approved in 1978 a Working Group on "Tropical Climatology and

M. Yoshino et al. (eds.), Climates and Societies – A Climatological Perspective, 3–17.

Human Settlements". Although the group's activities were formulated independently of WMO and UNEP initiatives, the aims and subjects persued by the three organizations were well coordinated because the chairman of the Working Group (this author) was also a member of the Science Advisory Committee of WCIP. There were two other IGU Working Groups, one on "Climate Change" and the other on "Topoclimatology Investigation and Mapping". These three Working Groups were unified into one "Commission on Climatology" in 1988. During the 10 year period (1978–1988), climatologists working mainly in geography met and discussed problems related to human activities and environmental change at conferences and symposia at least once a year. In addition, this author organized international symposia on climate and food production in 1976 (Takahashi and Yoshino 1978), socio-economic forces in global environment change (Takeuchi and Yoshino 1991a) and climate impacts on society and environment (Yoshino 1991b).

Results of these conference and symposia together with those from other established international programmes, are reviewed below.

2. Problems

The Intergovernmental Panel on Climate Change (IPCC) was established in 1988. Its first report was published in the proceedings of the Second World Climate Conference (SWCC) along with papers on climate change and socio-economic activities presented at the meeting (Jäger and Ferguson 1991). Specific topics considered in scientific sessions include water, agriculture and food, oceans, fisheries and coastal zones, energy, landuse and urban planning, human dimensions, environment and development, forest and integrated studies. In addition, several areas dealing with climate impacts, most notably impact on a regional scale, were covered. They include: (a) socio-technical systems: commercial/manufacturing, settlement/infrastructure, transportation and energy; (b) intermediate systems: water resources, agriculture/food, livestock/grassland and forests; and, (c) natural (less managed) systems: fisheries and natural reserves. Details of some of the components of these systems are discussed by Riebsame and Magalhaes (1991). Techniques and tools used in the impact work presented are: (a) historical reconstructions; (b) extreme events analysis and case studies; (c) statistical correlations and transfer functions; (d) sensitivity and longitudinal analysis; (e) simulation modeling; and (f) greenhouse projections. A schematic diagram of assessment skills across climate impact subsystems is shown in Figure 1.

The increase of global air temperature in the 1980s is quite striking in both hemispheres. In general, the observed rate of increase falls between 0.4 and 0.5°C over the last 100 years of meteorological observation. This coincides with both the global mean temperature increase calculated over the same period by Jones et al. (1991), and estimates of temperature rise obtained from model simulation for a doubled CO_2 climate (Houghton et al. 1990).

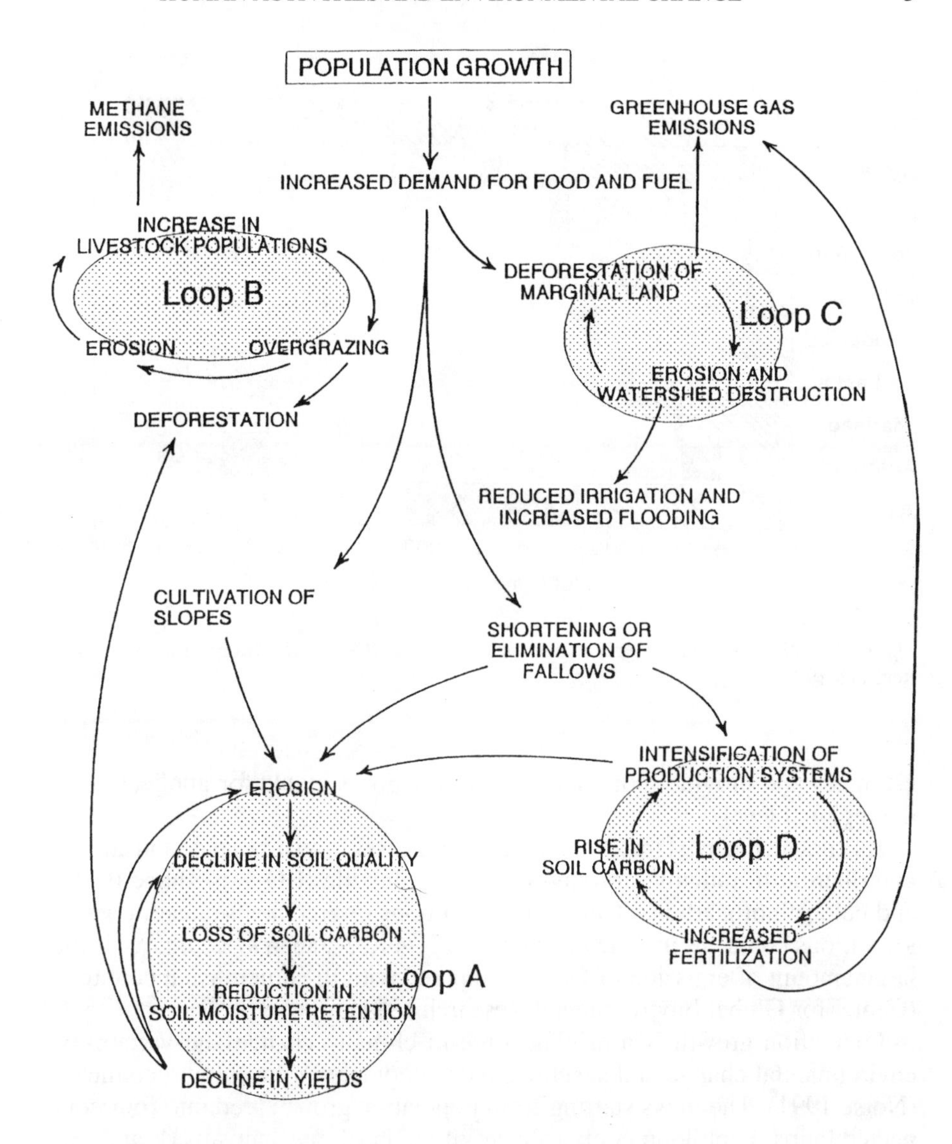

Figure 1. Schematic relationships between resource system and impact assessment skill (Riebsame and Magalhaes 1991).

The regional profiles of warming in East Asia estimated by various General Circulation Models vary (Yoshino 1991a). Roughly speaking, it ranges from 3–5°C for winter (mean of December, January and February), and 2–4°C for summer (mean of June, July and August). Estimates of changes in precipitation by those models vary more widely. Some project an increase while others predict a decrease. It is strongly hoped that future studies will refine these

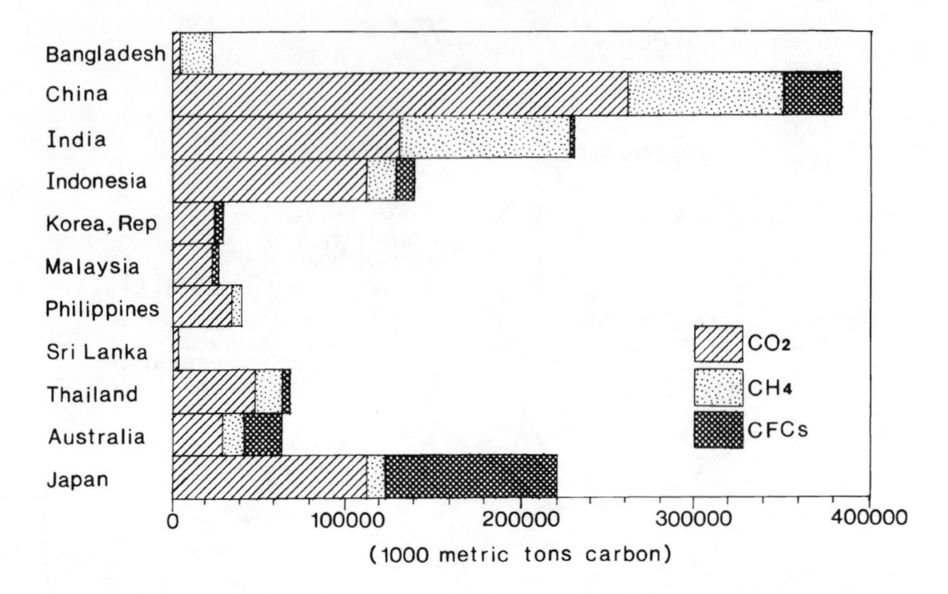

Figure 2. Net additions to greenhouse heating effect (Center for Global Environmental Research 1991).

estimates and provide more detailed information suitable for application at the regional level.

As known, the most important greenhouse gases in terms of past and current contribution to radiative forcing are carbon dioxide (CO_2), methane (CH_4) and chlorofluorocarbons (CFCs). Net additions to greenhouse heating gases for various countries in Asia are shown in Figure 2. Japan's peculiarity is the large amount of emission of CFCs compared with the other Asian countries (Center for Global Environmental Research 1991).

Population growth is a driving force in climatic change and variability, environmental change and sustainable development in developing countries (Norse 1991). The flows starting from population growth feed into four connected loops. Each loop contains a set of feedback mechanisms (Figure 3). The connections between greenhouse gas emissions and increased vulnerability to climatic change are well illustrated in the figure. Loop A highlights soil erosion processes. Loop B is dominated by processes and consequences of overgrazing. Loop C is the deforestation-marginal land loop; and Loop D emphasises responses and results which flow from the preceding processes.

The flow diagram and its component parts form a useful framework for further intensive studies on climate-environment interactions. The detailed plan of the Human Dimensions of Global Environmental Change Programme has been discussed, through symposium (Kosinski 1996), but the areas for which solutions are required are: (a) improvement of living standards and

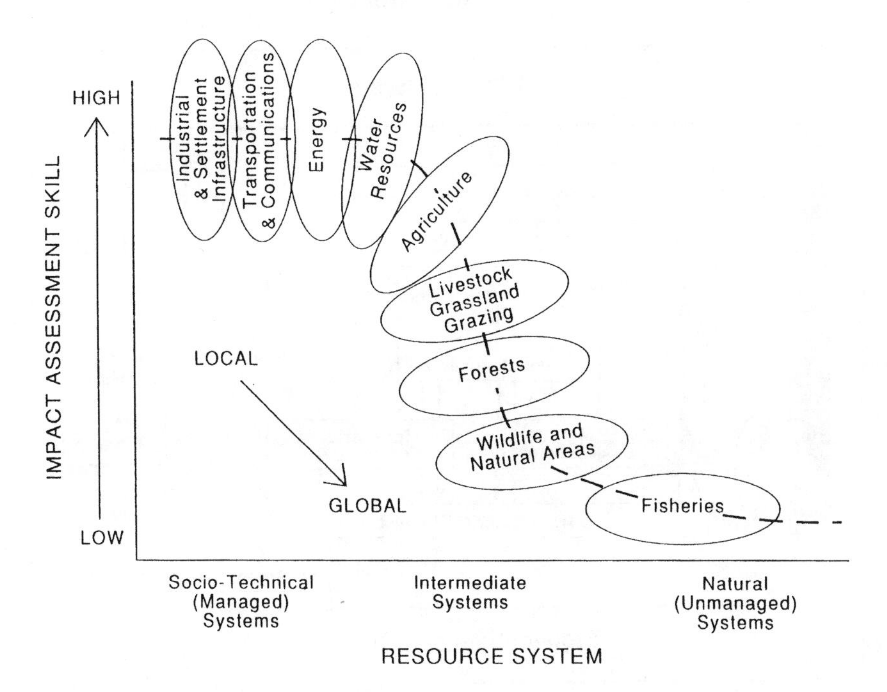

Figure 3. Impact of population growth on greenhouse gas emissions and methane emissions by agriculture and forestry (Norse 1991).

global environmental quality; (b) environmental education in primary and secondary schools, and universities; (c) present status of the cultural landscape from region to region and its evolution through history; (d) human behavioral studies from psychology to medical and educational sciences to sociology etc.; (e) methods in environmental and landscape conservation, planning, and policy development; (f) human driven biological processes including those linked to culture, adaptation and medicine; (g) regional studies in human, social and economic geography; (h) effects of regional economic development on global environmental change; (i) environmental sciences of agricultural chemicals; (j) security and sustainable development; (k) new concept on quality of life; (l) paddy field agriculture and global environment; (m) synthetic studies on desertification protection; (n) synthetic studies on land degradation; (o) farming systems based on land use hierarchy, and environmental conservation; (p) role played by economic, social and political factors in human-environmental relations; (q) studies on marine resources; (r) studies on economic and social developments in relation to policy; and (s) model-building.

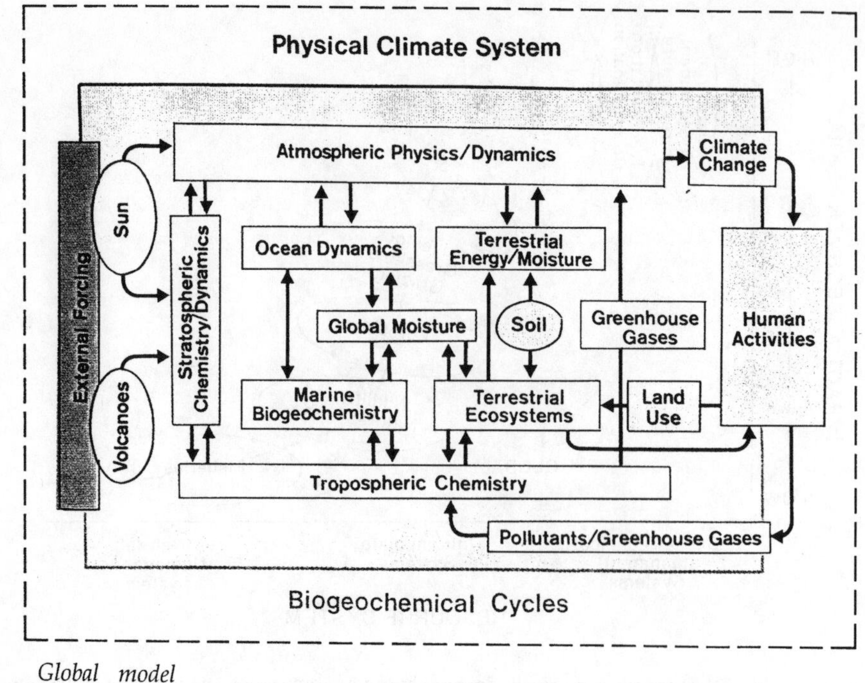

Figure 4. Global model of physical climate system and biogeochemical cycles (Williamson 1992).

3. Examples of "human activities and environmental change"

In a review article, Yoshino (1991a) provides examples of work undertaken in the following areas: (a) impacts of global warming on agriculture; in particular, effects of doubling CO_2 on the agricultural production; (b) sea level rise in coastal areas in Japan and East China; (c) acid rain in Japan, Korea and China; and (d) long-term fluctuation of paddy rice yield. Finally, the flow of impacts from human activities to agriculture, forestry and fishery was shown. In the present paper, land use, historical change of impact, some socio-economic aspects of response to global warming, and secular change in urban climates are used as examples.

3.1. *Land use*

Land use is the only component, which is basic to both IGBP and HDGEC. This is dealt with in another paper (Yoshino 1992). Human activities influence upon terrestrial ecosystems through land use is shown in Figure 4 (Williamson 1992). This diagram is a global model illustrating how biogeochemical cycles and physical climate systems are affected by human activities and external natural forcing. It provides a framework for improving our ability to predict changes in the behavior of Earth Systems as a whole.

The Environmental Agency in Japan distributed a questionnaire to the experts on global change studies in Japan and summarized the results (Environment Agency of Japan 1992). According to this report, land use problems should be solved particularly in the following areas: (a) feedback structure of changing land use pattern, including vegetation, local land-sea distribution pattern etc.; (b) climatic environment change during the historical period; (c) changes in NH_4 with respect to location of paddy fields and techniques used in rice production; (d) accurate estimation of change in levels of CO_2, NH_4 etc., from agricultural regions; (e) long-term observation of material balance and greenhouse gases in various vegetation and land use types; (f) population increase and sustainable food supply; (g) impacts of population increase and climatic fluctuation; for example, on sustainable agricultural system, limitation of soil resources etc.; and (h) impacts of climatic change and CO_2 increase on agriculture.

Molin (1986) discussed land use and agrosystem management in the humid tropics, stressing the importance of time and space. He dealt with flux exchange between plants and the atmosphere particularly the latent heat flux, which is a joint measure of the water consumption by plants and the direct evaporation from the soil. Burgos (1986) presented an idealized schema of several ecosystems and micro-agrosystems in the humid tropics with their respective energy and water balance as shown in Figure 5. Similar figures are available mostly for areas in the middle latitudes and for traditional land use techniques outside the humid tropics. In order to develop an appropriate land use such approaches are needed in the tropics.

3.2. *Historical change of impact on agriculture*

With respect to the impact of warming on agriculture, the problems are summarized as follows: (a) regional climate scenarios, with special emphasis on monsoon areas; (b) capability of farmers to respond to environmental change; and (c) tolerance of crops to changing environmental conditions.

Changes in land use are of importance. In particular, changes in: (a) cultivated area; (b) crop type; (c) crop location; and (d) crop calendar under the new conditions. Changes in management such as irrigation, fertilizer use, control of pest and diseases, soil drainage and erosion, etc. will be considered in relation to the problems mentioned above.

Other problems which agricultural production will face in the future are total production volumes and year to year fluctuations both due to a variety of changing natural and socio-economic conditions. An example drawn from Japan is given here.

The 100-year time series of paddy rice yield in Japan is divided into three periods (Figure 6). Period I started from the end of the 19th century and lasted

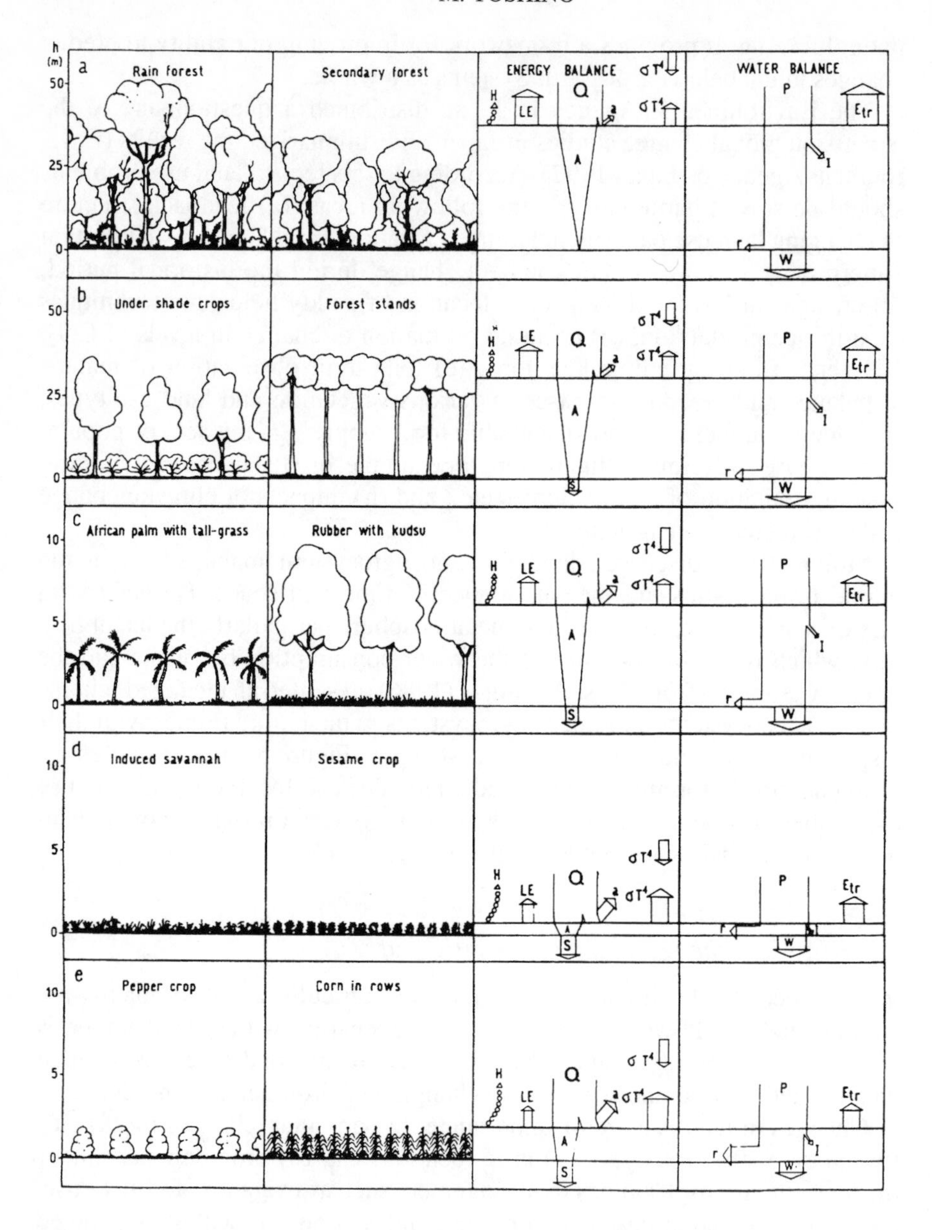

Figure 5. Ecosystems and micro-agrosystems and energy/water balance in the humid tropics (Burgos 1986).

till 1910–1915 when a trend towards a gradual increase in yield began. In Period II, 1915–1945, productivity was virtually constant. Period III which began in 1945 was marked by rising yields.

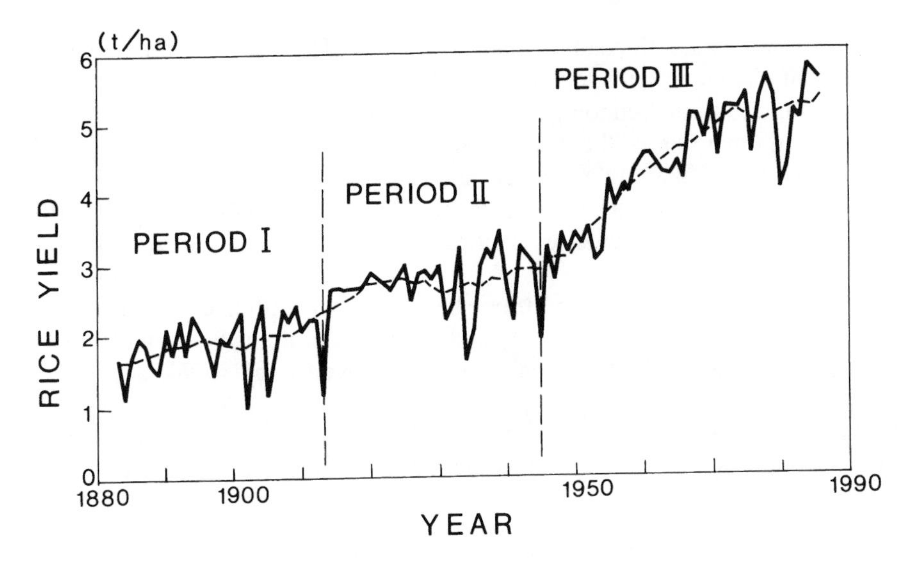

Figure 6. Secular change of paddy rice yield in Japan, 1883–1984, and its division from Period I to Period III (Yoshino 1991a).

Table 1. Fluctuation (defined by standard deviation) of paddy rice yield in each period in North Japan.

Period	Year	Standard deviation* (t/ha)
Period I	1885–1917	
1st half	1885–1900	0.118
2nd half	1901–1917	0.244
Period II	1918–1950	
1st half	1918–1933	0.122
2nd half	1934–1950	0.168
Period III	1951–1984	
1st half	1951–1967	0.063
2nd half	1968–1984	0.081

*Root-mean-square of deviation of each year from the 11-year-running mean.

It is noteworthy, however, that their standard deviation which represents the degree of fluctuation changed in a striking manner (Table 1): (a) it is one order smaller in Period III than in Periods I and II which may be attributed to technical developments in rice cultivation, including the introduction of new varieties; (b) further, it is quite interesting to note that the second half of each period shows greater fluctuation than the first half. This is of course not related to secular change in climate variability, but possibly to a step-wise development in agricultural technology and to trends in land use over the respective periods. In other words, the fluctuation of yields is affected by climatic variability while its magnitude is determined by the socio-economic conditions in place and changes in agricultural technology. Evidence of this type should be taken into account when estimating the effect of global environmental change.

Future land use potential in Europe has been dealt with in connection with climate change (Parry et al. 1991) and its potential impacts on world food supply have been studied by modelling (Rosenzweig et al. 1993).

3.3. *Japan's response to global warming: Socio-economic considerations*

(a) *Energy*

In Japan, electric energy used domestically is roughly 30% of the total supply. Commerce and industry use 20 and 50% respectively. Demand is higher in summer than in winter due to air conditioning. Air temperature rise of 1°C in daytime in summer results in an increase of 3 million kWh nationwide.

Kurosaka (1990) made clear, based on detailed analysis of domestic demand for electric energy supplied by Tokyo Electric Co., that an increase of 1°C in daily maximum temperature in July, August and September corresponds with an increase of 40 kWh. This value translates roughly into 1.5% of the total annual demand in 1987.

In addition to problems created by increasing demand, transmission difficulties occur, not infrequently, because of natural hazards. Meteorological phenomena account for 96–97% of such events. Earthquakes, landslides, etc. contribute only 3–4%.

(b) *Sea Level Rise*

Quarternary research done in recent years suggests that the rate of uplift since the last interglacial, about 120,000 years B.P. may not have been constant (Ohta et al. 1991). An accelerated rate in recent time is indicated by various pieces of evidence. A summary of all estimates of sea level change around Japanese islands during the last glacial maximum indicate a range from –80 to –140 m. Umitsu (1991) reviewed late studies on sea level changes in the Holocene and showed the temporal and regional variation in them during the period. From early to middle Holocene, (8,000–6,500 B.P.) the rate of sea level rise was very high attaining 1.5–2.0 cm/yr which is believed to be maximum for that period.

Due to heat expansion of sea water and melting of continental ice such as mountain glaciers, global mean sea level rise is believed to be 10 cm/100 yr during the last 100 years (Gornitz et al. 1982). Around 2030, sea level could rise by 5–17 cm due to heat expansion of sea water, 4–9 cm due to melting of land ice such as small mountain glaciers, 3–9 cm from melting of Greenland ice sheet, and 3–24 cm from contribution from Antarctic ice sheet. In Japan, the following figures were observed during 1956–1980 (Okada 1988); 0.0–0.5 cm/yr (subsidence) caused by crustal movement along the Pacific coast. In some parts of Tohoku district and Hokkaido, North Japan, subsidence was as high as 0.0–1.0 cm/yr. However, uplift of up to 0.25 cm/yr is believed to have occurred along the coast of Sea of Japan.

Given these facts, Japan should as a matter of urgency take the impacts of sea level rise into consideration when designing coastal management strategies and in construction planning. According to the IPCC Report, protection cost among Asian countries is highest in India ($22,500 million U.S.). At $22,300 millions U.S., Japan is second. These costs are almost twice the amount for the third and fourth ranked countries, China and Australia.

(c) *Water*

Masukura et al. (1991) studied precipitation variation in Japan during the last 100 years, analysing the occurrence of 2-day precipitation exceeding 300 mm. It is more frequent (twice the average) in the warmer period than in the colder period. However, at the same time, it was indicated that the frequency of the minimum rainfall for 60 consecutive days in the warmest decade is noticeably less than for the coldest decade. Therefore, it is suggested that frequent flood and drought can be expected as a result of global warming.

Yamada et al. (1991) proposed a feedback mechanisms between global warming and hydro-climatic systems in the Eurasian continent including Japan both in winter and summer. Because the mechanisms adopted differ from season to season and therefore more realistic, this approach is recommended for future studies, especially in the monsoon region, where Japan is located.

(d) *Japanese Task in Relation to Asian Countries*

The "Asian and Pacific Workshop" on global warming research (Center for Global Environmental Research 1991) provided the following summary of research projects on the environmental and socio-economic impacts of warmer climate coordinated by Environment Agency of Japan: (a) evaluation of the effects of global warming on plants: This is mainly a study on the prediction of changes in vegetation and survival of plants in conjunction with global warming; (b) clarification of effects of sea level rise which requires precise monitoring of sea level rise as well as other aspects of environmental change in the coastal areas; (c) assessment of global warming effects on the water balance designed to develop methods to predict changes in the hydrology of river basins; (d) assessment of the effect of global warming on urban environments and countermeasures; systematic and comprehensive assessment study of impacts; and (e) evaluation of the effects, and the risk of global warming on human health. In this regard, physiological impacts on

Table 2. Stages of secular change[a)] of urban climate (Yoshino 1990/1991).

Stage	I	II	III
Air temperature	+	+	+
Vapor pressure	0(–)	0(–)	–
Relative humidity	–	–	–
Number of days with fog	+	0(–)	–
Number of days with drizzle or total amount of drizzle +	+	0	
Number of days with heavy rain	0	0	+(0)
Falling soot	+	0(+)	–
SO_2		+	0(–)
Example of Tokyo	1920	1926–1961	Since 1961

*+ sign denotes "increasing", – sign "decreasing", 0 "no significant tendency" and ( ) denotes "or".

individuals due to global warming and those brought about by current life style and social group behaviour will be assessed.

3.4. *Secular change of urban climate*

Peculiar climate conditions in cities were recognized more than two thousand years ago. Vitruvius (75–26 BC) presented an outstanding account of city planning and climatic conditions in Roman cities. Evelyn (1620–1706) described the growth of air pollution in London due to a marked increase of population.

Modern urban climatology started in Europe in the 18th or early 19th century, based on instrumental records. By the end of the 19th century, city temperature (heat island) was observed in most big cities.

Other climatic elements are also affected by city growth. Although vapour pressure is about the same or slightly lower in the city than the country, relative humidity is less in the city due to higher temperatures. The number of days with fog or drizzle are higher in the city. Days with good visibility are fewer. Soot deposition, SO_x and NO_x also increase. These features are observed in many cities worldwide with varying degrees of severity (Geiger et al. 1995; Kratzer 1956; Oke 1988; Landsberg 1981; Yoshino 1975; Yoshino 1990/1991).

In Tokyo, records show that the city's climate has been changing during the last 70–80 years. It has gone through three different stages over that period. The standard features mentioned above occurred only in Stage I. Stage II is, on the whole, stable except for a decline in the number of days with fog. This stage is transitional between the classic urban climate of Stage I and a somewhat contradictory Stage III. Stage III features vapor pressure decreases, drastic decreases in relative humidity and number of days with fog. An increase in the number of days with good visibility is indicated while the number of days with drizzle and the total amount recorded show no clear change. Days with heavy rain increase. Stage I ended around 1920. The typical period for Stage II was 1926–1940. Stage III began in 1961 and intensified during the last 20 years. Stage divisions are given in Table 2.

Causes for differences between stages are attributable partly to regional or global climatic changes, and partly to urban effects which may change with developments in city function, number of automobiles, city size, population and building densities, etc. It is apparent that the change in air pollution levels is the result of legal enforcement of better air quality standards which took effect in the 1960s, hence the striking "improvements" in climate evident in Stage III.

Based on rainfall observation, the urban climate of Moscow, Russia, has been in Stage III since 1955. Also, Kiel, Germany, experienced Stage II by 1950 and Stage III after 1960. Toronto, Canada, shows a decrease in the number of drizzle days, indicating that Stage III may be starting.

These stages are related also to developments in investigation methods. In Stages I and II, horizontal distribution was the main focus. It is hoped that stage changes in urban climates will be studied further in other cities worldwide.

One more climatic problem for all cities is the potential consequences of doubling CO_2, perhaps by 2030. In such cases, more serious problems will occur in highly modernized major cities, which are more sensitive to the changes of environment. Numerical experiments will be one of the best approaches for identifying the scope of these problems and their solution. Architects, sociologists and psychologists should be involved in this project, because the future climate in cities could involve many areas of human life.

4. Conclusion

Human activities and environmental change interact in both directions. Even though, these interactions have been discussed almost continuously since ancient times, the problems which accrue from them are increasing markedly in importance because of a sharp rise in human activities such as industrial development, population growth, and urbanization.

Consistent with these developments, climates have been changing, but the nature of their influence (negative, positive, or neutral), and the degree of that influence, have varied from period to period.

There are many international programmes structured to study these changes. But it is necessary to push strongly for an interdisciplinary framework containing natural, human, sociological and economic sciences to conduct these studies.

Climatologists in the department/institutes of geography are in a good position to participate in this framework. Their active participation is encouraged.

References

Burgos, J.J. (1986): Equilibrium and extreme climatic conditions of the world's biomes and agrosystems. In: *Land use and agrosystem management under severe climatic conditions*. WMO Tech. Note, (184), 12–56.

Center for Global Environmental Research (1991): *Summary of the Asian and Pacific Workshop on Global Warming Research*, 18–20, 1991. 11 pp. + App.

Environment Agency of Japan (1992): *Methodology for the studies on the global warming and framework for IPCC*. Tokyo, 135 pp. (in Japanese).

Geiger, R., Aron, R.H. and Todhunter, P. (1995): *The climate near the ground*, 5th ed. Vieweg, Braunschweig, 528 pp.

Gornitz, V., Lebedeff, S. and Hansen, J., (1982): Global sea level trend in the past century. *Science*, 215, 1611–1614.

Houghton, J.T., Jenkins, G.J. and Ephraums, J.J. (eds.) (1990): *Climate change: IPCC Scientific Assessment*. Cambridge University Press, Cambridge, 362 pp.

Jäger, J. and Ferguson, H.L. (1991): *Climate change: Science, impacts and policy*. Cambridge University Press, Cambridge, 578 pp.

Jones, P.D. et al. (1991): Global temperature variations since 1981. In: *Long-term variability of pelagic fish populations and their environment*, ed. by T. Kawasaki et al. Pergamon Press, Oxford, 1–17.

Kosinski, L.A. (1996): Issues in global change research. Problems, data and programmes. ISSC, HDGEC Progr., Report No. 6, Geneva, 113 pp.

Kratzer, A. (1956): *Das Stadtklima*, 2nd ed. Vieweg, Braunschweig, 184 pp.

Kurosaka, H. (1990): Impact of the climate change on the demand of electric energy in the domestic purposes in Japan. *Ann. Rep. Fac. Education, Bunkyo Univ.* (24), 25–31 (in Japanese).

Landsberg, H.E. (1981): *The urban climate*. Academic Press, New York, 275 pp.

Masukura, K., Yoshitani, J. and Yoshino, F. (1991): Possible changes of hydrological cycle by greenhouse effect in Japanese rivers. In: *Proc. CIES*, Tsukuba, Jan. 27–Feb. 1, 1991. WMO TD-No. 435 B: 15–20.

Molin, L.C.B. (1986): Land use and agrosystem management in the humid tropics. In: *Land use and agrosystem management under severe climatic conditions*. WMO Tech. Note, 184, 114–137.

Norse, D. (1991): Population and global climate change. In: *Climate change: Science, impacts and policy*, ed. by J. Jäger and H.L. Ferguson. Cambridge University Press, Cambridge, 361–365.

Ohta, Y. and Omura, A. (1991): Late Quarternary shorelines in the Japanese islands. *Quarternary Res. (Japan)*, 30(2), 175–186.

Okada, M. (1988): Long-term fluctuation of mean tide level. In: *The 37th Meeting of Office of Climate Program, Japan Met. Agency*, 41–71 (in Japanese).

Oke, T.R. (1988): The urban energy balance. *Prog. Phys. Geogr.*, 12, 471–508.

Parry, M.L. and Carter, T.R. (1991): Climatic changes and future land use potential in Europe. In: *Climatic change and impacts*, ed. by R. Fantechi et al. Com. Europ. Comman., Luxembourg, 321–341.

Riebsame, W.E. and Magalhaes, A.R. (1991): Assessing the regional implications of climate variability and change. In: *Climate change: Science, impacts and policy*, ed. by J. Jäger and H. L. Ferguson. Cambridge University Press, Cambridge, 415–430.

Rosenzweig, C. and Parry, M.L. (1993): Potential impacts of climate change on world food supply: A summary of a recent international study. In: *Agricultural dimensions of global climate change*, ed. by H.M. Kaiser et al. St. Lucie Press, Delray Beach, 87–116.
Takahashi, K. and Yoshino, M. (eds.) (1978): *Climate change and food production*. University of Tokyo Press, Tokyo, 433 pp.
Takeuchi, K. and Yoshino, M. (eds.) (1991): *The global environment*. Springer-Verlag, Berlin, 257 pp.
Tamaki, M. and Koyama, I. (1991): The acid rain observed on ground level in Japan: A review of major recent issues and problems. *Jour. Japanese Soc. Air Pollution*, 26(1), 1–22 (in Japanese).
Umitsu, M. (1991): Holocene sea-level changes and coastal evolution in Japan. *Quarternary Res. (Japan)*, 30(2), 187–196.
Williamson, P. (1992): *Global change: Reducing uncertainties*. IGBP, Royal Swedish Academy of Sciences, Stockholm, 40 pp.
World Meteorological Organization (WMO) (1992): *World Climate News*, No. 1, June 1992, 12 pp.
Yamada, I. and Yamakawa, Sh. (1991): Effects on soil environment of global warming. Res. Rep., Div. Changing Earth and Agro-Environment, NIAES, (1), 29–44 (in Japanese with English abstract).
Yoshino, M. (1975): *Climate in a small area*. University of Tokyo Press, Tokyo, 549 pp.
Yoshino, M. (1990/1991): Development of urban climatology and problems today. *Energy and Buildings*, 15–16, 1–10.
Yoshino, M. (1991a): Impact of climatic change on agriculture from the viewpoint of East Asia. In: *The global environment*, ed. by K. Takeuchi and M. Yoshino. Springer-Verlag, Berlin, 16–41.
Yoshino, M. (ed.) (1991b): Proceedings of the "International Conference on Climatic Impacts on the Environment and Society" (CIES). WMO TD, No. 435, 416 pp.
Yoshino, M. (1992): The significance of rural land use study in the IGBP-HDGEC framework. A paper presented at the International Symposium on "Rural Land Use in Asia" held in Tokyo, 7–8 October, 1992.

Masatoshi Yoshino
Institute of Geography
Aichi University
Toyohashi-City
441 Japan

PART 2

CLIMATE CHANGES AND VARIABILITY

2.1. ON RELATIONSHIPS BETWEEN CLIMATE VARIABILITY AND CHANGE, AND SOCIETIES

ANNICK DOUGUÉDROIT

1. Introduction

Climate is only one of the many physical variables which impact life on Earth. Although humans have experienced the partial control exercised by climate on their economic activities and life style throughout the ages, investigations and wide development of theories on the relationship between climates and societies did not begin until the last several decades. This development gave climate a new perspective, which is being integrated into space and time organization of societies.

2. Two approaches to climate

Climate is defined as the average state of the atmosphere measured, usually, over a 30-year period – a normal – as recommended by the World Meteorological Organization (WMO). It includes, over the same period, the statistical variations of the components of the atmosphere which represent the climate variability. Recently, climate was described in the WMO Bulletin as "the statistical probability of the occurrence of various states of the atmosphere over a given region during a given calendar period" (Gibbs 1987).

2.1. *A physical approach to climate*

Climate describes a provisional state of a purely physical system with five components; atmosphere, land, ocean, ice and biosphere, all with strong linkages to one another (Kutzbach 1974; Houghton 1984; Schlesinger 1990). Climate is driven by the energy from the sun. The radiation balance of Earth

M. Yoshino et al. (eds.), Climates and Societies – A Climatological Perspective, 21–41.

changes under the influence of climate forcing agents such as changes in both the amount and seasonal distribution of solar radiation reaching the earth, changes in albedo, and changes in the composition of the atmosphere. The five components of the system react on different time-scales to changes in the forcing, from hours to days for the atmosphere, from days to months for lands, and for up to several millennia for the cryosphere and oceans. All the components interact, such that the system shows a considerable variance which are the climate fluctuations. The forcing agents have had a natural evolution since the existence of Earth's atmosphere and have undergone fluctuations. They have created climatic fluctuations at several time scales (Berger 1992). Even if a growing human influence affects two of these agents (the albedo may be modified by desertification or deforestation, pollution may increase the atmospheric aerosol loading), processes involved in climate are still purely physical. Results from physical models of climate systems are promising (Schlesinger 1988). However, their full potential are yet to be realised even among the few General Circulation Models (GCMs) of the atmosphere that are coupled with ocean models (see Gates et al. 1993; Le Treut, Chapter 2.5 of this volume).

Climate variability and change depend on the physical processes working in the climate system. Climate variability captures year to year fluctuations of climatic elements (temperature, precipitation, humidity, wind, etc.) at several time-scales, the most common are, one hour or less, one day, the month, the season and the year. Variability is defined generally as anomalies or deviations from the average and seems to depend on chance. Unlike climate change, climate variability assumes that the climate is stationary. If there is unanimity that long time-scale changes such as those between glacials and interglacials constitute real change, there is no similar agreement about climate during the instrumental period. Here, the distinction between climate variability and climate change has not been clearly defined. The usual and often confusing way of dealing with the difference between the two indicators is to compare the fluctuations of the suspect period against those of the reference period, namely, the standardized "normal" of 30 years as defined by the WMO (Berger and Labeyrie 1987; Mitchell 1990; Berger 1981).

2.2. *A social approach to climate*

The importance of climate to human activities appears evident but it was neither defined nor analysed until recently. Climate is one of the many interconnected variables, both physical and socio-economic, that may affect human activities. Activities affected directly or indirectly analysed by the climate have to be identified and analysed and the manner in which they are impacted have to be conducted. The values of climatic knowledge must be stressed at all levels; locally, nationally, regionally and internationally, thereby, increasing the ability of people to manage weather-sensitive activities (see Douguédroit et al., Chapter 2.6 of this volume).

Climate has many applications; some are favourable while others are not. On the positive side, climate can be viewed as a potential asset or a resource which can be exploited directly or indirectly. For example, the direct exploitation of the sun for holidays, and by business is evident. But in a more subtile way, many activities such as agriculture, water supplies, etc., benefit directly from several components of the climate system (precipitation, solar energy and others). Others such as transport and housing gain from a favourable climate. Climate is a special type of natural resource, but as is the case for all the other natural resources, it can be harnessed or ignored.

The physical limits of climate constrains its practical use in various areas of human endeavour (see discussion on agriculture and transportation below). They also determine the nature, magnitude, and timing including suddenness and duration, of those constraints, risks and natural hazards which are weather driven.

Benefits and constraints are two sides of the relationship between climates and societies. This relationship is summarized in the single notion of *climate impacts*. Climate impacts societies and societies impact climate(s). Lately, the impact of societies on climates has become a much discussed issue among scientists and in the media. The focus for this discussion is the potential effects of anthropogenic gas emission on the natural greenhouse effect. A social approach which deals with both types of impact (climates on societies and societies on climates) is required.

The knowledge about climate which societies need is a problem specific and in many ways different from the type of information packaged by professional climatologists. In this regard, people appear to be much concerned about climate variability which regulates climate sensitive activities. In light of its importance, climate data analyses including modelling must deal with present climate variability and also with the future one in a global warming perspective, even if people do not feel very much concerned.

3. Measuring, modelling and prediction of climate variability to estimate climatic impacts

Modelling and prediction of climate variability depend on the quality of the measures used and on the methods and techniques chosen. The description and discussion conducted here will focus on some features of climate fluctuations, mostly on temperature and precipitation.

3.1. *Data problem*

Data on climatic elements are obtained in different ways. Since the beginning of the instrumental period, most measurements have been and continue to be made at meteorological stations. Others are inferred from remote sensing.

Data are collected at meteorological stations at regular intervals. The stations themselves form an uneven mesh of networks whose size depends on the element sensed and geography. There are more rain gauges than thermometers, precipitation and temperature being most sensed elements. Data from oceans and upper atmosphere are much fewer in comparison with land surface measurements. The amount of data and their density vary from one continent to another, the Northern Hemisphere is better represented than the Southern Hemisphere. Europe and North America have the best data sets.

This uneven distribution leads to problems of interpolation between points when areal and global values and maps are desired. For global datasets, the problem is partially solved particularly for temperature, precipitation, pressure and wind, by dividing the globe into regular latitude-longitude grids and interpolating among stations in each grid. Such datasets are prepared in several world centres including the National Meteorological Centre (NMC) in the United States and the Meteorological Office and the ECMRWF in the United Kingdom. Inhomogeneity in datasets usually caused by a variety of non climatic factors (see list in Chapter 2.4), must be detected and corrected. The random characteristics of the data, which are necessary to estimate fluctuations, are isolated by tests (Sneyers 1975).

The problem of spatial representation of climatic data could be later on solved by remote sensing which produces common information for large areas at the same time. Except for a few cases, such as the shapes of clouds, climatic variables have to be reconstructed from data obtained from several spectral bands (visible, infrared, . . .) and the results compared with those obtained from field experiments on the ground. Earth and ocean temperatures are the best estimated climatic elements at present. Areas of drought or convective rainfall are fairly easily identified but the quantification of rainfall is still a very difficult problem. Soil moisture estimates are modelled successfully in cultivated areas but similar efforts in other areas, e.g. forests, are still at the experimental stage (Prosper-Laget et al. 1995). Very important investigations on climate are based on remote sensed data. Remote sensing represents an important tool for designing new datasets in future, especially in specific areas of applied climatology (see Douguédroit et al., Chapter 2.6 of this volume).

Another tool for future investigations are the numerical general circulation models (GCMs). Some of these are coupled with oceanic models. But there are questions about their accuracy including those raised by levels of uncertainties encountered in modelling atmospheric, cryospheric and oceanic interactions. As well, they are at present, less able to predict the characteristics of climate variability than they are to predict global changes in the climate system,

Table 1. Annual precipitations in Marseille in several 30-year periods (mm).

	Smallest value	First decile	First quintile	Median	Mean	Fourth quintile	Ninth decile	Highest value
1871–1900	323	349	407	540	558	698	788	1092
1901–1930	276	456	499	586	581	684	740	895
1931–1960	323	435	494	621	635	769	855	1020
1961–1990	261	359	418	550	565	706	777	836

even when using high resolution models or nested regional climate models (Hewitson 1994; von Storch 1994). It explains all the recent development on climate downscaling (von Storch et al. 1993; Hewitson 1995).

3.2. *Temporal structure of climate variability*

The time structure of the variability of data series is poorly described by special values, even if they are measures of central tendency such as means and medians. Variability is better represented by probabilities which are statements about the likelihood of an event rather than a prediction of its occurrence. "... the quality of a (rainy) season or a year as a whole, based on conditions preceding the season or year by more than a month" (Hastenrath 1990). Such statements are demanded by users for risk assessment in real time (see Douguédroit et al., Chapter 2.6 of this volume).

The fluctuations of a number of climatic elements, temperature and precipitation, for example, are widely studied. Their average range, space and time dimensions, changes in variability through time, and so on are estimated (Lamb 1977; Flohn and Fantechi 1984; Labeyrie 1985; Peguy 1989).

Differences in the fluctuations of temperatures derived for 30-year periods are published. As many climatic data series do not fit a Gaussian distribution, their characteristics have to be determined via percentiles (observed or calculated) rather than standard deviations. In Marseille (Table 1), characteristic values of 1871–1900 and 1961–1990 periods are close to each other except for the lowest and the highest ones, and different from those of 1931–1960. In that example, a 30-year period is not long enough to provide a good estimate of possible variability. Better estimates will be achieved with a longer period since the scale of the fluctuations and the size of the central values used to estimate the fluctuations both vary.

A question arises when fluctuations move far away from a previous central value: is this a signal of climate variation? The best example of such a problem is shown by the evolution of the Sahelian precipitation (Nicholson 1985;

Table 2. Annual precipitations in Niamey in different periods (mm).

	Smallest value	First decile	First quintile	Median	Mean	Fourth quintile	Ninth decile	Highest value
1912–1985	281	392	442	547	558	669	738	901
1912–1968	281	408	459	564	575	686	755	901
1912–1949	281	382	433	541	553	666	739	893
1950–1968	416	476	521	614	622	717	776	901
1969–1985	293	350	395	489	499	598	659	686

Janicot 1992) based on records from Niamey (Niger) and Kaedi (Mauritania) (Demaree 1990; Demaree and Nicholis 1990).

The characteristic values of precipitation of several series in Niamey are not significantly different from a statistical point of view (Table 2). They are lowest in 1968–1985 but not much lower than in 1912–1985, 1912–1949 and 1912–1968. It is wettest in 1950–1968. The only significant difference between the series occurred between the two last periods (1950–1968 and 1969–1985: 125 mm difference). Analysis of the Kaedi series (1904–1988) for annual precipitation reveals a highly significant downward trend since the sixties and an abrupt change in the estimated mean for the second half of the same decade (1904–1967; 413 mm mean, 1968–1988; 237 mm mean; 176 mm difference between the two means). Such differences between Niamey and Kaedi show that the statistical evidence for non-stationarity of the precipitation series in the Sahel area is not general, even if abrupt changes can be detected in many climatological series (using, for example, the Pettitt test (Pettitt 1979)).

Ideally, the probability calendar should correspond to the calendar year (Péguy 1976). This will make it easier to calculate the probability of occurrence of a determined level of climate variables such as solar radiation, temperature, and snow-cover etc. (Figure 1), at a fixed date in the year.

3.3. *Natural hazards and extreme events*

Many scientists and others have focused their attention on natural hazards and extreme events, because of their drama and costly impacts. Cyclones, droughts, floods . . . are often targets but other adverse weather types such as very cold or very hot days must not be neglected. We shall consider a few examples.

February 1956 and January 1985 constitute the most famous recent cold spells in Western Europe. Both events were caused by the sudden arrival of arctic air from the north-north east. In sequence (–16.8 and –18.2°C respec-

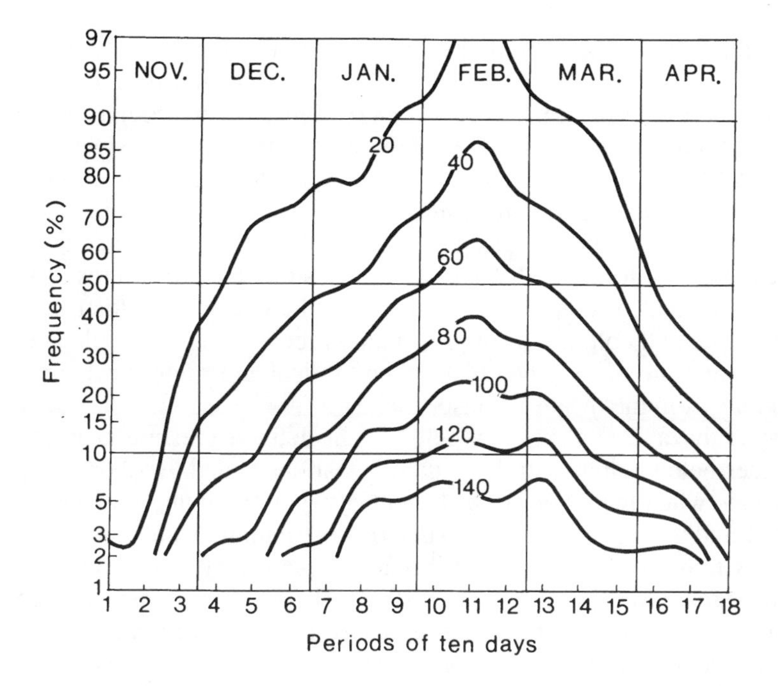

Figure 1. Frequency calendar of snow depths (in cm) in La Clusaz (French Alps) (in S. Martin, J. Mainguy, and M.F. de Saintignon, 1990–1991). For three years out of ten (i.e. 30%) snow depth is less than 20 cm at Christmas (period 6); for six years out of ten (i.e. 60%), snow depth is less than 40 cm at Christmas.

tively), they broke the minimum temperature record at Paris-Le Bourget established in 1931. But if we consider the longer series at Paris-Montsouris beginning in 1873, they are exceeded by the 10th December 1879 event (compare –11.1° in 1956, –11.5° in 1985, and –15.1° in 1879) which has a return period longer than a century (Besleaga 1991).

Precipitation falls as snow when temperatures are near or below zero. Extreme values in both small and large depths of snow cover arouse increasing interest with the development of skiing, and in road traffic.

Precipitation records are even more striking. At the monthly scale, for October and November 1961, with totals representing between 200 and 300% or more of the mean, was the most rainy year in East Africa between 1932 and 1983. During that period, the direction of the surface wind was opposite that of the circulation of the zonal cell that normally covers that area (Beltrando 1990). In Perpignan (France), 748 mm fell in October 1965 in comparison with monthly averages around 80 mm (1951–1980: 83 mm, 1865–1990: 80 mm). The difference was more than 900%. This convective rainfall due to atmospheric circulation was probably enhanced by the high surface temperature of the Mediterranean sea in the beginning of autumn.

But the most striking rainfall events occur in tropical areas affected by cyclones (Grazzuliz 1991), the highest being 3,240 mm over La Réunion island in three days in January 1980 when *Hyacinth* swept through the country, and 1,825 mm in 24 hours in January 1966 with *Denise* (Meleo-France, 1987). All those heavy rainfalls are, of course, associated with strong winds and very low pressure systems. Most of the heaviest precipitation occur on coasts of warm oceans or seas and not far inland during the autumn when the sea temperature is the highest. High sea surface temperatures speed up the development of convection. Because of the decrease of total humidity in temperate and cold air, local precipitation records in the middle and high latitude areas are much lower than in the tropical and subtropical zones and deviations from the means are also smaller.

Natural hazards and extreme events can be defined as events characterized by values below minus 2 or 3 standard deviations and above 2 or 3 standard deviations when the given series has a Gaussian distribution, or lower than decile 1 and higher than decile 9 when the distribution is not Gaussian.

All extreme events can be described and explained after the event but most cannot be forecast in advance (WMO 1990). The ability of forecasters depends on their size. Some, like tornadoes and severe storms are too small. Longer running events such as record frost events in cold spells can be forecasted a few days before they occur. But when the causes are not well understood, as in the case of cyclones, forecasting is more difficult. Cyclones grow out of depressions, but the forcing mechanisms at their origin are unknown. Nobody can predict if a depression formed over tropical oceans will become a cyclone before it is fully developed. And even after a cyclone's birth, its evolution and track are so uncertain that forecasts of the coasts which are likely to be affected by the cyclone is based on watch. The coasts are much more important than the effective disaster area, except during the last few hours before the arrival of the cyclone.

3.4. *Persistent weather*

Some climatic extreme events occur in persistent form. Extreme events of the persistent type have characteristics that are similar to those associated with the extreme events discussed in the preceding sections (the stronger wind blasts felt during a storm or the coldest day of a cold spell) but are generally less intense. The persistence of snow over several days or drought over several decades may, especially in the case of the later, impact adversely on agricultural production and water resources. Persistent extreme events which occur periodically as droughts are of great interest because of their possible influence on economic life, and therefore, have to be investigated (see Douguédroit et al., Chapter 2.6 of this volume).

In Paris (Le Bourget), 16th January 1985 (–11.4°) is included in a 5-day cold spell when daily temperatures averaged –6.3°C, and a 16-day one when

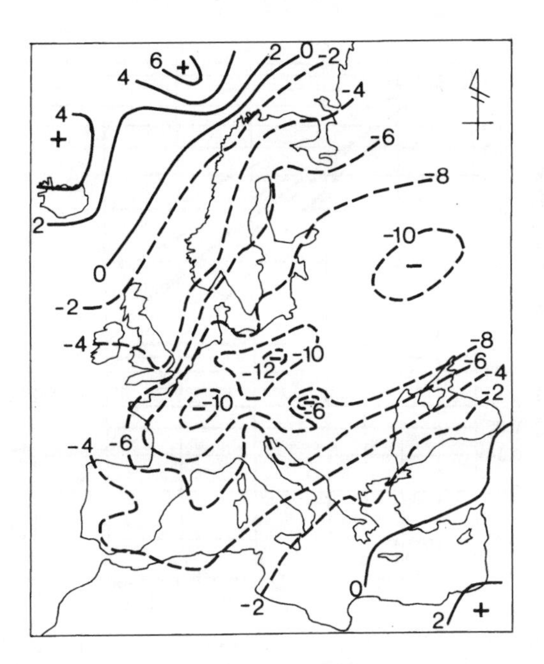

Figure 2. Deviations of monthly mean temperature (in °C) from the mean of the period 1931–1960 for February 1956 (after Flohn and Fantechi 1984).

the average reached –5.5°C (–5° in February 1956 and –4.2° during 24 days) (Figure 2). With respect to average monthly temperature, the coldest January in the 1951–1980 normal occurred in 1963. In Europe, cold spells can last several weeks due to strong cold advection followed by a period of radiative losses. Very low temperatures are sustained till the general high pressure field over the continent breaks. Their length is variable, the longest being linked with the most important modifications of the atmospheric circulation (Besleaga 1991). Characteristic values of persistent events depend both on the time scale involved and the reference period.

Statistical models of weather spells over successive days with similar characteristics have been investigated for a long time. The most usually used till Gabriel and Neumann (1962) are Markov chains. More recently, Galloy et al. (1983) proposed the shifted Negative Binomial Model, adapted unlike the previous ones, to increasing or decreasing persistence. All define probabilities of occurrence of spells of chosen lengths or even the recurrence interval of the same spells.

Modelling persistence was mainly applied to wet and dry spells of temperate and mediterranean areas. These spells are very short and not very different in length in oceanic countries as Belgium (Berger and Goossens 1983). In the mediterranean climate area, wet spells are shorter than those in more northern

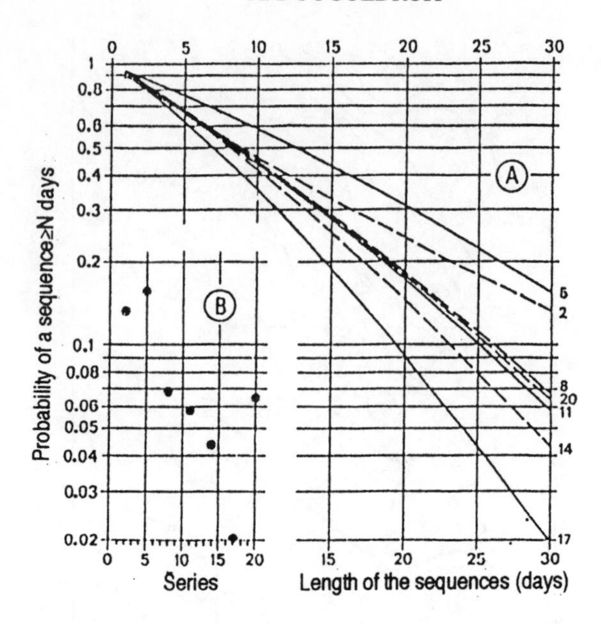

Figure 3. Distribution in Marseille of the probability of occurrence of a dry sequence beginning in August according to the SNB model during several 30-year periods, 1880–1899 (2), 1885–1914 (5), 1900–1929 (8), 1915–1944 (11), 1930–1959 (14), 1945–1974 (17) and 1960–1989 (20). A = All spells up to 30 days long; B = > 30 days long spell.

climates but dry spells are much longer, especially in summer. The probability of occurrence of a dry spell can be modelled for each month of 30-year periods. This enables us to estimate the return period for spells of defined lengths (Figure 3). In Marseille, during 1960–1989, the probability was 0.5 for a 10-day spell and 0.18 for a 20-day one. The comparison between the 20-overlapping 30-year series from 1865–1894 (No. 1) to 1960–1989 (No. 20) reveals, for spells of the same length in the different series, noticeable differences in probabilities that are statistically significant while similar differences in monthly precipitation are not significant (Douguédroit 1987a, 1991).

Results of recent investigations show that for some specific events remote sensing enables you to approach real time resolutions. The determination of the evolution of a drought makes possible the monitoring of future similar droughts (see Douguédroit et al., Chapter 2.6).

3.5. *Space repartition of temporal fluctuations*

Except for a small number of natural hazards distinguished by their small size, climatic elements display a space repartition consistent with physical processes (climate regionalization). Its determination is accomplished by methods such as correlations or component analyses.

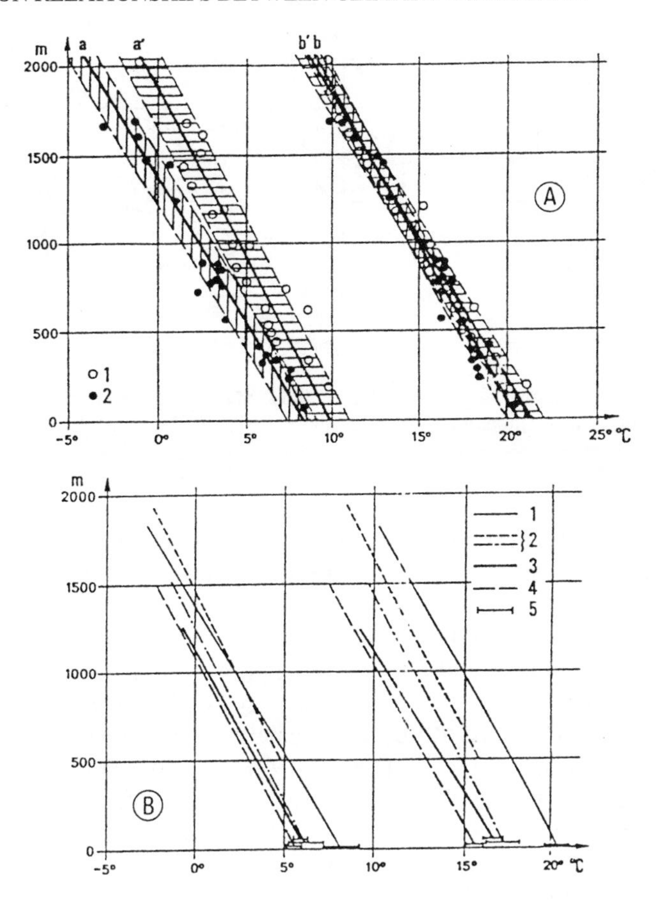

Figure 4. Decreasing of the mean annual temperature with the altitude: A = In the French Southern Alps and the inner Provence mean minimums (a, b) and maximums (a′, b′) in the bottom of the valleys (2) and on the sunny slopes (1). B = in the bottom of the Valleys in several mountains. 1 = Southern French Alps; 2 = Northern French Alps; 3 = Jura; 4 = Vosges and Black Forest (Germany); 5 = Confidence interval (after Douguédroit 1987b).

Temporal correlations in a climate parameter (temperature, precipitation, ...) between a given location and all others in the area reveal domains of significant relationship at different levels. Temporal linear regressions between monthly or yearly temperatures of stations in a location type (sunny or shady slopes or bottoms of valleys) and altitude give the most probable temperature (with confidence interval) at determined altitudes in the area where the stations are located (Douguédroit and de Saintingnon 1970; Douguédroit 1987b). The width of the confidence interval gives an estimate of the space fluctuation of the temperature of each type of location in the target area. Confidence bands are narrower in daytime ($< 2°C$) than night time. At night, they are narrower for valley bottoms (about 2°C) than for slopes (about 3°C). All

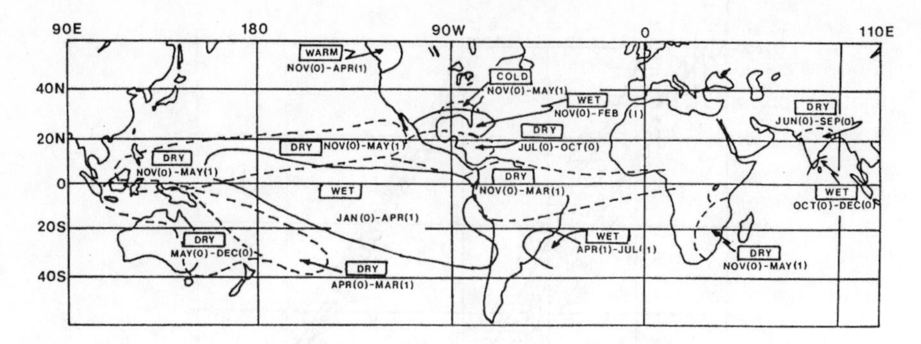

Figure 5. Climatic anomalies occurring during an ENSO event all over the globe. (0) = year when the Pacific Ocean temperature increases; (1) = the following year (after Ropolewski and Halpert 1987).

the mountains in Eastern France display similar temperature gradient, however, for the same altitude, temperature decreases with increasing latitude (Figure 4).

But since Steiner (1965), principal component analyses (PCA) are considered as more suitable and powerful tools for climate regionalization, the most used being the principal component analysis with varimax rotation. But several types of PCA exist, differing especially by the type of rotation used (Richman and Yarnal 1991; White et al. 1991). A physical interpretation of the regionalization produced is always a basic criterion for validity. PCA with rotation produces temporal series of each factor which can be used to estimate the fluctuations of climatic elements for each region (for geopotential heights, see (Barnston and Livezey 1987), for precipitation in North-Western Africa, etc., see (Douguédroit 1988)).

3.6. *Variability and teleconnections*

The multitude of interactions between the components of the climate system induces effects that are not only local, but regional and even global. Anomalies never occur in isolation as it might seem. Because of the scales of atmospheric and oceanic motions, they are associated, simultaneously or with a short delay, with similar or other anomalies in distant areas (Halpert and Ropolewski 1992). Such teleconnections are frequent, but only recently investigated and well described but less understood. Recent interest focussed on the El Niño–Southern Oscillation (ENSO), a major feature of the climate system which influences climate variability (Philander 1990; Glantz et al. 1991; Hastenrath 1991; Rasmusson and Carpenter 1982). Temporary interactions, especially volcanic eruptions, are also noticeable (Sear et al. 1987).

Cycles or quasi cycles shorter than the length of series used to estimate fluctuations are aspects of variability (see Schönwiese, Chapter 2.4 of this volume, for spectral tests). ENSO corresponds to a quasi cycle, as other phenomena which can be involved in climate variability. Among the many spectral analyses performed recently, the Quasi Biennual Oscillation (QBO) with a 40–50 day oscillation is the most common, even if some other cycles induce common fluctuations between areas far from each other. The causes of those kinds of cycles are not yet understood. Even with the QBO, its links with the stratospheric QBO are not established. So we shall neglect all the cycles except the ENSO. Spectrum of air-temperature oscillations in North Atlantic sector is shown in Figure 5.

3.6.1. *The influence of ENSO cycles on climate variability*

The El Niño–Southern Oscillation is a quasi cycle of the Pacific Ocean. Its occurrence is only well established since 1935, but the existence of previous ones are known but not quantified. Here, we shall not emphasize the development of the ENSO, nor detail all its influences.

The 1982–1983 ENSO characterized by an exceptional strength and timing provided an opportunity for the first intense investigation of the phenomenon during the event. A list of important climate anomalies occurring during the ENSO over lands and oceans was collected, most of them in the tropical zone (Figure 5). In general, they are studied by correlating the series of temperature and precipitation, with the Southern Oscillation index (which is the standardized difference of surface pressure anomaly between Tahiti and Darwin).

The tropical Pacific Ocean and countries around them are directly affected by a reversal of humidity conditions between the East Pacific and the coast of South America (Hastenrath and de Castro 1987) which become wet, and the western part with Indonesia and North-East of Australia which become dry during the summer of the Southern Hemisphere (Ropolewski and Halpert 1987; Ropolewski et al. 1992).

All these anomalies are caused by the movement of the Pacific zonal (Walker) cell during an ENSO event. Fluctuations of the Indian monsoon are also correlated with ENSO, but precedes it (Rasmusson and Carpenter 1985; Cadet 1985). According to Gregory (1990), the ENSO index explains the major part of anomalous rainfall conditions in the northwestern and northern Deccan.

The possible influence of the ENSO on the climate of other tropical areas was extensively investigated recently. Significant correlations exist but they are rather low for Africa (Nicholson and Entekhabi 1986) and North-East Brazil (Hastenrath and de Castro 1987). In the extra-tropical latitudes of the Northern hemisphere, ENSO events are mainly associated with a special pattern of pressure anomalies, the Pacific North America pattern or PNA over the Pacific Ocean and North America (Shukla and Wallace 1983; Henderson and Robinson 1994). It corresponds with dry and hot anomalies in the West of

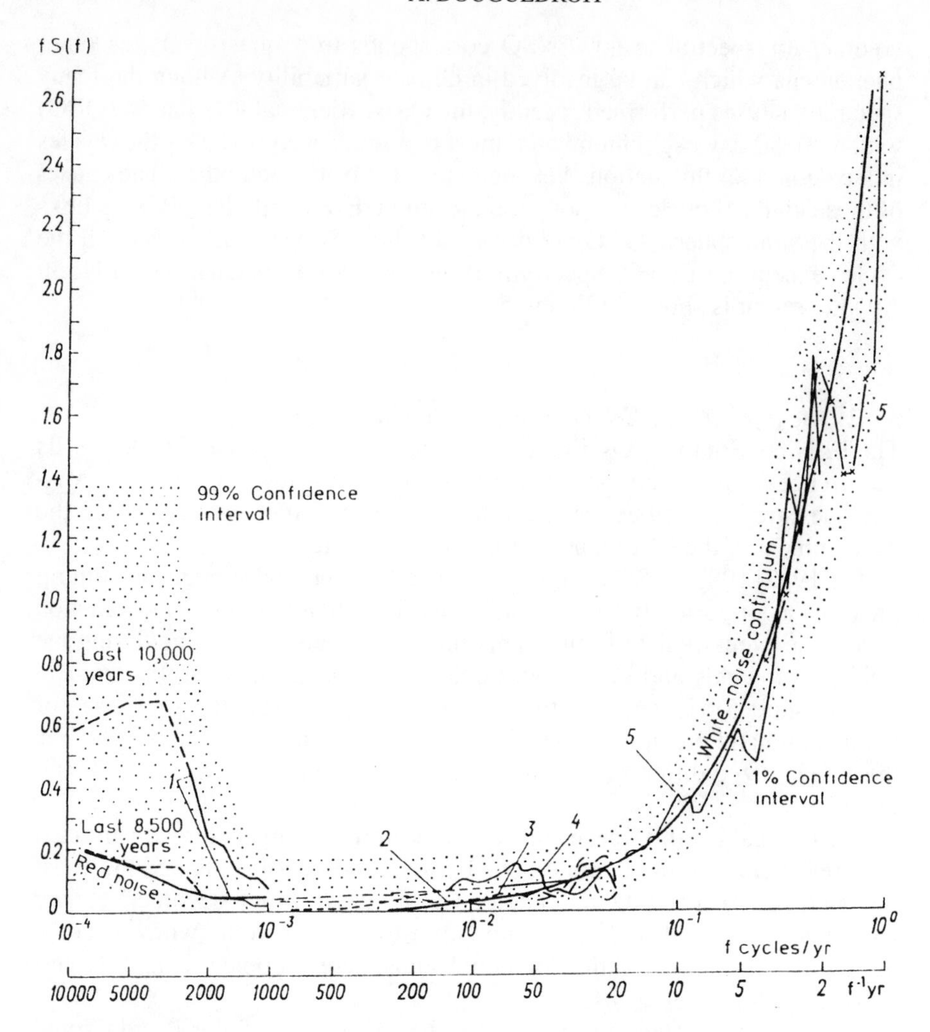

Figure 6. Spectrum of air-temperature oscillations in North Atlantic sector according to Kutzbach and Bryson. The ordinate is the amplitude squared $S(f)$ times the frequency f; the abscises are the frequencies f on a log scale and the corresponding periods. (1) Central England botanical data; (2) Central England, historical evidence; (3) Iceland, historical evidence; (4) Greenland, $d^{18}O$ data; (5) Central England, instrumental data.

the USA and opposite ones in the East. They are influenced by the anomalies of the high troposphere induced by the ENSO (Glantz et al. 1991).

3.6.2. *Other teleconnections and attempts of predictions*

Teleconnections can be demonstrated by correlations between series or indices of climatic elements, for instance, for correlations between temperature series for Greenland and Western Europe, the North Atlantic Oscillation

(NAO) is utilised (Van Loon and Rogers 1978). When correlations are calculated between the series of a climatic element and each grid point of a climatic field, they reveal areas of significant relationship with possible teleconnections and predictions. The geopotential heights in the atmosphere (700 and 500 hPa in general) and the surface ocean fields (sea surface temperature, wind, pressure) are the usual tools for that kind of research.

Investigations on relations between the fluctuations of oceanic climate fields and rainfall were recently carried out in different parts of the world, chiefly in the tropics. Anomaly patterns of sea surface temperature (SST) of the Atlantic Ocean are linked with anomaly patterns of precipitation in Western Africa without knowing if SST is a factor in precipitation or not. When rainfall is below (above) normal, SST displays warmer (colder) water in the south-east Atlantic in summer, which is the rainfall season (Lough 1986). This feature also corresponds to a shift of the Intertropical Convergence Zone (ITCZ) south (north) of its normal position. But such links are not evident every year.

As the fluctuations of the SST are slower than the fluctuations of the atmosphere, the produced anomalies are often established before the assumed corresponding fluctuations of precipitation. In the Atlantic Ocean they begin in spring while the wettest months occur in summer. That type of gap is the basis of various attempts at long-range predictions for the Sahel, Indian monsoon, Northeast Brazil, and Northern South America. These approaches and their results have been listed by Hastenrath (1990) who shows that 60 to 80% of the interannual variability can be predicted from antecedent departures in the large-scale circulation setting in Northeast Brazil, northern South America, and India.

Success appeared first more difficult in other areas, particularly in extra-tropical zones. When Namias (1990) studied a major break in circulation and anomalous temperature pattern which took place between autumn 1988 and the following winter in the USA, he concluded "Practically all conventional tools used in long-range forecasting failed to capture the big change in pattern". But recently new assessments give satisfactory results (Barnston 1993; CAC 1995).

4. Estimation of past climates by their impacts

Modelling and prediction of climate variability necessary for the understanding of the impacts imply reliable measures. But this is a relatively recent situation. The accuracy of the old gauges, the problems of interpreting the old units of measurement, instrument exposures and uses, and changes of location make it difficult to obtain long homogeneous series. This question was tackled first by Manley (1953) for measurements on land and by Folland et al. (1985) for measurements on boats. The series of temperature and precipitation which start earliest are respectively the monthly averages for Central

England and London–Kew in 1659 and 1697: but climatological networks spread worldwide only after World War II (see Schönwiese, Chapter 2.4 of this volume). Before the instrumental period, climate was not measured but estimated.

4.1. *Estimation of the climate by its impacts*

The estimation of past climates is supposed to reverse the significant problems encountered in determining the impacts of climates from knowledge of climate itself. Climate is inferred from knowledge of its impacts. This method could appear paradoxical since past climate impacts are poorly understood. Also, the method consists only in estimating qualitative or even quantitative departures from averages of temperatures and (or) precipitation from present values measured over the location of a given impact.

This procedure relies also on the axiom of time stability of the relations between climates and their impacts: equal impacts from equal climates. Also the impacts chosen to estimate past climates generally belong to the biological or to the physical environment which are determined by processes assumed to be time stable. The varied accuracy of sources and methodologies lead to estimates that a past period was warmer or colder, wetter or drier than present based on an empirical comparison. It may also be used to estimate the prevailing temperature or precipitation expressed in degrees Celsius or millimeters respectively. The second case relies on the use of a statistical relationship called a transfer function. It is computed by principal component analysis which determines sets of elements with similar relations with present day temperature or (and) precipitation. Thermal and rainfall conditions existing in the past are computed based on these relationships (see Serre-Bachet, Chapter 2.3 of this volume).

4.2. *Time scales of past climates*

The characteristic time scales of past climates are generally described loosely; there is no commonly accepted definition of climate variations from the statistical point of view. The hypothesis of cyclic or pseudo-cyclic rhythm has not been demonstrated. Periodicities were identified in some areas (see, for example, for South Africa (Tyson 1986)), but their transient or permanent features were not shown. Using instrumental, historical and paleobotanical data, the fluctuations of temperature on the scale 10^0–10^4 years over the North Atlantic sector reveals strong persistence for periods $>10^3$ years, randomness for decadal or shorter periods, and moderate persistence at intermediate time scales (Figure 6). Strong indications of abrupt changes also exist, such as those shown in precipitation in the Sahelian area about 1970. According to some researchers this should be connected with a non periodic behavior of climate depending on fractal attractors (Nicolis and Nicolis 1984; McCauley 1990; Lorenz 1991; Nicolis et al. 1993).

For our project, it was decided that the past climates of interest went back to the end of the Würm glaciation, limiting the period to the beginning of the present interglacial. The question of its division into periods characterized by their own time scale remains since statistical studies of the period of instrumental measurements do not provide any answer to that question. The only approach available is a division based on methods used for reconstructing climates and mainly on differences in the precision of the reconstructed climates.

The present climate is the only one which can be statistically characterized with regular diurnal and seasonal fluctuations and interannual variability within a 30-year (normal) series. Consequently, its impacts on economical activities can be evaluated (see Douguédroit et al., Chapter 2.6 of this volume).

The complete long series of temperature and (or) precipitation going back to the beginning of measurements give a comparison between the statistical characteristics of successive normals. The evolution of climates can be calculated. Temperatures are the only element which gives a general trend, an increasing one (see Schönwiese, Chapter 2.4 of this volume). Historical evidence of exceptionally favorable conditions on economic activities allows the separation of the influence of climate impacts from socio-economic factors, especially in agricultural production (see, for example, the study of the fluctuations of paddy rice yield in Japan during the last century (Uchijima 1981)).

Today, the main worry about climate concerns predictions about the future following the warming noticed for more than a century. If the debate concerning a partly anthropogenic origin for this warming persists, there is no doubt, according to the GCMs, that a greenhouse effect enhanced by a CO_2 doubling and an increase in the emission of other gases such as methane will occur. In the next decades, warming of the low atmosphere will increase, however, and most importantly, uncertainty about its magnitude remains (see Le Treut, Chapter 2.5 of this volume).

Before instrumental measurements, estimates of periods under a variety of climates were based on climate impacts. Current climate reconstructions yield estimates of mean temperature and precipitation with a precision dependent on the period involved. Precision decreases with the passage of time measured from present. It also varies with the types and sometimes the number, of sources used for the reconstruction. The time scales of the dating increase in the same way.

Also, the present interglacial can be divided into two large periods preceding the instrumental one: the postglacial and the historical periods. The historical period goes back to the beginning of the first millennium A.D., but records are more in number and their precision better in the second millennium especially its second half. Reconstructions can be specified to the tenth of a Celsius degree for average temperature, for annual, seasonal or even monthly temperatures and very scarcely to the millimeter for average annual or seasonal precipitation. They rely mainly on tree-ring method, even though the historical, isotope ratio and pollen analyses methods are quickly developing. From the standpoint of the spatial extent of knowledge, Europe is still favoured, but investigations are rapidly increasing in North America (see Serre-Bachet, Chapter 2.3 of this volume).

Beyond the historical period, climates are reconstructed over smaller time-scales in the order of decades or centuries because of the recent rapid improvement in the precision of reconstruction. Two methods are mainly used: pollen analysis and more recently isotopic ratio (see Zhang et al., Chapter 2.2 of this volume).

References

Barnston, A.G. (1993): Linear statistical short-term climate predictive skill in the Northern Hemisphere. NOAA Technical Report NWS 49, US Dept. of Commerce, NOAA, National Weather Service, 137 pp.

Barnston, A.G. and Livezey, R.E. (1987): Classification, seasonality and persistence of low-frequency atmospheric circulation pattern. *Mon. Wea. Rev.*, 115, 1083–1126.

Beltrando, G. (1990): Space-time variability of rainfall in April and October-November over East Africa during the period 1932–1983. *J. of Climatol.*, 10(7), 691–702.

Berger, A. (ed.) (1981): *Climatic variations and variability. Facts and theories*. NATO ASI Series C, Vol. 72. Reidel, Dordrecht.

Berger, A.L. and Goossens, C. (1983): Persistence of wet and dry spells in Uccle (Belgium). *J. of Climatol.*, 3, 21–24.

Berger, A. (1992): *Le climat de la terre. Un passé pour quel avenir?* De Boeck Université, Bruxelles, 479 pp.

Berger, W.H. and Labeyrie, L.D. (1987): *Abrupt climatic change. Evidence and application.* NATO ASI Series, Series C, Vol. 216. Reidel, Dordrecht.

Besleaga, N. (1991): Phenomènes remarquables n° 2. Vagues de froid sur la France et les pays voisins. Direction de la Météorologie Nationale, 41 pp.

CAC (from 1991): *Climate Diagnostics Bulletin*. NOAA/NWS/NMC, Washington, DC (Monthly issue).

Cadet, D.I. (1985): The Southern Oscillation over the Indian Ocean. *J. of Climatol.* 5, 189–212.

Demaree, G.R. (1990): Indication of climatic change as seen from the Rainfall Data of Mauritanian Station. *Theor. Appl. Climatol.* 42, 139–147.

Demaree, G.R. and Nicolis, C. (1990): Onset of Sahelian drought viewed as a fluctuation induced transition. *Quart. Jour. Roy. Met. Soc.*, 116, 221–238.

Douguédroit, A. (1987a): The variations of dry spells in Marseilles from 1865 to 1984. *J. of Climatol.*, 7, 541–551.

Douguédroit, A. (1987b): Les topoclimats thermiques de moyenne montagne. Agrométéorologie des régions de moyenne montagne. In: *Les Colloques de l'INRA*, 39, 197–213.

Douguédroit, A. (1988): The recent variability of precipitation in North-Western Africa. In: *Recent climatic changes*, ed. by S. Gregory. Belhaven Press, London, 130-137.

Douguédroit, A. (1991): Drought in the French Mediterranean area. 5th Conference on Climate Variations. Denver (USA), Oct. 1991 (in press).

Douguédroit, A. and de Saintignon, M.F. (1970): Méthode d'étude de la décroissance des températures en montagne de latitudes moyennes: example des Alpes du Sud. *Rev. Geo. Alpine*, 453–472.

Flohn, H. and Fantechi, R. (1984): *The climate of Europe: Past, present and future*. Reidel, Dordrecht, 356 pp.

Folland, C.K., Parker, D.E. and Newman, M.R. (1985): Worldwide marine temperature variations on the season to century time-scale. In: *NOAA Proc. 9th Annual Climate Diagnostics Workshop*, Corvallis, Oregon, USA, Washington, D.C. 22–26 October 1984, 70–85.

Gabriel, K.R. and Neuman, J. (1962): On the frequency distribution of dry and wet spells at Tel Aviv. *Quart. J. R. Met. Soc.*, 88, 90–95.

Galloy, E., Lebreton, A. and Martin, S. (1983): A model for weather cycles based on daily rainfall occurrence. *Lect. Not. Bimatic*, 49, 303–318.

Gates, W.L., Cubash, U., Meehl, G.A., Mitchel, J.F.B. and Stouffer, R.J. (1993): An intercomparison of selected features of the control climates simulated by coupled ocean-atmosphere General Circulation Models. WCRP – 82, WMO/TD – No. 574, 46 pp.

Gibbs, W.J. (1987): Defining climate. *WMO Bulletin*, 36, 316–322.

Glantz, M.H., Katz, R.W. and Nichols, C. (1991): *Teleconnections linking worldwide climate anomalies, scientific basis and societal impacts*. Cambridge University Press, Cambridge, 535 pp.

Grazzuliz, T.P. (1991): *Significant tornadoes, 1880–1989. Vol. 1: Discussion and analysis*. Environmental Films, St Johnsbury, VT, 526 pp.

Gregory, S. (1989): Macro-regional definition and characteristics of Indian summer monsoon rainfall, 1871–1985. *Intern. J. of Climatol.*, 9, 465–483.

Halpert, M.S. and Ropolewski, C.F. (1992): Surface temperature patterns associated with the Southern Oscillation. *J. of Climatol.*, 5, 577–593.

Hastenrath, S. (1990): Tropical climate prediction: A progress report, 1985–90. *Bull. of the Am. Met. Soc.*, 71(6), 819–825.

Hastenrath, S. (1991): *Climate dynamics of the tropics*. Kluwer Academic Publishers, Dordrecht, 488 pp.

Hastenrath, S. and de Castro, L.C. (1987): The Southern Oscillation in the tropical Atlantic Ocean. *Quart. J. Roy. Meteor. Soc.*, 103, 77–92.

Henderson, K.G. and Robinson, P.J. (1994): Relationship between the Pacific/North American teleconnection patterns and precipitation events in the Southeastern USA. *Intern. J. of Climatol.*, 14, 307–324.

Hewitson, B. (1994): Regional climates in the GISS general circulation model = surface air temperature. *J. Climate*, 7, 283–303.

Hewitson, B. (1995): The development of climate downscaling: Techniques and applications. In: *Proceedings of the 6th International Meeting on Statistical Climatology*. University College, Galway, 33–36.

Houghton, J.T. (ed.) (1984): *The global climate*. Cambridge University Press, Cambridge.

Janicot, S. (1992): Spatio-temporal variability of West African rainfall. Part I: Regionalizations and typings. Part II: Associated surface air mass characteristics. *J. Climate*, 5, 489–497 and 499–511.

Kutzbach, J.E. (1974): Fluctuations of climatic monitoring and modelling. *WMO Bulletin* 23, 47–54.

Labeyrie, J. (1985): *L'homme et le climat*. Denoël, Paris, 281 pp.

Lamb, H.H. (1971): *Climate: Present, past and future*, Vol. 1. Methen, London, 835 pp.

Lamb, H.H. (1977): *Climate: Present, past and future*, Vol. 2. Methuen, London, 835 pp.

Lorenz, E.N. (1991): Dimension of weather and climate attractors. *Nature*, 353, 241–244.

Lough, J.M. (1986): Tropical Atlantic sea surfaces temperatures and rainfall variations in sub-Saheran Africa and northeast Brazil. *Mon. Wea. Rev.*, 114, 560–570.

Manley, G. (1953): The mean temperature of Central England, 1698–1952. *Quart. J. Roy. Meteor.*, 79, 242–261.

McCauley, J. (1990): Intruduction to multifractals in dynamical systems theory and fully developed fluid turbulence. *Physics Reports. A Review of Section of Physics Letters*, 189, 5.

Meteo-France, (1987): La Réunion. Dépression tropicale Clotilda du 9 au 16 février, 1987, 33 pp.

Mitchell, Jr., J.M. (1990): Climate variability: Past, present and future. *Climate Change*, 16(2), 231–246.

Monin, A.S. (1980): *An introduction to the theory of climate*. Reidel, Den Haag, 261 pp.

Namias, J.(1990): Basis for prediction of the sharp Reversal of Climate from Autumn to Winter 1988–1989. *Intern. J. of Climatol.*, 10, 659–678.

Nicholson, S.E. (1985): Subsaharian rainfall 1981–1984. *J. of Climate and Appl. Meteo.*, 24, 1388–1391.

Nicholson, S.F. and Entekhabi, D. (1968): The quasi-periodic behavior of rainfall variability in Africa and its relation to the Southern Oscillation. *Arch. Met. Geop. and Bio. Ser.*, 34A, 311–348.

Nicolis, C. and Nicolis, G. (1984): Is there a climatic attractor? *Nature*, 311, 829–832.

Nicolis, G., Nicolis, C. and McKernan, D. (1993): Long-term climatic transitions and stochastic resonances. *J. of Stat. Physics*, 70(1/2).

Palmer, T.N. and Anderson, D.L.T. (1994): The prospects for seasonal forecasting – a review paper. *Q. J. R. Meteorol. Soc.*, 120, 755–793.

Peguy, C.P. (1976): Une nouvelle expression graphique de la variabilité interannuelle des climats: les calendriers de probabilités. *Bull Assoc. de Géogr. Franç.*, 431–432, 5–6.

Peguy, C.P. (1989): *Jeux et enjeux du climat*. Masson, Paris, 254 pp.

Pettitt, A.N. (1979): A non-parametric approach to the change-point problem. *Appl. Statis.*, 28, 125–135.

Philander, S.G. (1990): *El Niño, La Niña, and the Southern Oscillation*. Academic Press, New York, 293 pp.

Prosper-Laget, V., Douguédroit, A. and Guinot, J.P. (1995): Mapping the risk of forest fire occurrence using NOAA satellite information. *EARTSeL*, 4(3), 30–38.

Ramage, C.S. (1979): *Monsoon meteorology*. Academic Press, New York, 296 pp.

Rasmusson, E.M. and Carpenter, T.H. (1982): Variations in sea surface temperature and surface wind fields associated with Southern Oscillation (El Niño). *Mon. Wea. Rev.*, 110, 354–384.

Rasmusson, E.M. and Carpenter, Th. (1985): The relationship between eastern equatorial Pacific sea surface temperatures and rainfall over India and Sri Lanka. *Mon. Wea. Rev.*, 111, 517–528.

Ropolewski, C.F. and Halpert, M.S. (1987): Global and regional scale precipitation and temperature patterns associated with El Niño/Southern Oscillation. *Mon. Wea. Rev.*, 115, 1606–1626.

Ropolewski, C.F., Halpert, M.S. and Wang, X. (1992): Observed tropospheric biennal variability and its relationship to the Southern Oscillation. *J. Climate*, 5, 594–614.

Schlesinger, M.E. (1988): *Physically-based modelling and simulation of climate and climate change*. Kluwer Academic Publishers, Dordrecht, 2 vols., 1084 pp.

Schlesinger, M.E. (1990): *Climate–ocean interaction*. Kluwer Academic Publishers, Dordrecht, 385 pp.

Sear, C.B., Kelly, P.M., Jones, P.D. and Goodess, C.M. (1987): Global surface temperature responses to major volcanic eruptions. *Nature*, 330, 365–367.

Shukla, J. and Wallace, J.M. (1983): Numerical simulation of the atmospheric response to equatorial Pacific sea surface temperature anomalies. *J. Atmos. Sci.*, 40, 1613–1630.

Sneyers, R. (1975): Sur l'analyse statistique des séries d'observations. OMM. Note technique n° 143, 192 pp.

Steiner, D. (1965): A multivariate statistical approach to climatic regionalization and classification. *Tijdschrift van Het Koninkklijk Nederlandsch Aardrijkskundig Genootschap*, LXXXII, 4, 329–347.

Tyson, P.D. (1986): *Climatic change and variability in Southern Africa*. Oxford University Press, Cape Town, 220 pp.

Uchijima, Z. (1981): Yield variability of crops in Japan. *GeoJournal*, 5(2), 151–163.

Van Loon, H. and Rogers, J.C. (1978): The seasaw in winter temperatures between Greenland and Northern Europe (Part I. General description). *Mon. Wea. Rev.*, 106, 296–310.

Von Storch, H. (1994): Interdecadal variability in a global coupled model. *Tellus*, 46A, 419–432.

Von Storch, H., Zorita, E. and Cubasch, U. (1993): Downscaling of global climate change estimates to regional scale, application to Iberian rainfall in wintertime. *J. Climate*, 6, 1161–1171.

White, D. and Yarnal, B. (1991): Climate regionalization and rotation of principal components. *Inter. J. of Climatol.*, 11, 1–25.

Wigley, T.M.L., Ingam, M.J. and Farmer, G. (eds.) (1991): *Climate and history*. Cambridge University Press, Cambridge.

WMO (1990): The role of the World Meteorological Organization in the International Decade for Natural Desaster Reduction, WMO No. 745, 32 pp.

Annick Douguédroit
Institute of Geography and CNRS/CAGEP-URA 903
University of Provence (Aix-Marseille 1)
29, av. Robert Schuman
13621 Aix-en-Provence Cédex
France

2.2. CLIMATE CHANGE IN POST GLACIAL PERIOD IN MONSOON ASIA

JIACHENG ZHANG, YASUNORI YASUDA AND
MASATOSHI YOSHINO

1. Introduction

The general trend of climate change in East Asia is discussed in this chapter in relation to world trends. To understand climatic change in this region, Asian Monsoons are the most important. The air coming from the polar regions predominates over East Asia as winter monsoon and, on the other hand, the air transporting moisture and heat from the tropical regions predominates in Monsoon Asia as summer monsoon. The main rainfall or cloud belt is located at the interacting zone of these two monsoons, while the year-to-year fluctuation or changing pattern of these two monsoons control climate change in the region.

This paper starts with a world survey of the general features of post glacial climate change. This is followed by a discussion of the results of climatic reconstruction for Monsoon Asia (Southeast and East Asia). It ends with a similar discussion for China and Japan. Originally, the part for China was written by Jiacheng Zhang and for Japan by Yasunori Yasuda. Masatoshi Yoshino compiled them in a final form, adding some recent literature.

2. General tendencies

In the post glacial period, the climate became gradually warmer with some fluctuation. Generally, three epochs are observed (Heusser 1966; Jelgersma 1966; Lamb 1988; Wiseman 1966; Yoshino 1976): 11,000 to 8000 yrs. B.P. The first epoch which began just after the ending of glacial age was marked

M. Yoshino et al. (eds.), Climates and Societies – A Climatological Perspective, 43–60.

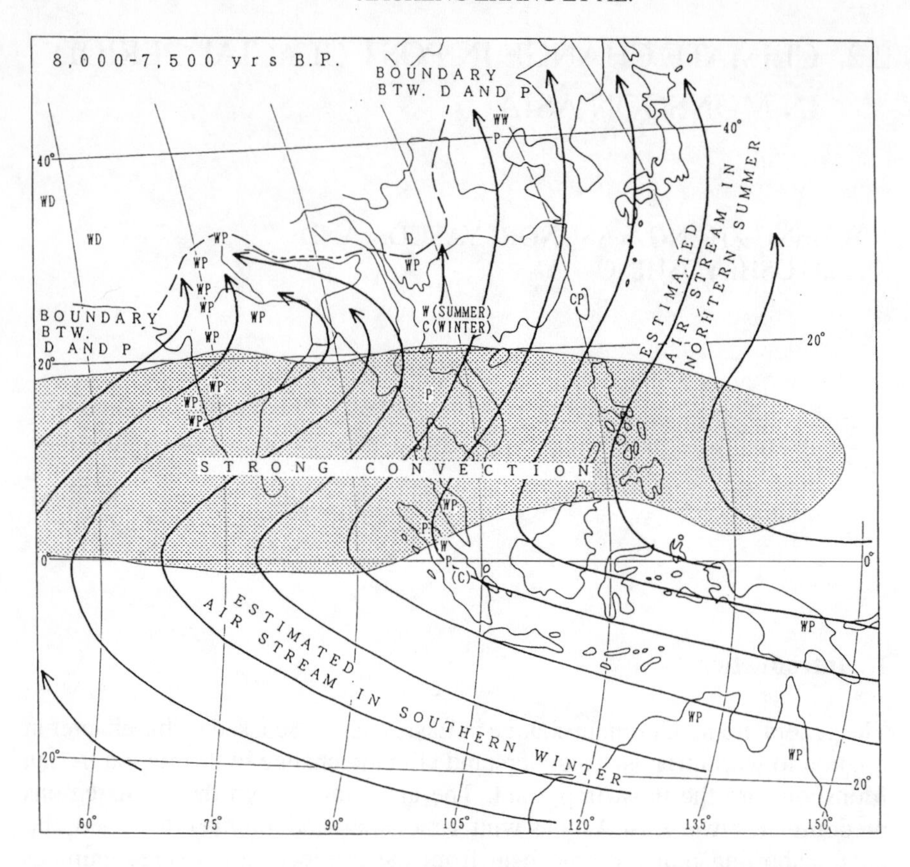

Figure 1. Paleoclimatic conditions in the period 8000–7500 yrs. B.P. and estimates of air streams in monsoon and strong convection areas in northern summer (Yoshino et al. 1993). WW = very warm or hot, W = warm, C = cold or cool, P = humid, and D = dry.

by a transition from the cold and dry climate of the Glacial Period to a warm and moist climate.

8000 to 3000 yrs. B.P. was the second epoch; warm and moist climate prevailed in most parts of the world including East Asia. This epoch is called "Hypsithermal" or "Climatic Optimum", because conditions were generally favourable.

By plotting the results of analysis, mainly palynological, distribution of wet/dry and warm/cold conditions was mapped as shown in Figure 1 (Shi 1992; Yoshino et al. 1993). Estimated monsoonal circulation pattern is included in the map. It is based on recent knowledge and can be explained rationally. From Figure 1, the following facts are noteworthy: (a) the boundary between the dry (D) area and wet (W) area, roughly coincides with NITC in summer as observed at present. This suggests that even though temperature was higher and precipitation generally good in Monsoon Asia, the inland limit of the summer monsoon was almost at the same position as today; (b) only some locations such as Yunnan and Taiwan showed cold condition in winter. It also remained cold in the central part of Sumatra. Such pieces of evidence suggest that warming in the post glacial period initially began in summer, perhaps caused by the intensification of the monsoon in northern summer rather than a weakening of the wind in northern winter; and (c) it is very interesting to note that the areas with cold conditions in winter are located just outside that part of tropical Asia which experienced strong convection during the warming phase (Huang and Sun 1992). This implies that the Southwest summer monsoon with strong convection played an important role in stipulating the warm conditions which prevailed between 8000–7500 yrs. B.P.

In the third epoch, 3000 yrs. B.P. to present, the climate became cold and dry, but was still much warmer and wetter than the first epoch. The coldest phase during this epoch is called "Little Ice Age" from the 18th century to the 19th as described later. From the middle 20th century the climate began to change under the stronger anthropogenic influences discussed in the other chapters of this book.

The relationship between air temperature and precipitation can be positive as seen for the post glacial period. Sarnthein (1978) has presented the global dryness/wetness distributions for the present, the last glacial maximum of 18,000 yrs. B.P., and the hypsithermal of 6000 yrs. B.P. It was indicated that drier conditions prevailed during the glacial maximum in almost all parts of the middle latitudes. In contrast, wetter conditions prevailed in the middle and lower latitude regions during the climatic optimum.

The reason for this may be attributable to moist instability (Zhang 1989). For example, frontal activities are much more active in winter than in summer, but winter precipitation is much less than that in summer. This prominent difference is mainly caused by the large difference in the atmospheric water content in the two seasons due to the close dependence of moisture holding capacity on air temperature. Moist instability plays a very important role in rain formation in summer, but its role is insignificant in winter.

3. China

After the last glaciation, the climate in China became gradually warmer in accordance with general trends in the world (Ye et al. 1994). A study on vegetation and climate change in the southern Liaoning region in China (Labor. Quart. Palynol. and Labor. Radiocarbon, Kweiyang Inst. 1978), estimated that the annual mean air temperature was 3°C warmer than present between 8000 and 3000 yrs. B.P., the climate was more humid than present from 8000 to 5000 yrs. B.P., but semi-humid to semi-arid from 5000 to 2500 B.P. (Figure 2). Analysis of palynological records in the Yangtze River Delta indicated that the climate in the period 10,300–7500 yrs. B.P. was 1–2°C colder than

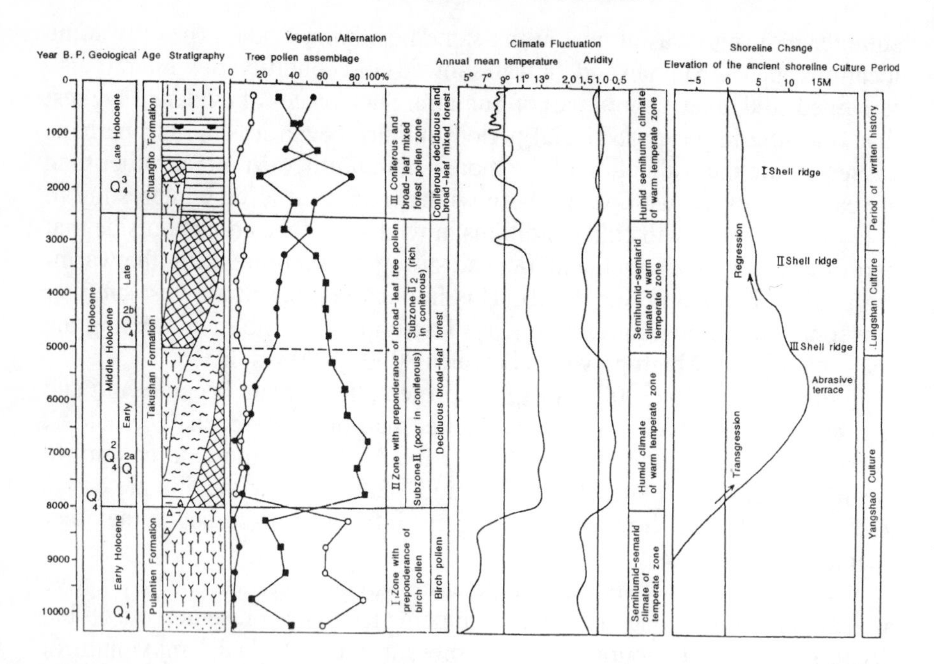

Figure 2. Holocene vegetational and climatic changes in the southern Liaoning region in China (Lab. Quarternary Pal. and Lab. Radiocarbon, Kweiyang Inst. Geochemistry, Acad. Sinica 1978).

present mean annual temperature. The climate was also drier (Shi et al. 1990). It was also indicated that the region around Qinghai Lake on the northeastern part of Tibetan Plateau had shrubs or forest-shrubs vegetation, reflecting transitional conditions from a cold-dry climate to a warm-moist interglacial one from 11,000 to 8000 yrs. B.P. Duan et al. (1981) estimated that before 8000 yrs. B.P., the annual mean air temperature in Shanxi, Liaoning and Inner Mongolia was about 5–6°C.

In Henan Province, the climate was dry in 9000–8000 yrs. B.P. (Wang et al. 1991) and after 8500 yrs. B.P. a change to a moist climate began. Wang and Jian (1989) reported that the air temperature in the Beijing area was about 4–5°C lower than in about 12,000 yrs. B.P., but a cooler-drier trend was established in 10,000–8000 yrs. B.P. In conclusion, just after the last glaciation, 10,000–8000 yrs. B.P., temperature was lower and precipitation smaller than present. The magnitudes differ from place to place, from 5–6°C in some places, and at least 1–2°C lower than present.

Many archaeological finds indicate a flourishing ancient civilization in China in the period 8000–3000 yrs. B.P.

A famous work on the reconstruction of past climate by Zhu (1973) revealed that the snow line change in Norway coincides roughly with temperature change in China for the last 5000 yrs. (Figure 3). Small fluctuations

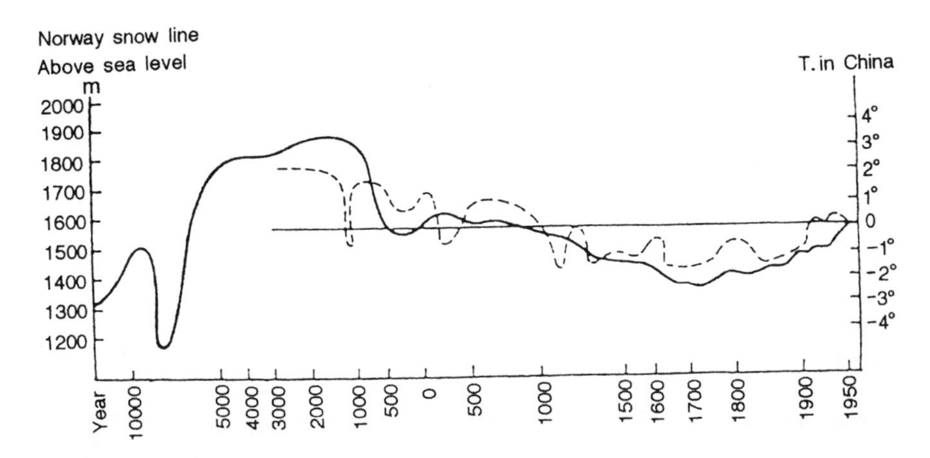

Figure 3. Snow line change in Norway from 12,000 yrs. B.P. (solid line) and temperature fluctuation in China since 5000 yrs. B.P. (broken line) (Zhu 1973).

are shown on the line for China. These results are summarized in Table 1, which shows evidence of a 3–5°C warmer period from roughly 8000–5000 yrs. B.P., and 2–3°C from 5000–3000 yrs. B.P. Drier conditions dominated before 8000 yrs. B.P., but moist climates prevailed thereafter. Cool and dry conditions started about 4000 yrs. B.P., but as a whole, the period was still relatively warmer and moister. Estimated temperature curve, as shown in Figure 3, reveals that these warm trends continued to A.D. 1000. There were striking, but short, cold periods around 3000 yrs. B.P. (1000 B.C.) and A.D. 300. The cold period of A.D. 300 continued to about A.D. 500 as shown in Figure 4 with a severity equivalent to that of the "Little Ice Age". It is noteworthy that the warm period from 7th–10th centuries, which coincided with the peak of the Tang Dynasty, was the most outstanding climatic feature of the last 2000 years. It had strong impacts on the establishment of prosperous dynasties supported by stable agricultural production.

The trend of climatic change in western part of China is similar to that of eastern part, as studied recently (China Society of the Qinghai-Tibetan Plateau Research 1995).

4. Japan

The climate of Japan is characterized primarily by the contrast between the climates on the Pacific side and the Japan Sea side as well as the difference between the lower and higher latitudes. As shown in Figure 5, the contrast in winter is caused by the northwesterly monsoon winds, which bring heavy snow accumulation on the Japan Sea side of Honshu. In the area with precipitation more than 150 mm at least one month under the present climate

Table 1. Climate change in China during the period 12,000 to 3000 yrs. B.P.

Years B.P.	Temp.* (in °C)	Precip.	Region	Literature
12,000	−4 to −5	–	Beijing	Wang, Jian (1989)
10,500	−8 to −11	–	Inner Mongolia	Shi et al. (1990)
10,300	−1 to −2	–	Yangtze Ri. Delta	Wang, Keifa (1979)
9000	transitional	–	Qinghai Lake	Shi et al. (1990)
9000	–	drier	Henan Prov.	Wang, Chun et al. (1991)
8000	+5 to +6	–	Shanxi, Liaoning, Inner Mongolia	Duan et al. (1981)
8000	+3 to +4	–	Hobei, N-China Plain	Zhang, Zibin et al. (1981)
8000	+3 to +4	humid	Southern Liaoning	Lab. Quart. Pal. et al. (1978)
7500	+2 to +4	moist	NW-China	Kung, Zhoazheng et al. (1990) Shi et al. (1990)
5600	+2 to +3	moist	Xian	Zhu (1973)
5040	+4 to +5	–	Zhenghou	Wang, Chung et al. (1987)
5000	+2 to +3	semi-humid to semi-arid	Southern Liaoning	Lab. Quart. Pal. et al. (1978)
4800	warmer	dry cond. started with frequent flood	Dahe Village in Zhengzhou	Zhang, Zibin et al. (1984)
4000	warmer, but started	–	Sha-Capital	Wang, Chun et al. (1987) Rei and Li (1981)
3400				
3100	+2 to +3	–	Henan	Zhu (1973)
2500	warmer	moist	Qinghai Lake	Shi (1992)
500	+1 to +2	humid to semi-humid	Southern Liaoning	Lab. Quart. Pal. et al. (1978)

*+ denotes warmer than present, − cooler than present.

conditions coincides with the zone of mixed forests comprising *Fagus crenata* and *Cryptomeria japonica*, representing one of the typical vegetation types. Because of this representation, palynological analysis can be done using this vegetation as an indicator of winter monsoon activities in past climates.

On the other hand, the summer (warmer-half year) precipitation is striking on the Pacific side in south to central Japan. This condition results in the contrast of vegetation distribution between the Pacific side and the Japan Sea side. One of the typical patterns is shown in the distribution of *Tuga diversitoria* and *T. Sieboedii*, of which the former occupies relatively lower elevations (400–1600 m a.s.l.) of the mountains and the latter the higher parts (800–2300 m) and the northern areas of Honshu (see the right hand part of Figure 5). In other words, the *Tsuga* distribution area shown on the right side

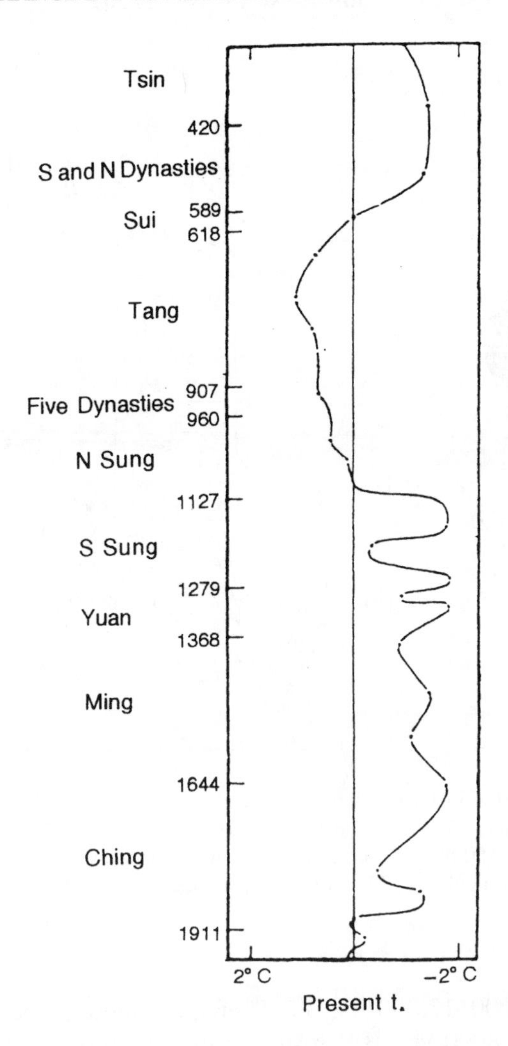

Figure 4. Temperature fluctuations for the last 1700 years in China based on the phenological studies (Zhu 1973).

of Figure 6 coincides quite well with the light snow accumulation area shown on the left side of Figure 5. Therefore, using these conditions as climate indicators, results from the palynological studies can be interpreted and used to reconstruct the past climates of Japan.

The climate of Japan during the maximum glaciation, especially from 21,000 to 15,000 yrs. B.P., was very cold and dry. It has been indicated that the mean annual temperature in Hokkaido, North Japan, was 9°C lower than present, 7–8°C lower in central Japan, and 6–7°C lower in Kyushu. This cold and dry climate turned warmer after 15,000 yrs. B.P., but precipitation did not

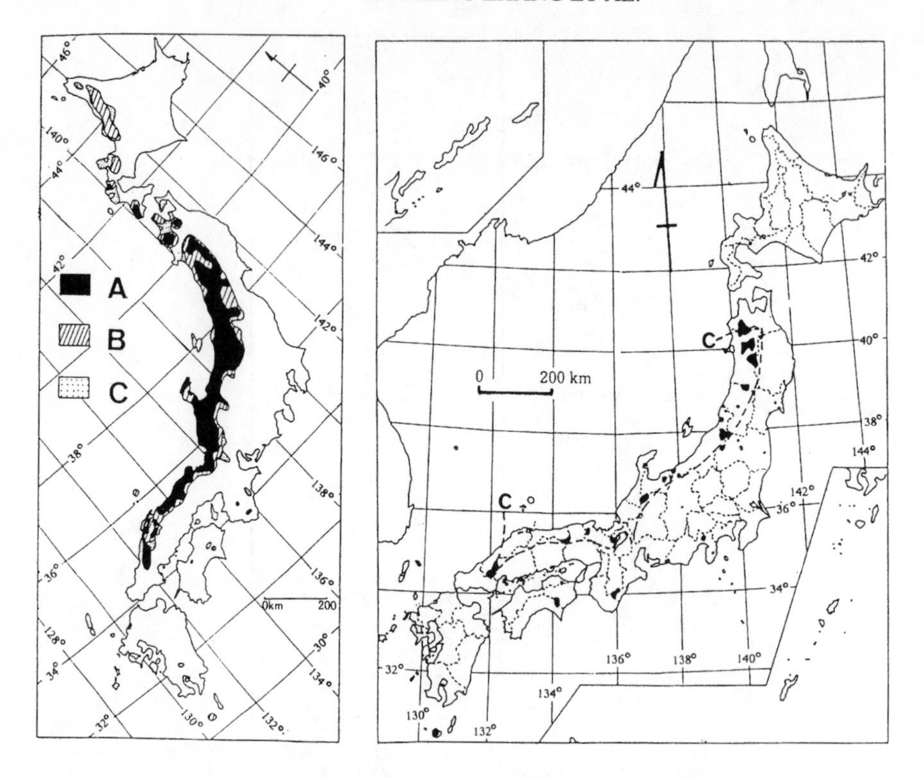

Figure 5. Precipitation in winter (left), and the distribution of the mixed forest of *Fagus crenata* and *Cryptomeria japonica* and the sites of pollen analyses discussed in this paper (right). A: More than 150 mm in each of the three winter months; B: More than 150 mm in each of the two winter months; C: More than 150 mm in the one winter month (Yasuda, orig.).

increase until 13,000–12,000 yrs. B.P. Thereafter, snow on the Japan Sea side increased around 8000 yrs. B.P. resulting in a maritime climatic condition and suitable for *F. crenata* growth. On the other hand, drier climate continued on the Pacific side over a longer period.

The native forest of *Cryptomeria japonica* is adapted to a wet climate. At present, it is found on the Japan Sea Coast, up to about 41°N (Figure 5). This fact indicates that the summer precipitation also increased after 10,000 yrs. B.P., especially the frequency of heavy rain in the Bai-u season, an early summer rainy season (Yasuda 1990b). For the period between 10,000–8000 yrs. based on the present state of knowledge, it is believed that a somewhat treeless vegetation developed in central and southwest Japan between 10,000 and 8000 B.P., while the lowlands of north Japan were covered by a shrubby forest of *Betula*. But we cannot envisage the abundant growth of the temperate broad-leaved forests in either of these two regions. Probably, after 10,000 yrs. B.P., the climate lacked stability in both space and time scales. Climates in

Photo 1. Beech forest on the Japan Sea side (Mt. Daisen, Tottori Prefecture).

Japan were transitional during 10,000–8000 yrs. B.P., corresponding to those in China given in Table 1.

Since 8000 yrs. B.P., the climate became warm as confirmed by the spread of deciduous broad-leaved forests composed mainly of *Quercus* in Hokkaido,

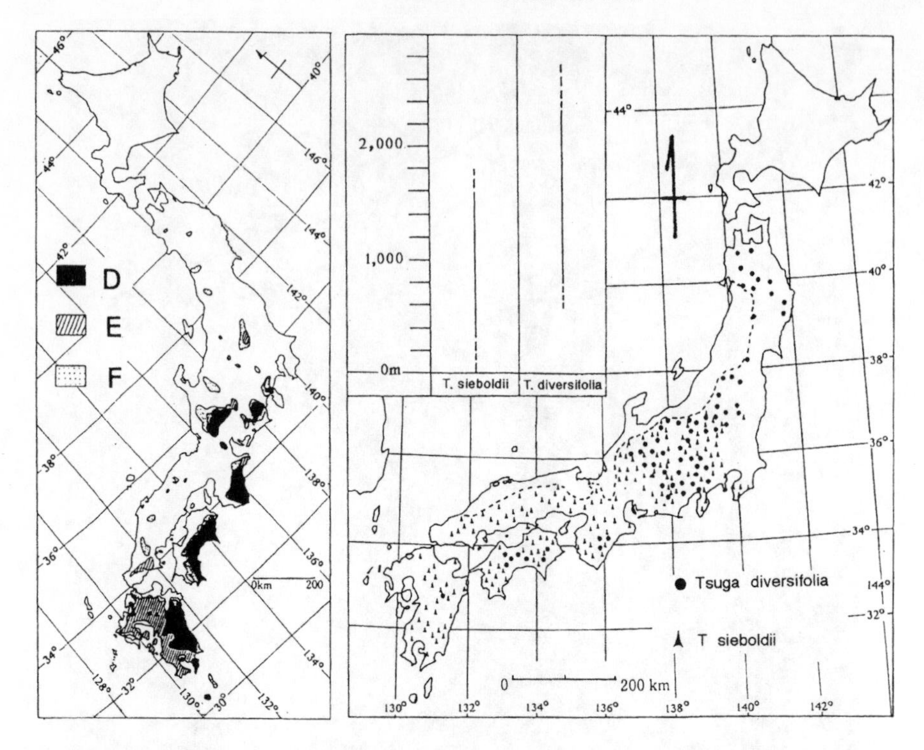

Figure 6. Precipitation in summer (left) and the distribution of Tsuga in Japan (right) (Yasuda 1994, orig.) D: More than 400 mm, E: More than 300 mm, and F: More than 200 mm per summer month.

by the expansion of *F. crenata* to the north of 40°N (Yasuda 1992), and by the establishment of an archaeological life style of the Jomon age (Nishida 1989).

Mean annual temperature during 6500–5000 yrs. B.P. was 2–3°C warmer in North Japan and 1–1.5°C in Southwest Japan than equivalent temperatures in the North, and East Japan was warm and dry. In particular, the Japan Sea side was dry, because of decreasing snow fall.

On the other hand, the climate of Southwest Japan was warm and moist. *Fagus crenata* forest arrived in Hokkaido at 6000 yrs. B.P. and evergreen broad-leaved forests covered the lowlands of Southwest Japan. The sea level during this period was estimated as 0–5 m (Umitsu 1991).

Evidence for a short cold period around 7000–6500 yrs. B.P., found in the fluctuations of lake levels in Africa (Owen et al. 1982; Gillespie et al. 1983; Hassan 1986) is rare in the pollen diagrams in Japan, which reveal only a temporary cold epoch.

Paleotemperature curve since 5900 B.C. shown in Figure 7, was obtained by analyzing pollen diagram from the Ozegahara moor in Gumma Prefecture,

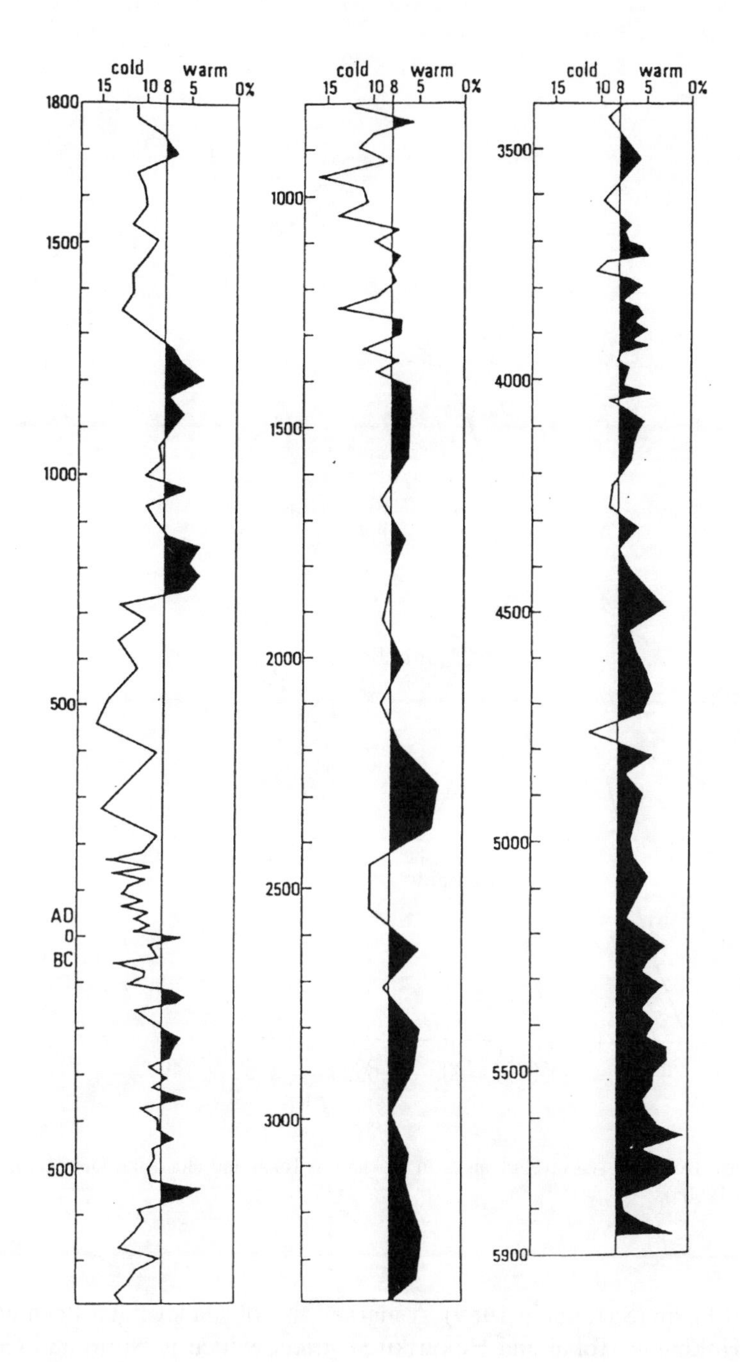

Figure 7. Paleotemperature curve since 5900 yrs. B.C. from the Ozegahara moor in Central Japan (Sakaguchi 1989).

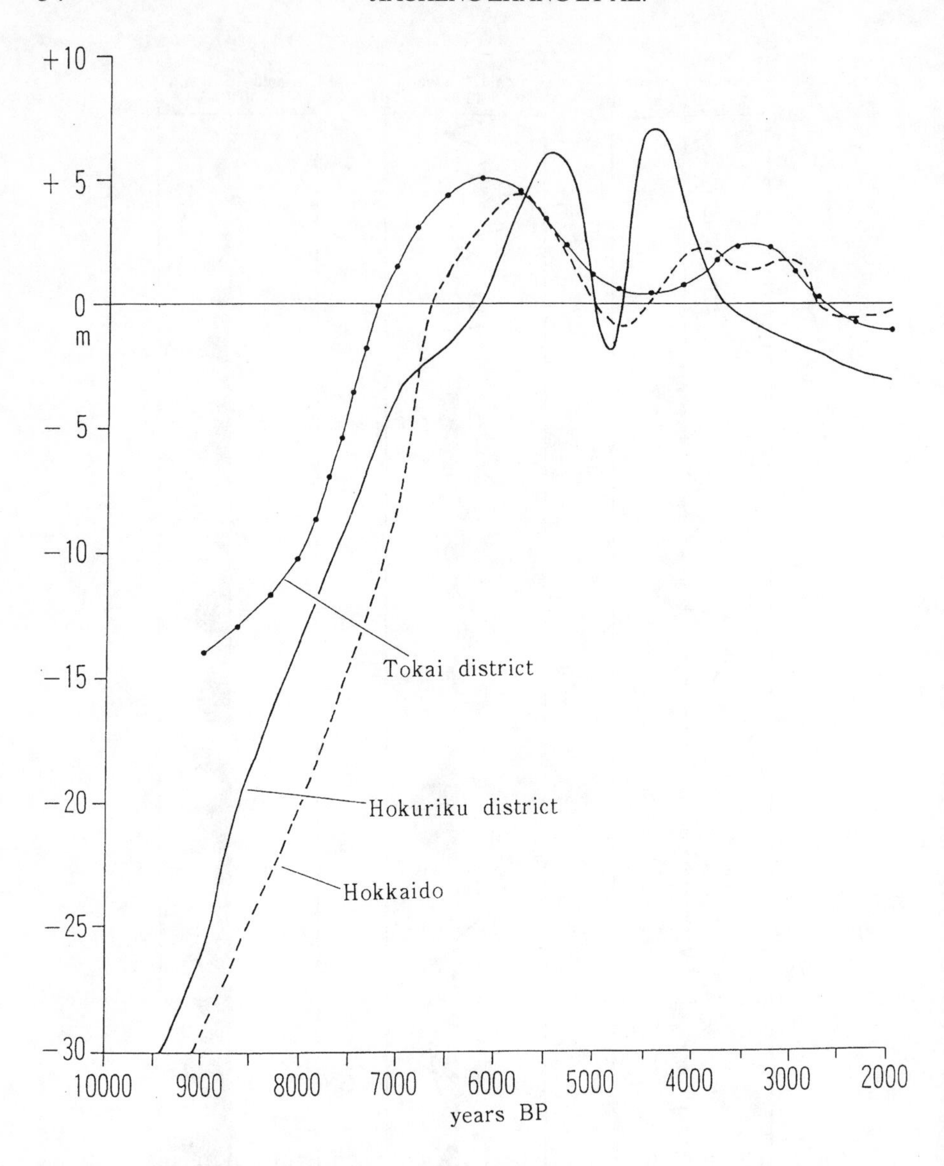

Figure 8. Holocene sea-level changes in Hokkaido, Tokai and Hokuriku Districts in Japan (Yasuda 1990a).

Central Japan (Sakaguchi 1989). A marked rise of sea level has been detected in Hokkaido, Tokai and Hokuriku Districts, which is North and Central Japan after the post glacial period about 6000 yrs. B.P. (Yasuda 1990a), as shown in Figure 8. It is noteworthy that, even though there were some local differences, a minimum occurred about 4500–4000 yrs. B.P. followed by a

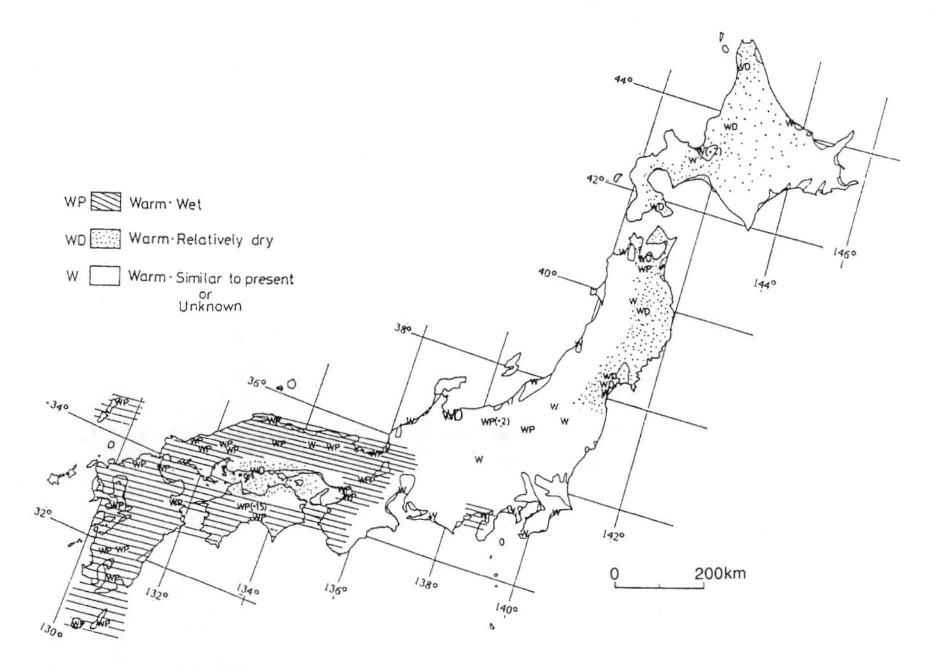

Figure 9. Paleogeographical and paleoclimatological map of Japan at 6500–5000 yrs. B.P. (Yasuda 1983).

maximum during 4000–3000 yrs. B.P. After this maximum, sea level had fallen, corresponding to the arrival of a cool climate.

One of the interesting spatial patterns of paleogeographical and paleoclimatological conditions in Japan during 6500–5000 yrs. B.P. is shown in Figure 9 (Yasuda 1983). All over Japan, warm climate prevailed, but the southwestern part was wet. In contrast, the northeastern part was dry. The wet condition was caused by the summer monsoon related to the stream lines shown in Figure 1, but the dry condition was caused by the winter monsoon activity mentioned earlier in this chapter, and elsewhere (Yoshino et al. 1993).

5. Climate change and human activities in the historical period

During the historical period (Zhang, Jiacheng et al. 1976; Zhang 1989a; Yoshino 1978), the warm period between the 8th and 10th centuries and the cold period between the 17th and 19th were striking over Monsoon Asia, corresponding to hemispheric fluctuations. There is considerable evidence linking population changes in China to changes in climate. Table 2 gives population (%) in the various regions of China and the corresponding climatic condition (Yoshino 1994). During the period of Little Climatic Optimum (8–10th century), the Tang Dynasty, one of the most distinguished periods

Table 2. Population (%) in various regions of China and the climatic conditions.

Year (A.D.)	2	140	742	1102	1491	1820
North China Plain	69.7	57.9	43.2	25.6	25.7	23.9%
Northern China[a]	13.1	5.6	17.8	16.8	15.8	11.2%
Southern China	17.2	36.5	38.9	57.6	58.5	64.9%
Total population	59.60 mill.	49.15	48.91	43.82	50.50	353.38
Climate[b] Temperature	Warm	Cold	Very warm	Cold to very cold	Very cold	Coldest
Precipitation	Wet	Dry	Wet	Dry	Very dry	Very dry

Notes: Data of population (Fang 1990).
[a] The area including the present China's territories north of the Huihe River, except the North China Plain.
[b] The climate was estimated from various sources (Zhu 1973; Zhang et al. 1976; Yoshino et al. 1993) by the present writers.

in history, flourished in China. Bohai Country, an active country, was established in the area occupied by today's Northeast China and its surrounding during this period. Also, Heijokyo, the new capital in Central Japan, and the "Hakuho" and "Tempyo" culture, flourished in Japan. It is believed that all these countries were supported by stable and favourable agricultural economy under the influence of a warm and humid climate. In Table 2, it is clear that population increased in northern China in A.D. 742.

In parallel with such observable developments in countries of East Asia, Viking enterprise was conspicuous from the 8th to the 9th century. They reached Shetland Island in A.D. 700, Faeroes in 800, and Greenland and Central Europe by the end of the 9th century. Further, they reached the Mediterranean Sea and the Black Sea coast. Trade by the Vikings was supported not only by their high ship building and sailing technology, but also by good conditions for forest, farm and sea products in the supply regions and by favourable economic conditions in the market regions. The period of "Little Climatic Optimum" had a climate good enough to support these conditions in North, Northwest and Central Europe, and in North Atlantic. It should be stressed that these conditions occurred concurrently on both side of the Eurasian Continent.

Roughly speaking, the "Little Ice Age" in the 17–19th centuries occurred also hemispherically. In China, the cold climates during those centuries resulted in a series of great famine which were more serious in North China than in South China. The number of death frequently exceeded millions. So the migration to the South was intensified. As a result, the population (%) was higher in South China than in North China (Table 2).

Important results obtained from paleoclimate reconstruction are as follows: (a) most of the migrations related to the southwards nomadic invasions from Central Asia and the Mongolian grasslands coincided with prolonged droughts, frequent dust storms and winter thunderstorms, weak floods on the Yellow River, lake level drops, and increased desertification rates; (b) scarce precipitation periods must have had a great influences on food supply; (c) there are indications that oases in the desert, now found as abandoned settlements, suffered from gradual declines in water supply before their abandonments; and (d) according to some historical literature, the migrations of the Hans were caused, actually, by climatic disasters.

In densely populated North China Plain Yellow River plays a particular role in amplification of disasters owing to the large silt content in her flow that deteriorates the river bed and causes the flooding of the low plain (Zhang 1992).

Climatic impacts on human activities are more important in the marginal areas, because they are more sensitive to minor changes in conditions. Human societies, which cannot adapt to the new conditions, in other words, cannot establish new cultivation and production systems, historically decline.

In considering the time scales of century to several centuries, the relationship between climatic conditions and human enterprise is displayed in Figure 10. The space scales are shown on the left side of the figure. In this flow chart, the decline of civilization is connected to human conditions such as economic deterioration, political instability, nomad invasion, decrease of urban activities, disease epidemics and accelerated negative environmental change. These occur at the regional or local scale. Practically speaking, they might coincide roughly with a region or an area controlled by a country or a dynasty during the historical period. The decline of civilization could occur at a regional scale as large as the Mediterranean Area, the Middle East, and the Huang Ho (Yellow River) Tributary. Conversely, reverse effects driven by good conditions can occur under the influence of favourable human activities.

It can be concluded that the impact of climatic conditions was sensitively reflected in the distribution of population in ancient times.

6. Conclusions

In the post glacial period, the climate became warmer and wetter reaching a maximum rather quickly. "Hypsithermal" or "Climatic Optimum", the maximum, prevailed between 6500 and 5500 yrs. B.P. during the warm-moist period of 8000 to 3000 yrs. B.P. These features were generally seen in Monsoon Asia, but the inland limit of strong summer monsoons was almost at the same position as present. Some locations showed cold and dry conditions in winter even in these period. Therefore, it is believed that the post glacial warming began initially in the summer caused by the intensification of the monsoon in northern summer.

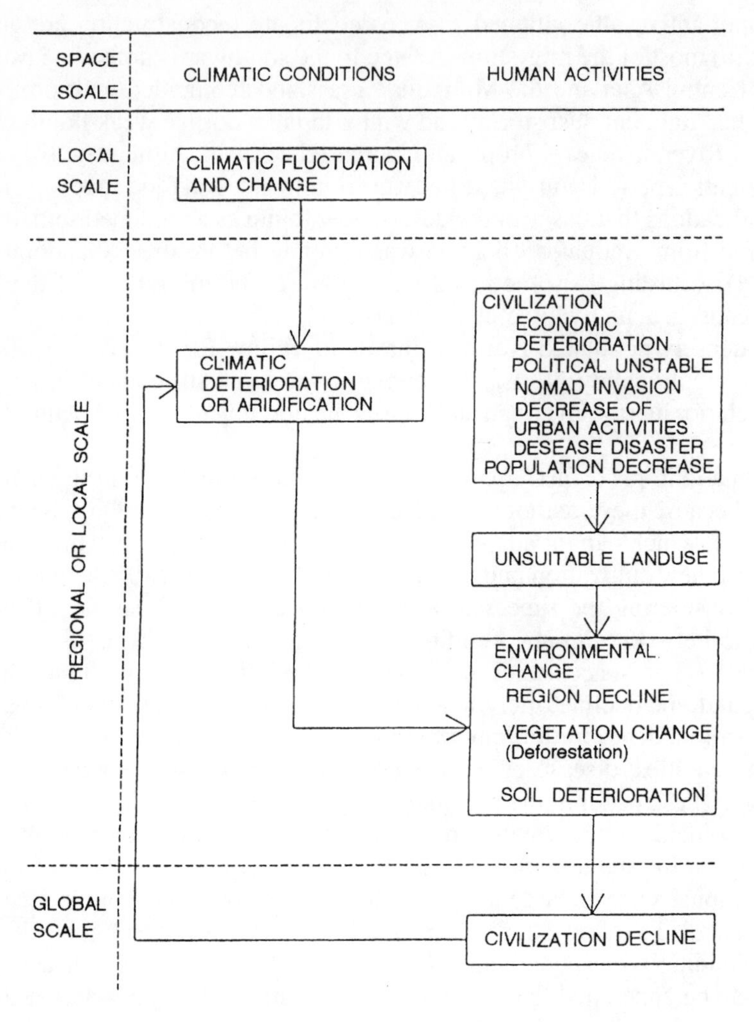

Figure 10. Relationship between climatic conditions and human activities considering the time scale of a century to several centuries during the historical period (Yoshino 1994).

7. Epilogue

The "Little Climatic Optimum" which appeared in the 8th to 10th centuries in China and Japan corresponds to the period of Viking enterprises in Europe and the Atlantic. The "Little Ice Age" in Monsoon Asia occurred concurrently with those in Europe. Societal decline marked by indicators such as economic deterioration, political instability, nomads immigration, decrease of city functions, and disease are believed to have a strong relationship to climatic change.

References

China Society of the Qinghai-Tibetan Plateau Research (1995): *Qinghai-Tibetan Plateau and global variations*. China Meteorological Press, Beijing.

Duan, Wanzhou et al. (1981): A preliminary study on climate fluctuation in Quarternary for China. In: *Proc. China Symp. Climate Change, 1978*. Science Press, Beijing.

Fang, Jinqi (1990): The impact of climatic change in the Chinese migrations in historical times. In *Global change and environmental evolution in China*, Proc. Sec. 3, Regional Conf. in Asian Pacific Countries of IGU, 96–103.

Gillespie, R., Street-Perrott, F.A. and Switsur, R. (1983): Postglacial arid episodes in Ethiopia have implication for climate prediction. *Nature*, 306, 680–683.

Hassan, F.A. (1986): Holocene lakes and prehistoric settlements of the western Faiyum, Egypt. *Jour. Archeological Sci.*, 13, 483–501.

Heusser, C.T. (1966): Polar hemispheric correlation: Palynological evidence from Chile and the Pacific Northwest of America. In *World Climate from 8,000 to 0 B.C.*, Proc. Intern. Symp. held at Imperial College, London, 124–141.

Huang, Ronghui and Sun, Fengying (1992): Impacts of the tropical western Pacific on the East Asia Summer Monsoon. *Jour. Met. Soc. Japan*, 70, 243–256.

Igarashi, Y. (1986): The distribution of Quereus in the Holocene period in Hokkaido. *Hoppo-Ringyo*, 38, 10–14.

Jelgersma, S. (1966): Sea level changes during the last 10,000 years. In: *World Climate from 8,000 to 1 B.C.*, Proc. Intern. Symp. held at Imperial College, London, 54–71.

Kung, Zhoazheng et al. (1990): A preliminary investigation on the vegetation and climate in the warm period of Holocene for some places of China. *Ocean Geology and Geology of Quarternary*, No. 1.

Laboratory of Quarternary Palynology and Laboratory of Radiocarbon, Kweiyang Institute of Geochemistry, Academia Sinica (1978): Development of natural environment in the southern part of Liaoning Province during the last 10,000 years. *Scientia Sinica*, 21, 516–532.

Lamb, H.H. (1988): *Weather, climate and human affairs*. Routledge, London and New York, 364 pp.

Nishida, M. (1989): *Palaeoecology of the Jomon culture*. The University of Tokyo Press, Tokyo, 104 pp.

Owen, R.B., Barthelme, J. W., Renaut, R.W. and Vincens, A. (1982): Palaeolimnology and archaelogy of Holocene deposits north-east of lake Turkana, Kenya. *Nature*, 298, 523–529.

Rei, Zhenqiu and Li, Zhisen (1981): The influence of planets movement on the climate change in the last 5,000 years for China. In: *Proc. of Symp. on Climate Change, 1978*. Science Press, Beijing.

Sakaguchi, Y. (1989): Some pollen records from Hokkaido and Sakhalin. *Bull. of Geogr., Univ. Tokyo*, 21, 1–17.

Sarunthein, M. (1978): Sand deserts during glacial maximum and climatic optimum. *Nature*, 127.

Shi, Yafeng et al. (1990): *Studies in climatic and sea level changes in China, I*. China Ocean Press, Beijing, 146 pp.

Shi, Yafeng (ed.) (1992): *The climates and environments of Holocene megathermal in China*. China Ocean Press, Beijing, 213 pp.

Umitsu, M. (1991): Holocene sea-level changes and coastal evolution in Japan. *The Quarternary Research, (Tokyo)*, 30, 187–196.

Wang, Chun et al. (1987): The annual precipitation change in central China for the last 500 years. *Chinese Science (B)*, (1), 104–112.

Wang, Chun et al. (1991): *The precipitation change in last 10,000 years and its trend in future*. The Main Hydrological Station of Henan.

Wang, Keifa (1989): *Proc. Symp. on Quarternary*.

Wang, Jian (1989): A qualitative study of climate variation since 16Ka, B.P. in North China. *Resources and Environment in Arid Region*, 3(3).

Wiseman, J.D.H. (1966): Evidence for recent climatic changes in cores from the ocean bed. In: *World Climate from 8,000 to 1 B.C.*, Proc. Intern. Symp. held at Imperial College, London, 84–98.

Yasuda, Y. (1983): Climatic changes since the last glacial age: Based on various methods. *Kisho-kenkyu Note*, 147, 47–60.

Yasuda, Y. (1990a): *Climatic changes and the rise and fall of civilizations*. Asakurashoten, Tokyo, 358 pp.

Yasuda, Y. (1990b): Monsoon fluctuations and cultural changes during the last glacial age in Japan, *Japan Review*, 1, 113–152.

Yasuda, Y. (1992): *The climate and Japanese culture*. Asakurashoten, Tokyo, 206 pp.

Ye, Duzheng and Chen, Panqin (1994): *Global change report*, No. 1. China Met. Press, Beijing, 1–62.

Yoshino, M. (1978): Regionality of climatic change in Monsoon Asia. In: *Climatic change and food production*, ed. by K. Takahashi and M. Yoshino. University of Tokyo Press, Tokyo, 331–342.

Yoshino, M. (1994): Climatic change and its impact on the population during the historical age. In: *Contemporary climatology*, ed. by R. Brazdil and M. Kolar. Brno, Czech Rep., 600–609.

Yoshino, M. and Urushibara-Yoshino, K. (1993): Monsoon change and paleoenvironment in Southeast Asia. In: *Proc. International Symp. on Global Change (IGBP), 27–29 March, 1992*. Tokyo, 700–705.

Zhang, Jiacheng et al. (1976): *Climatic fluctuation and its causes*. Science Press, Beijing, 212 pp.

Zhang, Jiacheng (1989a): *The reconstruction of climate in China for historical times*. Science Press, Beijing.

Zhang, Jiacheng (1989b): The greenhouse effect and the precipitation in North China. *Jour. Meteorology*, (3) (in Chinese).

Zhang, Jiacheng (1992): The vulnerability of socioeconomical development on climatic change in China. In *Nature and human kind in the age of environmental crisis*. International Research Center for Japanese Studies, 161–192.

Zhang, Zibin et al. (1981): The evolution of natural environment since 13 Ka B.P., in Beijing region. *Geological Science* (3).

Zhu, Kezhen (1973): A preliminary study on the climate fluctuations during the last 5,000 years in China. *Scientica Sinica*, 16, 226–256.

Jiacheng Zhang
Chinese Academy of Meteorological Science
State Meteorological Administration
46 Baishiqiao Road Beijing 100081
P.R. China

Yasunori Yasuda
International Institute for Japanese Studies
3-2 Goryo-oheyama, Saikyo-ku
Kyoto
610-11 Japan

Masatoshi Yoshino
Institute of Geography
Aichi University
Toyohashi-City
441 Japan

2.3. APPROACHES TO CLIMATIC VARIATIONS DURING THE HISTORICAL ERA: A FEW EXAMPLES

FRANÇOISE SERRE-BACHET

1. Introduction

"Climate, even under its natural development alone, varies continually. Each year, each decade, each century, each millennium, since long before any question of impact of human activity, has produced a somewhat different record" (Lamb 1984, p. 25). To this temporal variation we may add the spatial variation which complicates any attempt at synthesis in the course of a given period of time.

If the study of climatic variations is relatively easy for the period covered by instrumental data, which, nevertheless, vary greatly depending on the region (the first public networks of measurement stations were set up back in the 19th century), it is particularly difficult when inquiry is extended beyond it. When the study reaches into the very beginning of the historical era, as this chapter does, the difficulty is compounded. The documents, particularly for the first millennium, are still too few in number despite the attention given to them over the last few years.

Perhaps it is worthwhile specifying that rainfall, temperature and, to a lesser extent, atmospheric pressure, are practically the only climatic elements with sufficiently reliable data for variability studies. This is due to the frequency of observations or measurements made over the time period, or estimations based on indirect observations. In fact, over the last twenty years, increasingly sophisticated techniques have been used to reconstruct climatic data as homogeneously and accurately as possible from all types of sources, including proxies (Rotberg and Rabb 1981).

M. Yoshino et al. (eds.), Climates and Societies – A Climatological Perspective, 61–76.

2. Climatic information sources and climate reconstitution methods

2.1. *Climatic information sources*

Pfister (1988) distinguished two sources or record types from which climatic information can be reconstituted namely, anthropogenic and natural.

2.1.1. *Anthropogenic sources*

Anthropogenic sources are linked to human literary, scientific and agricultural activities. The more curious humans are about their environment and their dependence on it, the more observant of their natural surroundings they become. Their observations, put into writing, include fairly detailed descriptions of climatic elements or meteorological changes on different time scales. These descriptions were accompanied by temperature and pressure data in the 17th and 18th centuries following the invention and subsequent widespread use of the barometer and thermometer (Pichard 1988). To these direct observations we may add indirect ones which provide information based on physical or biological processes that are controlled to a large extent by meteorological parameters: the size and duration of snow cover, the freezing over of stretches of water, the movement of glacier fronts and their effects on habitats and cultures, flowering and ripening dates for cultivated plants: cereals, vines, and fruit trees. Administrative or private records (statements of damages, dates of grape harvests, auctions of tithe farms, cereal prices), parish registers, great famines and epidemics are also sometimes indirect indicators of meteorological conditions in the short, medium or long term.

2.1.2. *Natural sources*

The natural sources or records are made up of the set of elements from the vegetable, animal and mineral kingdom, which are subjected, in the same way as the human, to climatic factors that they have recorded in their own manner. Among these elements can be cited charcoal trapped in archaeological and pedological layers, pollen and spores, insects or mollusca of peaty or loamy sedimentological horizons, i.e. all the elements whose presence, extent or absence provide indications of the climate required for the biological processes and evidence represented. The annual growth rings in the wood of living trees or their remains, whether they have been reconstructed or not, can also be cited. Their thickness and texture provide information on the climatic conditions which prevailed during their formation. The ratios of stable carbon, oxygen or hydrogen isotopes present in ice or organic matter are again valuable indicators of climatic factors.

2.2. *Climatic reconstitution methods*

The analysis of the set of anthropogenic and natural sources depends on numerous varied scientific disciplines. The deciphering of the climatic significance of these sources use methods whose complexity depends on the nature of the sources themselves. Besides, the accuracy of these sources is highly variable as is the type of sequential or discontinuous data they are liable to provide. The principles of some of these methods are briefly described below.

2.2.1. *The historical method*

The historical method which is applied to sources of anthropogenic origin, is based on comparisons and continuous critical reviews of these sources. All climatic historians agree on the need to emphasize the complexity of the records at their disposal and the extreme care that must be taken in examining and using them. Since the work of Le Roy Ladurie (1967), a pioneer in the field and a starting block for later work, traditional historical critical examination has been applied by Alexandre (1987) in his study of narrative sources of the Middle Ages (A.D. 1000 to 1400) for the continental Europe and by Pfister (1984, 1988) in his study of the Swiss climate for the period 1525–1975. The latter has furthermore described a computerized method he developed for climate reconstruction based on a specific software program which uses "jusqu' aux plus petits éléments disponibles . . . et les condense en évaluations numériques des températures et des précipitations" (down to the smallest elements available . . . and condenses them into digital evaluations of temperatures and precipitations) (Pfister 1988, p. 314).

In other disciplines involved with changes in the climate, reference is sought to validate the results of the historical methods used to assess the validity of conclusions.

2.2.2. *The tree-ring method for climate reconstruction (dendroclimatology)*

This method involves the relationships between ring-width or wood-density and the climatic factors which contribute, directly or indirectly, to their formation. The existence of these relations allows the reconstruction of some climatic factors from long tree-ring series which predate the instrumental period. This climatic reconstruction of the unknown period uses climate-ring relationships that are calibrated then checked over the period for which meteorological records exist (Fritts 1976; Guiot 1989). It generally uses sophisticated statistical procedures such as transfer functions (Cook and Kairiukstis 1989). Climatic reconstruction involves differing annual periods: year, season, growing period, month. Its reliability depends on the correlations of observed and estimated data during the calibration period. Climatic reconstruction based on the calculation of transfer functions rarely goes beyond the last millennium. In spite of its annual specificity, which brings it close to the historical method, its interpretation is generally global. It is particularly applied to temperatures but there are also reconstructions of atmospheric pressure (Briffa et al. 1987) and rainfall (Till and Guiot 1990; Serre-Bachet 1988).

Other than calculations, the close relationship between the latewood maximum density of tree-rings and the average summer temperatures, demonstrated many times (Schweingruber et al. 1987), makes it possible to reconstitute such temperatures during the Holocene. These reconstitutions have particularly been compared to glacial oscillations and used in the interpretation of ^{14}C dates provided by the soils and fossilized wood of the period preceding the historical era: 3900–2600 BP (Rothlisberger et al. 1980).

Due to the necessity of working with annual and variable rings, dendroclimatology has so far mostly been applied to the extra-tropical regions.

2.2.3. *Methods relative to stable hydrogen, oxygen and carbon isotopes*

Generally speaking, the different isotopes of the systems in view are selectively distributed in thermodynamic transformations. This distribution is called isotope fractionating.

The D/H and $^{18}O/^{16}O$ ratios in precipitation in a region depend on geographical position and climate, in particular, average annual temperature. Monthly measurements carried out by the International Atomic Energy Agency show that "heavy isotopes are depleted in precipitation when water vapor is distilled from cold oceans in the winters and enriched in precipitation when water vapor is distilled from warm oceans in the summers" (Libby 1987, p. 82). "Rainwater retains the signature of the isotope content caused by the temperature of evaporation from the sea surface" (ibid., p. 84).

The isotopic ratios are used as climatic indicators in the analysis of ice caps in the polar regions and high mountains as in the analysis of land or marine deposits (Rotberg and Rabb 1981; Lowe and Walker 1984; Berger and Labeyrie 1987). They are also studied in organic matter of vegetable origin and, in particular in the two components, lignin and cellulose, of the wood in annual tree-ring which therefore provide information on the variation in these ratios with time (Libby 1987).

The main sources of ^{18}O in plants are ground water, the carbon dioxide and the oxygen in the air used in photosynthesis and respiration. The isotopic composition of organic oxygen implies significant isotopic fractionating effects in the plant and, more so, in its leaves, in conjunction with evapo-transpiration conditions (atmospheric water vapor, temperature and relative humidity of the air), which are linked to geography (Yapp and Epstein 1982; Förstel and Hützen 1983). The organic matter produced from water by photosynthetic processes therefore particularly gives an account of the availability or scarcity of water with respect to the plant and the vegetal cover in general. However, there is no certainty that this can be attributed to purely thermal or purely pluviometric effects, or both.

The $^{13}C/^{12}C$ ratios measured in the cellulose of wood are also considered to be indicators of drought (Freyer and Belacy 1981; Saurer and Siegenthaler 1989). The reduction of the available CO_2 during the closing of the stomata, brought about by a lack of moisture, results in less discrimination of the $^{13}CO_2$

with respect to the $^{12}CO_2$during photosynthesis than when the CO_2 is present normally; an increase in the ratio thus shows a lack of water. The linking of high ratios and narrow rings, evidence of drought in the semiarid regions west of the United States, strengthens the use of ^{13}C as a proxy indicator of this drought (Leavitt and Long 1989). Thus, wood is found to provide a record of the climate of the past not only through the variations in its ring-width or its density but also in its chemical composition.

2.2.4. *Pollen analysis*

This analysis deals with pollen grains and spores which are deposited and preserved in lakes or in present or fossil peat bogs (Lowe and Walker 1984; Reille 1990). As the form and size of these elements as well as the sculpture patterns of their outer coat, the exine, are characteristics of the genus and often even of the plant species from which they have been released, their identification and number in sediments allows a reasonably exact reconstitution of the flora and the vegetation in the region surrounding the area of the deposit. The analysis of cores of sediment is generally made at intervals of 1 to 10 cm and results in the construction of diagrams which trace the evolution in time of the relative abundance of each of the plants identified. The dating of such pollen diagrams is based on ^{14}C analysis of the organic matter at one or several levels. Indications of the climate which can be derived from the diagrams are generally broad, in fact, it is said that periods are more or less cold or more or less wet depending on the vegetal associations shown. It is quite obvious that this calls for a good knowledge of the present ecology of the species in question. Recently, the search for analogues of fossil pollen spectra among the greatest variety of reference modern spectra has made it possible to obtain calibrations which are more accurate than before and to estimate more reliably, past rainfall and temperature values (Guiot et al. 1989; Guiot 1990; Pons 1990).

Pollen analysis is far less used as a source of climatic information for the historical era than it is between 2000 and 15000 BP and beyond. Several examples of climatic reconstruction for the historical era can, however, be cited for the United States: Minnesota, Michigan (Webb III 1981); Africa (Nicholson 1980); Senegal (Feller et al. 1981) and Cameroon (Medus 1991).

2.2.5. *Sedimentological analysis*

Sedimentological analysis, which is based on stratigraphy, makes use of the chemical and physical composition of sediments and involves carrying out granulometric and mineralogical studies as well as analysis of organic matter. It is used in land areas (Brochier 1983) as well as in lacustrine and coastal environments (Oldfield and Clark 1990). Like pollen analysis, it has not been widely used in studies of the historical era although it is likely to provide a detailed and continuous sequence of climatic events. "With the development of appropriate dating techniques, the methods developed for long-term studies are now being applied to the recent past on a time scale of a few hundred years or less and with a temporal resolution at the level of individual events to decades" (Oldfield and Clark 1990, p. 203).

In land environments, caves in particular ("ce sont les évènements climatiques exceptionnels, et surtout leur accumulation dans une période donnée, qui sont à l'origine des phénomènes de sédimentation [fossilisés]") are exceptional climatic events, and more particularly their accumulation in a given period, marks of the beginning of a (fossilized) sedmimentaton process (Brochier 1983, p. 426). Thus, the estimation of a cold climate, synonym for a cold winter in this case, is based on the analysis of its sedimentological effects. Because of the human factor, the reconstruction of a wet climate based on vegetation change is much less accurate than one based on sediments alone (1988) and underscores the need to evaluate the anthropogenic factor when climate is being reconstituted from proxies.

Finally, the presence of charcoal in sediments ensures the application of ^{14}C dating.

3. Some examples

Climatic variation during the historical era will be illustrated here with examples based on the application of three of the methods reviewed above. These examples are intentionally limited to the European continent with small references to events in Africa.

3.1. *Pollen analysis*

As an example, we will consider an analysis of sporo-pollen material from silt interstratified in the sands of a sub-present river terrace at the foot of the Mandara Mountains in the Cameroons, Africa (Medus 1991). The six levels analyzed, ranging from the oldest to the most recent, together highlight several features: the decline of trees in the wooded savanna and dry forest, the increase in *Poaceae* in relation to the development of cereal cultivation, probably millet, and the replacement of the spores of the *coprophilous fungi* by parasitic *Poaceae* fungi spores. The changes in the entire sporo-pollen material show a somewhat irregular reduction of rainfall accompanied by a cessation of pastoralism, the development of cereal cultivation and an increase in human settlements in the region. The sequence, which has not been dated by ^{14}C but has been compared with historical sources, can be placed between the 11th and 12th centuries specifically at the beginning of the dry phase of the 13th–14th century, whose traces are still found in oral folklore.

3.2. *Sedimentary analysis*

The sedimentary analysis of two caves in the South of France (Brochier 1983) enables us to distinguish two well-defined cold periods. The beginning of the first period is around the 7th century while the second is in 14th century. In each case, the maximum occurs much later in the period. The two periods are separated by a mild and wet climatic phase, with a distinctive regional character. The "pause" ended at the beginning of the 14th century. According to the author, these results are generally in line with results of other climatic sequences such as those from the analysis of the Fernau peat bog in Austria (Le Roy Ladurie 1967). In the two sites in question, the beginning of the second cold wave could easily mark a first sign of the "Little Ice Age".

3.3. *Climatic reconstructions based on tree-rings*

The reconstructions analyzed are limited to the European continent (Scandinavia and Western Europe) and North Africa.

3.3.1. *The first millennium*

It is difficult to carry out a climatic reconstruction of the first millennium of the historical era based on the calculation of transfer functions. This is due to the rarity of adequate tree-ring series, to the frequently heterogeneous character of the series and the short length of calibration and verification periods. Such an attempt has nevertheless been made in Scandinavia for tree-ring series of Scots pines (*Pinus silvestris* L.) of nearly 1,500 years (A.D. 500–1975) (Briffa et al. 1990). An analysis of the same type in the USA for tree-ring series of more than 1,600 years can also be cited (A.D. 372–1985) (Stahle et al. 1988).

In Western Europe, a small number of long tree-ring series cover all or part of the first millennium. The longest, from Irish, English and German oaks, covers more than 1,000 years (Pilcher et al. 1984). These long series, however, have not yielded systematic reconstructions of the climate as yet. But the presence of characteristic rings (narrow or wide) common to several series, therefore useful for cross-dating, show the same sensitivity to a common factor: climate, and can be used for climatic reconstruction (Kelly et al. 1989). A limit to the use of ring-width lies, however, in the ambiguity of the climatic significance of the narrow or wide ring which can be, in a complex way, due both to rainfall and temperature fluctuations.

In Scandinavia, summer temperature reconstruction based on the calculation of a transfer function (Briffa et al. 1990) identifies at least three periods approximately as warm as the period 1930–1949 (A.D. 750–780, 920–940 and 960–1000) and two periods at least as cold as the latter decades of the 19th century (A.D. 780–830 and 850–870) in the first millennium. It appears, from the results obtained for the second millennium with the same tree-rings series and their comparison to other series, that these results may not be extended to the rest of Europe.

3.3.2. *The second millennium*

The second millennium is in all respects much better documented than the first. We have selected 18 tree-ring reconstructions and summarized the most striking phases in Figure 1, including rough estimates of their time and space dimensions. In this graph a group of calculated reconstructions (series 1 to 14) is distinguished from a group of reconstitutions (series 15 to 18) based on the variations of the maximum latewood density or on those of ring width indices interpreted from historical or palynological data. The graph enables an assessment of the concomitance of the climatic variations from Scandinavia (1–3) to Morocco (14) for the reconstructed series and from the Alps to Spain for the reconstituted series. With the exceptions of series 14, 16 and 18 relating to rainfall, these involve series of temperatures, especially summer temperatures which correspond closely to the temperatures of the tree-ring growth period.

The variations in the reconstructed climatic factor, whatever it is, and however established, are often compared by the authors to the variations in other proxy data, such as the glacial fronts whose movements in the Alps are well known (Le Roy Ladurie 1967). They can also be compared with the longest thermometric observations known, such as Manley's series for central England (Manley 1974) or Schüpp's series for Switzerland (Schüpp 1961). Furthermore, some reconstructions are based only partly on tree-rings and rely on other proxy sources (see on Figure 1, series 8, 9 and 13). Only such a comparison can produce a more accurate indication of climatic variation over a given period and region. We hope that Figure 1 will be of some help.

We will avoid detailed analysis of the climatic variations summarized in Figure 1 and instead focus attention on two major contrasting periods of the millennium: The Medieval Warm Epoch or Little Climatic Optimum and the Little Ice Age.

3.3.2.1. *The Medieval Warm Epoch.* According to Lamb (1977, 1988) and Alexandre (1987) the Little Climatic Optimum appears to have lasted from at least A.D. 1100 to A.D. 1300 in Europe, but according to Le Roy Ladurie (1967) a short glacial thrust recorded in the peat bog at Fernau in Austria marks the period as A.D. 1150–1200 to A.D. 1300–1350. From A.D. 1100 to A.D. 1300, the reconstruction of summer temperatures in the Scandinavian countries (1) shows along with the other series some similarities in Western and central Europe (6 and 13) and even with the Moroccan series (14) where the period is rather dry. On the other hand, the Lauenen alpine series (15) which shows definite warming over the 12th century is consistent with the conclusions reached by Lamb and Alexandre.

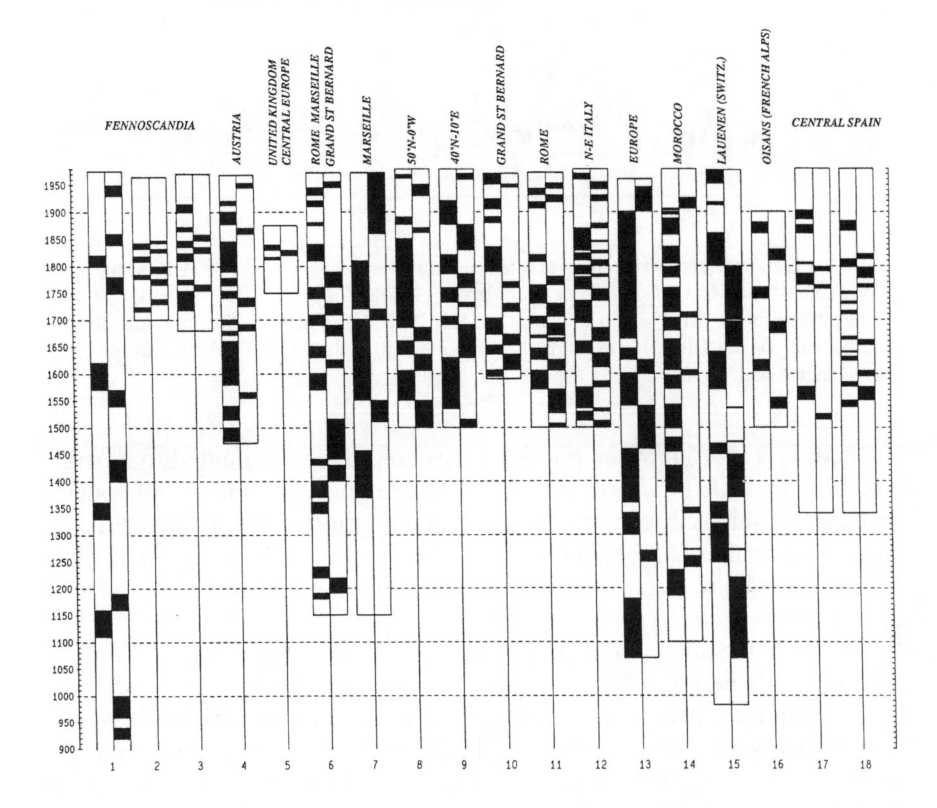

Figure 1. Recapitulation of the 18 climatic reconstructions based on tree-rings selected for Scandinavia, Western Europe and North Africa. There are as many vertical blocks as summarized climate reconstructions. These reconstructions have an annual to multi-decadal resolution. The central vertical line of each block corresponds to the average of the climatic factor for the calibration period. The coldest or driest episodes (to the left of the central vertical line) and the warmest or wettest ones (to the right of the central vertical line) are in black. The length of each episode is specified by the author(s) of the reconstruction and, when not given, from our own interpretation. The white parts of blocks correspond to climatic values close to the average or with high frequency variations. References and characteristics of the 18 reconstructions concern Figure 1. (Ref.: p. 75)

The deterioration in the 14th century pointed out by Lamb is perceptible everywhere (1, 6, 7, 13, 15) but a warming towards the turn of the century is highlighted in the Lauenen Series (15); the first half of the 15th century itself is rather warm as indicated in series 1, 6 and 15.

3.3.2.2. *The Little Ice Age.* There is considerable controversy (Grove 1988) over the duration of the Little Ice Age in which temperatures are believed to have fallen to their lowest levels since the end of the last glaciation. In Europe, particularly on the basis of the Fernau peat bog (Le Roy Ladurie 1967), it is timed approximately between A.D. 1550 and 1850. For Scandinavia, as

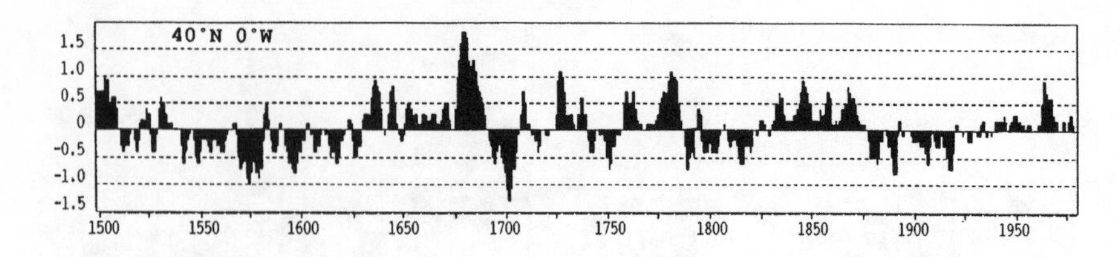

Figure 2. Detailed reconstructed annual temperature variations since 1500 A.D. at the 40°N 0°W point of the Jones et al. temperature grid (1985), cf. Figure 1 No. 9; temperature in °C expressed as departure from the mean of the 1950–1970 period. At this point, there is no evidence for the Little Ice Age such as reported in central and Northern Europe (from Serre-Bachet et al. 1990).

Briffa et al. (1990) have emphasized, this cold period is limited to a period around A.D. 1600 (1570–1620) and might even have been less cold than the first half of the 12th century. Series 2 and 3 (much shorter than series 1) based on a smaller number of tree-ring series, nevertheless reveal some slight differences with respect to the distribution of cold periods since A.D. 1700 in the region.

The set of the other temperature series effectively show a definite predominance of cold periods over warm periods between A.D. 1550 and 1850 with the exception of three series (9, 15 and 17) where the period would appear to be rather warm (15) or, at the very least, regularly alternated between cold and warm (9) (see also the corresponding period of this series in Figure 2). On the whole, except in series 15, the 17th century would appear to be warmer than the 18th century, however, warm periods are not excluded in the latter.

The second half of the 19th century and the 20th century appear rather warm in all regions, and the variations that they contain are, on the whole, in line with results obtained from analyses of instrumental data (Jones and Kelly 1983; Jones et al. 1986).

Hints of pluviometric conditions from the Moroccan (14) and Spanish (18) series would appear to indicate, for the Little Ice Age, a predominance of dry periods in 14 and wet periods in 18. In 14, Till and Guiot (1990) show that, in year-by-year comparisons, there is not, strictly speaking, any correspondence with Europe (cold winters in Europe and dry winters or wet years in Morocco; mild winters in Europe and dry or wet winters in Morocco). These authors do not find any correlations either between the droughts in Morocco and those reported for the Sahel (Nicholson 1980), but point out some anticoincidences. In 16, the principal phases with a more or less large formation of snow cover which allow us to define the principal stadial and interstadial phases of the Little Ice Age in a region of the French Alps, correspond to the majority of the temperature series.

4. Conclusion

As we have just seen, climatic variations during the historical period can be studied with a wide range of methodological tools that are increasingly efficient and useful in highlighting a wider range of climatic features. However, systematic development of their application is compounded by wide differences in materials available from one geographical area to another.

We must also emphasize that at present, accuracies in temporal and numerical specification of climatic information likely to be obtained from these methods are unequal. Also at present, conclusions are best when they are based on results from a combination of sources including those of anthropogenic origin (Pfizer 1988), elements from the vegetable and mineral kingdoms which have a well defined annual or intra-annual rhythm, tree-rings in particular, and instrumental data. When possible, the number attached to the meteorological factor, whatever its source, depends on the calibration used which itself is based on present data. It is clear then that this calibration should be made over sufficiently long periods and should ideally be carried out using standardized data so as to reduce, if not eliminate, the individual and random point variations. The temperature grid expressed as deviations from the 1951 to 1970 average and adopted by the group at Norwich for the Northern Hemisphere for the period 1851–1980 (Jones et al. 1985) is a valuable resource in this field.

A real limitation to the use of the majority of natural sources is still the difficulty of identifying, without ambiguity, the forcing climatic factor, rainfall or temperature particularly, when biological remains are being analysed. In this case, more complex are relationships between living species and their environment. In isotopic analyses we have noted the doubts about the real hydrous or thermal causes to their behaviour. The same is true for tree-rings with respect to the climatic factors that are actually responsible for their width, etc. Resort especially to anthropogenic sources and comparison with information provided by other proxy data in a multidisciplinary analysis most often make it possible to best discern the underlying climatic factor or factors, and specify both their temporal and spatial variations (Oldfield and Clark 1990).

Among major objectives of climate research as we approach the end of this century is the need to improve our understanding of those variations which occurred during the historical era as well as the global scale changes. The instrumental and historical periods are short compared to the geological era but both are distinguished largely by human interventions.

Acknowledgements

I thank J. Medus for his special contribution and J. Medus and G. Georgelin for their comments on the manuscript. I am also grateful for the technical support of C. Goeury.

References

Alexandre, P. (1987): *Le climat en Europe au Moyen Age*. Ed. Ecole des Haute Etudes en Sciences Sociales, Paris, 827 pp.

Anior, R.W. and Eckstein, D. (1984): Dendroclimatological studies at the Northern timberline. In: *Climatic changes on a yearly to millenial basis*, ed. by N.A. Morner and W. Karlen. Reidel, Dordrecht, 273–279.

Berger, W.H. and Labeyrie, L.D. (eds.) (1987): *Abrupt climatic change. Evidence and implications*. Reidel, Dordrecht, 425 pp.

Briffa, K.R., Bartholin, T.S., Eckstein, D., Jones, P.D., Karlen, W., Schweingruber, F.H. and Zetterberg, P. (1990): A 1,400-year tree-ring record of summer temperatures in Fennoscandia. *Nature*, 346, 434–439.

Briffa, K.R., Jones, P.D. and Schweingruber, F.H. (1988a): Summer temperature patterns over Europe: A reconstruction from 1750 A.D. based on maximum latewood density indices of conifers. *Quarternary Research*, 30, 36–52.

Briffa, K.R. Jones, P.D. and Schweingruber, F.H. (1988b): Reconstructing summer temperatures in Northern Fennoacandinavia back to A.D. 1700 using tree-ring data from scots pine. *Arctic and Alpine Research*, 20(4), 385–394.

Briffa, K.R., Wigley, T.M.l., Jones, P.D., Pilcher, J.R. and Hughes, M.K. (1987): Patterns of tree-growth and related pressure variability in Europe. *Dendrochronologia*, 5, 35–58.

Brochier, J.E. (1983): Deux mille ans d'histoire du climat dans le midi de la France: Etude sédimentologique. *Annales*, 2, 425–438.

Cook, E.R. and Kairiukstis, L.A. (eds.) (1989): *Methods of dendrochronology: Applications in the environmental sciences*. Kluwer Academic Publishers, Dordrecht, 247–258.

Eckstein, D. and Aniol, R.W. (1981): Dendroclimatological reconstruction of summer temperature for an alpine region. Mitt. Forstl. Bundesvers. Anst. Wien, No. 142, 391–398.

Feller, C., Medus, J., Paycheng, C. and Chavane, B. (1981): Etude pédologique et palynologique d'un site protohistorique de la moyenne vallée du fleuve Sénégal. In: *Palaeoecology of Africa and the surrounding islands*, ed. by J.A. Coetzee and E.M. Van Zinderen Bakker Sr. Balkema, Rotterdam, 235–248.

Förster, H. and Hützen, H. (1983): $^{18}O/^{16}O$ ratio of water in a local ecosystem as a basis of climate record. In: *Palaeoclimates an palaeowaters: A collection of Environmental isotope studies*. International Atomic Energy Agency, Vienna, Austria, 67–81.

Freyer, H.D. and Belacy, N. (1981): $^{13}C/^{12}C$ record in Northern hemispheric trees during the past millennium anthropogenic impact and climate superpositions. In: *Analysis and interpretation of atmospheric CO^2 data*. WMO/ICSU/UNEP Sci. Conf. Geneva, 209–215.

Fritts, H.C. (1976): *Tree rings and climate*. Academic Press, New York, 567 pp.

Grove, J.M. (1988): *The Little Ice Age*. Methuen, London, 498 pp.

Guiot, J. (1984): Deux méthodes d'utilisation de l'épaisseur des cernes ligneux pour la reconstitution de paramètres climatiques anciens, l'exemple de leur application dans le domaine alpin. *Palaeogeography, Palaeoclimatology, Palaeoecology*, 45, 347–368.

Guiot, J. (1988): The climate of Central Canada and South-Western Europe reconstructed by combining various types of proxy-data: A detailed analysis of the 1810–1820 period. In: *The year without a summer? World Climate in 1816*, ed. by C.R. Harrington. Canadian Museum of Nature, Ottawa, 291–308.

Guiot, J. (1989): Methods of calibration. In: *Methods of dendrochronology: Applications in the environmental sciences*, ed. by E.R. Cook and L.A. Kairiukstis. Kluwer Academic Publishers, Dordrecht, 165–178.

Guiot, J. (1990): Methodology of the last climatic cycle reconstruction in France from pollen data. *Palaeogeography, Palaeoclimatology, Palaeoecology*, 80, 49–69.

Guiot, J., Pons, A., de Beaulieu, J.L. and Reille, M. (1989): A 140,000-year climatic reconstruction from two Europian pollen records. *Nature*, 338, 309–313.

Jones, P.D. and Kelly, P.M. (1983): The spatial and temporal characteristics of Northern Hemisphere surface air temperature variations. *Journal of Climatology*, 3, 243–252.

Jones, P.D., Raper, S.C.B., Bradley, R.S., Diaz, H.F., Kelly, P.M. and Wigley, T.M.L. (1986): Northern Hemisphere surface air temperature variations, 1851–1984. *Journal of Climate and Applied Meteorology*, 25, 161–179.

Jones, P.D., Raper, S.C.B., Santer, B.D., Cherry, B.S.G., Goodess, C., Bradley, R.S., Diaz, H.F., Kelly, P.M. and Wigley, T.M.L. (1985): A grid point surface air temperature data set for the Northern Hemisphere, 1851–1984. DOE Tech. Rep. TRO22 (US Department of Energy, Washington DC).

Kelly, P.M., Munro, M.A.R., Hughes, M.K. and Goodess, C.M. (1989): Climate and signature years in West European oaks. *Nature*, 340, 57–60.

Lamb, H.H. (1977): *Climate: Present, past and future*. Vol. 2, *Climate, History and the Future*. Methuen, London.

Lamb, H.H. (1984): Climate in the last thousand years: natural climatic fluctuations and change. In: *The climate of Europe: Past, present and future*, ed. by H. Flohn and R. Fantechi. Reidel, Dordrecht, 25–64.

Lamb, H.H. (1988): *Weather, climate and human affairs*. Routledge, London and New York, 364 pp.

Leavitt, S.W. and Long, A. (1989): Drought indicated in CARBON-13/CARBON-12 ratios of Southwestern tree rings. *Weather Resources Bulletin*, 25(2), 341–347.

Le Roy Ladurie, E. (1967): *Histoire du climat depuis l'an mil*. Flammarion, France.

Libby, L.M. (1987): Evaluation of historic climate and prediction of near-future climate from stable-isotope variations in tree-rings. In: *Climate, history, periodicity and predictability*, ed. by M.R. Rampino, J.E. Sanders, W.S. Newman and L.K. Konigsson. Van Nostrand Reinhold, New York, 81–89.

Lowe, J.J. and Walker, M.J.C. (eds.) (1984): *Reconstructing Quarternary environments*. Longman, London, 389 pp.

Manley, S. (1974): Central England temperature: monthly means 1659 to 1973. *Quart. Jour. Roy. Meteorol. Soc.*, 100, 389–405.

Medus, J. (1991): Human consequences of an historical climatic change evidenced in Northern Cameroon by the pollenanalytical study of river terrace deposits. Personnal communication.

Nicholson, S.E. (1980): Saharan climate in historic times. In: *The Sahara and the Niles*, ed. by M.A.J. Williams and H. Faure. Balkema, Rotterdam, 173–200.

Oldfield, F. and Clark, R.L. (1990): Lake sediment-based studies of soil erosion. In: *Soil erosion on agricultural land*, ed. by J. Boardman, I.D.L. Foster and J.A. Dearing. John Wiley, 201–228.

Pfister, C. (1984): The potential of documentary data for the reconstruction of past climates. Early 16th to 19th Century. Switzerland as a case study. In: *Climatic changes on a yearly to millenial basis*, ed. by N.A. Morner and W. Karlen. Reidel, Dordrecht, 331–337.

Pfister, C. (1988): Une rétrospective météorologique de l'Europe. Un système de reconstitution de l'évolution du temps et du climat en Europe depuis le Moyen Age central. *Histoire et Mesure*, 3(3), 313–358.

Pichard, G. (1988): Les météorologistes provençaux aux XVIIe et XVIIIe siècles. *Provence Historique, Fasc.*, 153, 249–184.

Pilcher, J.R., Baillie, M.G.L., Schmidt, B. and Becker, B. (1984): A 7,272-year tree-ring chronology for Western Europe. *Nature*, 312, 150–152.

Pons, A. (1990): Analyse pollinique et reconstitution quantitative de climat. In: *Leçons de palynologie et d'analyse pollinique*, ed. by M. Reille. Ed. C.N.R.S., Paris, 187–206.

Reille, M. (1990): *Leçons de palynologie et d'analyse pollinique*. Ed. C.N.R.S., Paris, 187–206.

Richter, K. (1988): Dendrochronologische und Dendro-klimatologische Untersuchungen an Kiefern (*Pinus* sp.) in Spanien. Ph.D. Dissertation, Hamburg, 296p.

Richter, K. and Eckstein, D. (1986): Estudio dendro-cronologico en Espana. Dendrochronologia, 4, 59–74.

Rotberg, R.I. and Rabb, T.K. (eds.) (1981): *Climate and history. Studies in interdisciplinary history*. Princeton University Press, Princeton, NJ, 280 pp.

Rothlisberger, F., Haas, P., Holzhauser, H., Keller, W., Bircher, W. and Renner, F. (1980): Holocene climatic fluctuation-Radiocarbon dating of fossil soils (fAh) and woods from moraines and glaciers in the Alps. *Geographica Helvetica*, 35(5), 21–52.

Saurer, M. and Siegenthaler, U. (1989): $^{13}C/^{12}C$ isotope ratios in trees are sensitive to relative humidity. *Dendrochronologia*, 7, 9–13.

Schüpp, M. (1961): *Klimatologie der Schweiz. C: Lufttemperatur*, 2, Teil Beih. Annalen Schw. Meteo. Zentral., Zürich, 15–62.

Schweingruber, F.H., Bartholin, T., Schar, E. and Briffa, K.R. (1988): Radiodensitometric-dendroclimatological conifer chronologies from Lapland (Scandinavia) and the Alps (Switzerland). *Boreas*, 17, 559–566.

Schweingruber, F.H., Braker, O.U. and Schar, E. (1979): Dendroclimatic Studies on conifers from Central Europe and Great Britain. *Boreas*, 8, 427–452.

Schweingruber, F.H., Braker, O.U. and Schar, E. (1987): Temperature information from European dendro-climatological sampling network. *Dendrochronologia*, 5, 9–33.

Serre-Bachet, F. (1988): La reconstruction climatique à partir de la dendroclimatologie. *Publications de l'Association Internationale de Climatologie*, 1, 225–233.

Serre-Bachet, F. and Guiot, J. (1987): Summer temperature changes from tree-rings in the mediterranean area during the last 800 years. In: *Abrupt climatic change. Evidence and implications*, ed. by W.H. Berger and L.D. Labeyrie. Reidel, Dordrecht, 89–97.

Serre-Bachet, F., Guiot, J. and Tessier, L. (1992): Dendroclimatic evidence from South-Western Europe and North-Western Africa. In: *Climate since A.D. 1500*, ed. by R.S. Bradley and P.D. Jones. Harper Collins, London, 349–365.

Serre-Bachet, F., Martinelli, N., Pignatelli, O., Guiot, J. and Tessier, L. (1991): Evolution des températures de Nord-Est de l'Italie depuis 1500 AD. Reconstruction d'après les cernes des arbres. *Dendrochronologia*, 9, 213–229.

Stahle, D.W., Cleaveland, M.K. and Hehr, J.G. (1988): North Carolina climate changes reconstructed from tree rings: A.D. 372 to 1985. *Science*, 240, 1517–1519.

Tessier, L., Couteaux, M. and Guiot, J. (1986): An attempt at an absolute dating of a sediment from the last glacial recurrence through correlations between pollenanalytical and tree-ring data. *Pollen et Spores*, 28(1), 61–76.

Till, C. and Guiot, J. (1990): Reconstruction of precipitation in Morocco since A.D. 1100 based on Cedrus atlantica tree-ring widths. *Quarternary Research*, 33, 337–351.

Webb, T. III (1981): The reconstruction of climatic sequences from botanical data. In: *Climate and history. Studies in interdisciplinary history*, ed. by R.I. Rotberg and T.K. Rabb. Princeton University Press, Princeton, 169–192.

Yapp, C.J. and Epstein, S. (1982): Climatic significance of the hydrogen isotope ratios in tree cellulose. *Nature*, 297, 636–639.

Figure 1. References and characteristics of the 18 reconstructions

1. Briffa et al. (1990). April-August mean "summer" temperature in Fennoscandia from 500 to 1975 AD. Sixty-five standardized mean ring-width series and sixty-five maximum latewood densities series of living and remnant Scots pine (*Pinus silvestris* L.) as predictors, averages of Jones et al.'s (1985) gridded monthly temperature data as predictands.
2. Briffa et al. (1988b). July–August mean temperature for northern Fennoscandia from A.D. 1700 to 1964. Twenty-one ring-width and maximum latewood density chronologies of Scots pine as predictors, gridded Jones et al.'s (1985) Northern hemisphere monthly-mean temperature as predictands.
3. Aniol and Eckstein (1984). July temperature for Swedish Lapland back to A.D. 1680 (1680–1980). Four ring-width series of Scots pine as predictors, monthly temperature data in Kiruna as predictands.
4. Eckstein and Aniol (1981). Summer temperature (May-August) for Tyrol (Austria) from A.D. 1471 to 1968. Mean tree-ring widths of larch (*Larix decidua* Mill), spruce (*Picea abies* Karst.) and stone pine (*Pinus cembra* L.), living trees and building timbers, as predictors, temperature data in Vent as predictands.
5. Briffa et al. (1988a). April to September mean temperature over United Kingdom and central Europe (A.D. 1750–1875). Network of maximum latewood density chronologies of coniferous trees as predictors, subset of Jones et al. (1985, 1986) Northern hemisphere compilation as predictands.
6. Serre-Bachet and Guiot (1987). Mean summer temperature (June–September) common to Rome, Marseille and Grand St Bernard from A.D. 1972 back to 1150. Four mean ring-width series of larch, fir (*Abies alba* Mill.) and pine (*Pinus leucodermis* Ant.) as predictors, temperature data of Rome, Marseille and Grand St Bernard pass as predictands.
7. Serre-Bachet and Guiot (1987). Main climatic periods for Marseille since A.D. 1150. Same data as 6.
8. Serre-Bachet et al. (1990). Mean annual temperature at 50° N 0° W gridpoint back to A.D. 1500 (1500–1979). Eleven mean ring-width series of deciduous and coniferous trees plus nine other proxy series derived from historical archives (grape harvesting dates, thermal indices, temperature decadal estimates, winds frequency) and isotopic data (^{18}O in the Arctic ice) as predictors, average of Jones et al.'s (1985) gridded monthly temperature data as predictands (cf. Guiot 1988).
9. Serre-Bachet et al. (1990). Annual temperature at 40° N 0° W gridpoint back to A.D. 1500 (1500–1960). Same predictors and predictands as before.
10. Serre-Bachet et al. (1990), Guiot (1984). June to August temperature for Northern Alps (1590–1970). Five pine (*Pinus cembra*) and larch ring-width chronologies from Switzerland and France as predictors, temperature at Grand St Bernard pass as predictand.
11. Serre-Bachet et al. (1990), Serre-Bachet and Guiot (1987). June to September mean temperature from A.D. 1500 to 1972 in Rome. Same data as 6.
12. Serre-Bachet et al. (1991). April to September mean temperature for Veneto, North-Eastern Italy (A.D. 1500–1979). Seven pine (*P. leucodermis*), fir, spruce and larch ring-width chronologies from France and Italy as predictors, averaged temperatures of Veneto region and at 45° N 10° E gridpoint as predictands.
13. Guiot (1988). Annual mean temperature in Europe from A.D. 1070 to 1960, trend of 16 series at latitudes ranging from 40° N to 55° N. Same data as 8.
14. Till and Guiot (1990). Annual (October to September) main characteristics of rainfall from A.D. 1100 to 1979 in Morocco. Forty-six mean ring-width series of cedar (*Cedrus atlantica*) as predictors, monthly precipitation records for the humid, subhumid and semiarid regions as predictands. (see also Serre-Bachet et al. 1990).
15. Schweingruber et al. (1979, 1988). Summer (August–September) temperature in the Swiss Prealps since A.D. 982 (982–1975) derived from latewood mean maximum densities of coniferous recent wood and construction wood from old buildings (Lauenen series). In the

sub-alpine regions of the Alps, the high temperatures in August and September influence cell wall development in the latewood.

16. Tessier et al. (1986). Main stadial and interstadial phases during the last glacial recurrence in Oisans, French Alps (A.D. 1500–1900). Phases evidenced both by the relationships between some spores and snow cover and by tree-rings response of larch to climate.
17. Richter (1988). Winter (November–March) mean temperature of central Spain from A.D. 1340 to 1980. Main climatic periods derived from mean ring-width series of recent wood and construction wood of Pine (*Pinus nigra, Pinus silvestris* L.) (Teruel series, Richter and Eckstein 1986), winter temperature reconstructions from A.D. 1690 to the present and historical archives.
18. Richter (1988). Summer (May–August) rainfall of central Spain from A.D. 1340 to 1980. Same data as 17.

Françoise Serre-Bachet
ERA 1152 de CNRS
Laboratoire de Botanique Historique et Palynologie, case 451
Université d'Aix-Marseille III
Faculté des Sciences et Techniques de St Jérôme
Avenue Escadrille Normandie-Niemen
13397 Marseille Cédex 13
France

2.4. SOME STATISTICAL ASPECTS OF OBSERVED REGIONAL AND GLOBAL CLIMATE CHANGE WITHIN THE INSTRUMENTAL PERIOD

C.-D. SCHÖNWIESE

1. Introduction

Normally, the instrumental period in climatology is said to have begun when information based on direct instrument measurements of climatic elements (air temperature, precipitation, humidity, air pressure, wind etc.) became available. The quality of these instruments should be consistent over time so that earlier measurements are approximately comparable. Such comparability will advance analysis of the long-term records (time series) which could reveal the typical variations and variability of climate in space and time.

Apart from a number of earlier and primitive measurements e.g. of precipitation (India, Chile etc.; see also Chapter 2.3, e.g. Nile river level, Egypt) the invention of the thermometer (Galilei 1611) and the barometer (Torricelli 1643) enabled the development of quasi-modern methods used in experimental physics and climatology. Probably the longest record within this instrumental period for climatology is the monthly means of surface air temperature in "Central England" which dates back to 1659 (Manley 1974). Because this record is not really continuous, Manley used measurements from the Netherlands to fill some of the gaps.

In addition to time continuous records, a satisfactory data coverage in space is an important requirement in modern climatology. Following the "Academia del Cimento" (1654–1670) which successfully compared measurements from a number of Italian stations, the "Societas Meteorologica Palatine" (Germany, 1780–1792) organized the first international network of climatological stations. Some of these stations survived such that, in addition to the "Central England" record, a few time series of climatological data (mostly daily temperatures) which cover more than two centuries exist. Figure 1 adopted from

M. Yoshino et al. (eds.), Climates and Societies – A Climatological Perspective, 77–97.

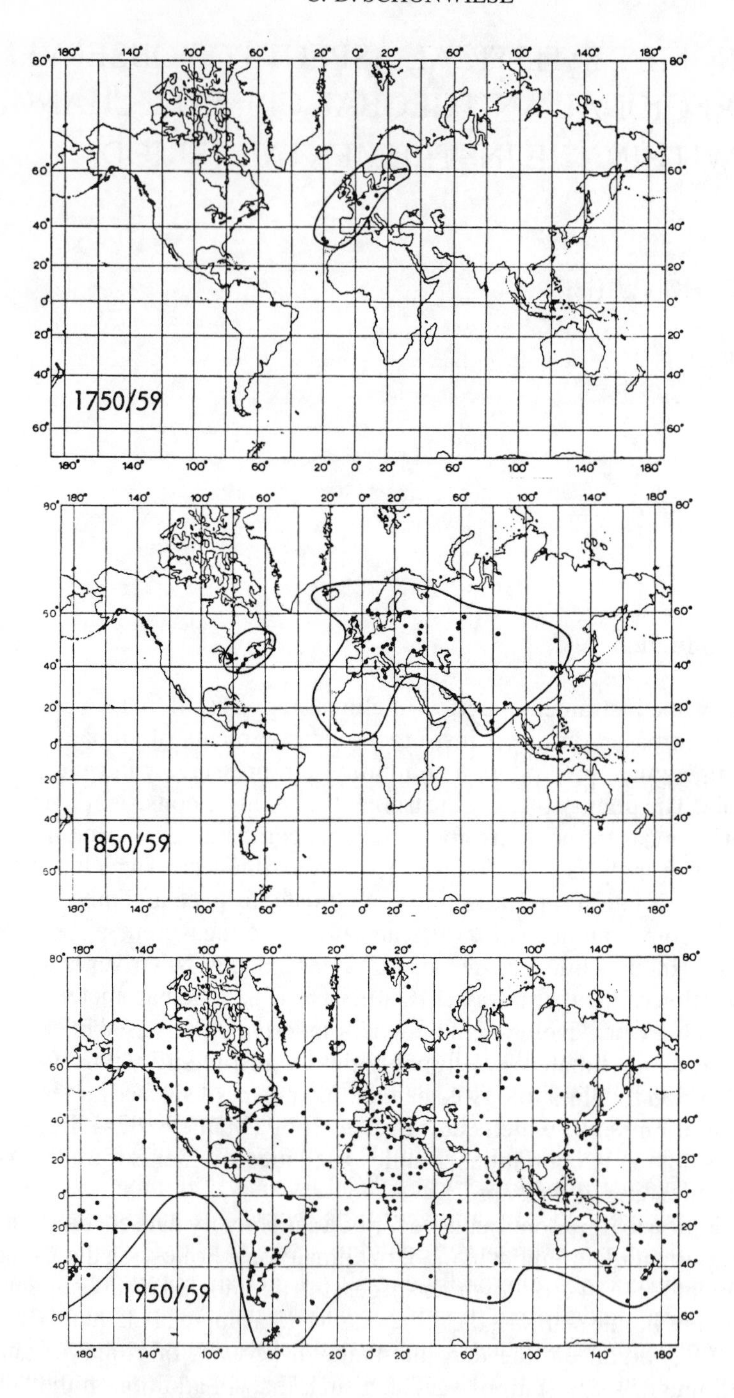

Figure 1. Growth of the world network of meteorological observation stations. The heavy lines indicate the area which may be covered by reliable field analysis (isotherms etc.); after Lamb (1969), reproduced from Schönwiese (1987).

Lamb (1969) shows a progressive overview of the growth of climatological networks since the 1750/1751 decade.

Today, approximately 1,200 stations operate with procedures established by the World Meteorological Organization (WMO). The U.S. Carbon Dioxide Information Analysis Centre (CDIAC) (1985) reported that there were more than 171 stations with records dating back to 1859; more than 361 to 1879; and 701 to 1899. All the stations were in operation till 1950 and some, beyond. The climatic elements observed are surface air temperature, precipitation, and surface air pressure. Occasionally, wind, humidity and sunshine duration (cloud information is not based on instrument measurements in those records), are also available. Upper air data which are necessary for any understanding of the atmospheric circulation are available only for recent decades. They are not discussed here. A very important milestone in data preparation and analysis is the evaluation of hemispheric or global gridded data sets based on more recent station records (surface air temperature, and less successfully, precipitation). These led, for example, to assessments of global mean temperature change, including sea surface temperatures, since 1861 (IPCC 1990).

In the following brief summary some data problems are addressed and some selected fluctuation indicators such as trend characteristics, including some spectral (power spectrum) features of observed climate variations are described. In addition, to a general consideration of the global scene, some results from Europe are discussed. The climate elements reviewed include a few aspects of precipitation. Finally, the most important causes of global climate change are listed and some concluding remarks concerning large-scale anthropogenic forcing (the so-called "greenhouse" effect) are presented.

2. Data problems

The first question which arises with respect to measurements usually is accuracy. This strictly depends on the climate element considered, the method of measurement itself, and the behavior of the particular element. For example, in the case of temperature, the accuracy of any measurement taken 100 or 200 years ago was roughly +0.5 or +1 °C compared with +0.1 °C today. The fluctuations at any stations from minute to minute are relatively small so that measurements performed every hour lead to a fairly reliable assessment of the daily average value. In contrast to that, wind fluctuations are very pronounced, sometimes even from second to second. As a consequence, many measurements are not temporally correct which may compromise results of analyses based on daily and monthly means. In the case of precipitation, the accuracy depends very strongly on location, temperature (snow or rain), and wind. In an ideal setting the accuracy of any precipitation measurement may amount to c. +20% or even c. +10% decreasing to c. +100% in mountain regions and at low temperatures if strong and gusty winds allow only a small and varying part of the snow to fall into the gauge used. Unfortunately the shape of these gauges varies from country to country. Sevruk (1989) refers to the frequent occurrence of astonishingly high error levels in precipitation measurements. Growth of the world network of observation stations is shown in Figure 1.

Another problem is that of representativeness in space. Schönwiese et al. (1986); Malcher and Schönwiese (1987) computed station-to-station correlations of secular annual time series of surface air temperature and precipitation. They found that in the mid-latitudes (Europe and USA) on average, the temperature correlation dropped to 0.8 over a distance of 100 km; to 0.4 over 2000 km; and to 0 over 3000 km. For precipitation, however, it is likely that at a distance of only 50 km the annual data time series correlation is smaller than 0.8; and at 100–200 km distance smaller than 0.6. For monthly or daily data the situation is even worse. So a relatively small area like the former West Germany today maintains c. 500 stations for temperature measurements but 2,700 stations for precipitation. In general, any precipitation station record represents only a very small area often significantly smaller than for temperature.

Data homogeneity may be the most serious problem encountered in the analysis and interpretation of long records. Many data reliability discussions with respect to the instrumental period reflect this problem. According to Conrad and Pollack (1962) "a numerical series representing the variations are caused only by variations of weather and climate". It seems that homogeneity is a basic assumption underlying all climate analyses but in reality one that is seldom met. In practice, the question is: Is the inhomogeneity involved so large that it is better to discard the record or are the inhomogeneities relatively small such that the results from analysis of the data are only marginally compromised? The most frequent causes of inhomogeneity are:

- station replacements (position and/or elevation);
- changes in instruments and exposure;
- changes in other observational procedures (e.g. time);
- changes in methods used to calculate averages (e.g. daily);
- changes in the environment of the station (e.g. urban growth or deforestation) (modified from CDIAC 1985).

All these non-climatic influences may introduce abrupt discontinuities. Such discontinuities may be detectable in the data even if the cause of the inhomogeneity is not documented. For this detection a variety of homogeneity tests is available which compare the climatic variations at any station with related variations at neighboring stations whose records are assumed to be homogeneous (relative tests). There are, however, also tests available which identify, based only on one station record, any discontinuities to be probably not climatic (absolute tests); see e.g. WMO (1966), and Schönwiese et al. (1986).

Inhomogeneities may, however, also introduce gradual (long-term) trends or fluctuations, e.g. increasing temperature caused by the increasing heat island effect of growing cities. This is a much more complicated situation because it is not often possible to efficiently separate such a long-term trend which may be due to the urban heat islands from a real and large-scale (possibly global) trend due to enhanced "greenhouse" effect (human activities) to some long-term natural forcing of climate. It may be useful, when this problem is taken into account, to perform global temperature trend analyses first including and thereafter excluding city stations. This has been done, for example, by Hansen and Lebedeff (1987) in the context of "greenhouse" detection studies. It may also be possible to correct for the urban inhomogeneity, especially where the station under consideration is relocated from the city center to the surrounding countryside (e.g. airport) and some years of overlapping measurements exist. In the case of precipitation, and depending on the weather condition, the urban effects may be realized leeward of the city because the urban heat increases the lapse rate of the advancing air, instability is increased, and with some time lag, convective precipitation is increased.

One should keep in mind all the data problems mentioned or not mentioned in this text, in selecting data for analysis and interpreting the results. Similarly, results presented in the literature should be checked carefully. Not seldom, data problems, in particular non-representativeness and inhomogeneities, are underestimated such that results may be over interpreted.

3. Some remarks on observed fluctuations

Keeping in mind data restrictions and possible errors, a few aspects and selected results of observed climate fluctuations within the instrumental period will be examined. Analyses on long time scales are often based on particular long-term station records such as the surface air temperature at Hohenpeissenburg Germany (1781–1991) (Figure 2). This station is situated on a mountain (983 m) just north of the Alps (47.8 N 11.0 E), roughly 100 km south of Munich. The observations were started by the "Societas Meteorologica Palatina" and it is fortunate that they are continued today and that the record is not influenced by any urban effects. Based on a cascade of homogeneity tests and comparisons with 36 station records in the vicinity, Malcher and Schönwiese (1987) found that at least within the 1880–1980 interval there is a very high probability for homogeneity. This temperature record is just one example of an excessively large number of long-term climate series (despite this fact climatologists always require more and better observational information). Three typical features can be identified in the time series:

- a pronounced year-to-year variability which seems to be irregular (random);

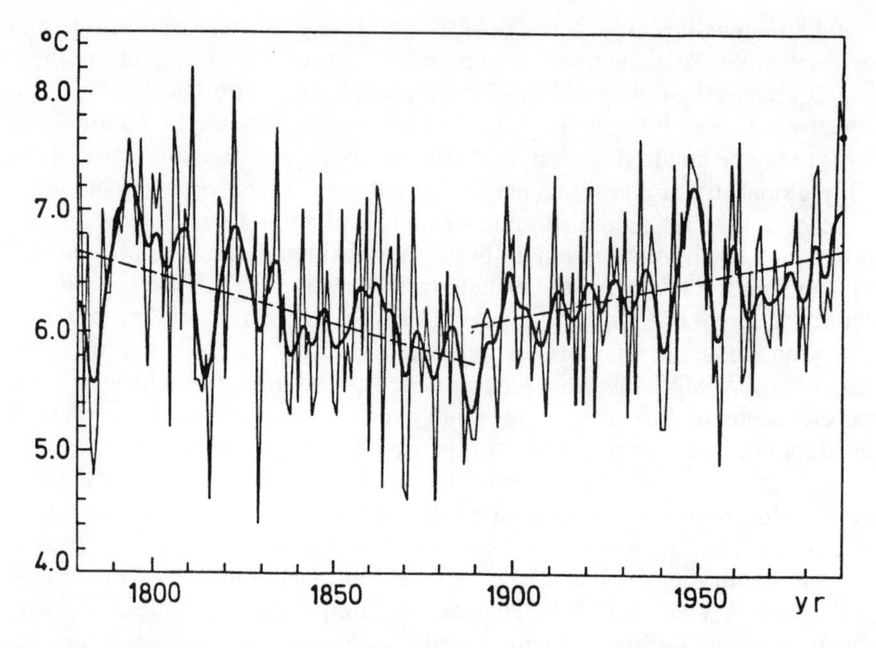

Figure 2. Hohenpeissenberg (Germany, 47.8 N 11.0 E, 983 m) annual mean surface air temperature time series 1781–1990 (data source: German Weather Service), Gaussian 10 yr low-pass filtered data (smoothed, heavy line) and linear trends (dashed lines) 1781–1890 (–0.90 K, trend-to-noise ratio 1.1) such as 1890–1990 (+0.66 K, trend-to-noise ratio 1.0).

- some relatively long-term fluctuations, more or less cyclical, which seem to be a superposition of different periods (the smoothed plot is based on 10 yr low-pass filtered data);
- long-term trends, more or less linear, which may be part of an ultra-long fluctuation.

Before the question of forcing and causes can be discussed it is reasonable to define, describe, and test the statistical characteristics of the series. In this regard, one may compute the total interval statistical moments (average, variance; skewness etc. to characterize the frequency distribution). It is often the case, however, that the moments and other statistical characteristics are not stable in time (non-stationarity). The existence of fluctuations and trends, for example, points to the fact that the record average varies in time. In the case of Hohenpeissenberg temperature, the following linear trends can be calculated: 1781–1890, –0.90°C; 1890–1989, +0.57°C.

The next question addresses the confidence or significance of the statistical features. Without discussing all the statistical measures and tests suitable for this analysis (see textbooks, e.g. Panofsky and Brier 1958; Sneyers 1975; Schönwiese 1985), the simplest method for judging trends will be used namely, trend-to-noise analysis in which the trend is divided by the standard deviation of the underlying data. For the data in Figure 2, these ratios are 1.1

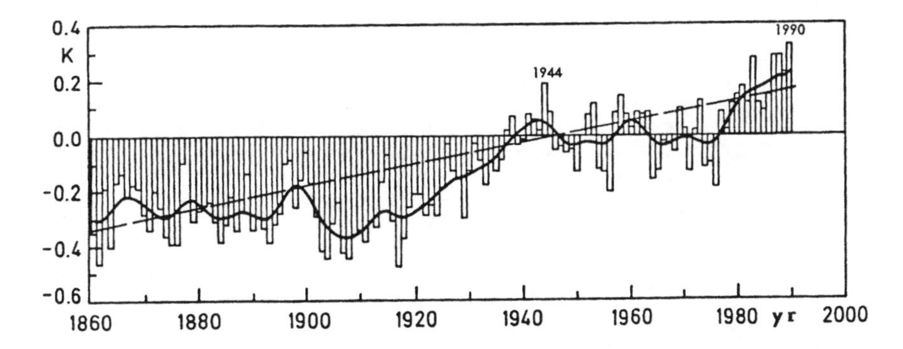

Figure 3. Global mean surface air temperature (land and marine data) annual mean data time series 1861–1990 (data source: IPCC 1990; Jones et al. 1991), Gaussian 10 yr low-pass filtered data (smoothed, heavy line), and linear trend (dashed line; 0.5 K, trend-to-noise ratio 2.6).

(1781–1890), and 0.9 (1890–1989) respectively. It has to be concluded that even though these trends are detectable by calculation, they are not significant because trends and standard deviation are approximately of the same magnitude. (If Gaussian frequency distributions are assumed, ratios of 2, or better still 3, are required for significance at the 95 or 99% confidence levels; error probability 5 or 1%, respectively.) A confidence analysis of fluctuations should be combined with a power (variance) analysis (see the next section).

One may consider and discuss some hundreds or thousands of additional climatic records within the instrumental period; examples are IPCC (1990), Rudloff (1967), Schönwiese (1987); as well as other contributions on regional and seasonal scales in this book. Here we will try to obtain a global overview of temperature fluctuations and trends. Figure 3 presents the IPCC (Houghton et al. 1990) assessment of annual mean surface air temperature variations for land and sea surface for the period 1861–1990, in terms of anomalies (deviations from the reference period indicated). Again we see some year-to-year variability which is, however, considerably smaller than seen in Figure 2. This is a typical feature of time series which are large-scale spatial averages. The trend-to-noise ratio is 2.6 and statistically significant. The trend amounts to 0.5°C. The ten-year smoothed data also plotted in Figure 3 indicate, however, that there are only two intervals where a temperature increase can be identified: 1917–1944 and 1976 to present. The fact that the five warmest years (since 1944) all occurred in the 1980s is a consequence of the long-term trend or fluctuations and not a series of isolated dramatic events. Similarly, neither the trend nor any fluctuations can be *a priori* attributed to the "greenhouse" effect because there are many mechanisms, natural, anthropogenic, or both, which may cause climate change. Global surface air temperature data sets, grid or box data, are provided by Jones et al. (1991), and Hansen and Lebedeff (1987, 1989). These data sets have been used to assess hemispheric and global mean monthly and annual data; see also Vinnikov et al. (1990).

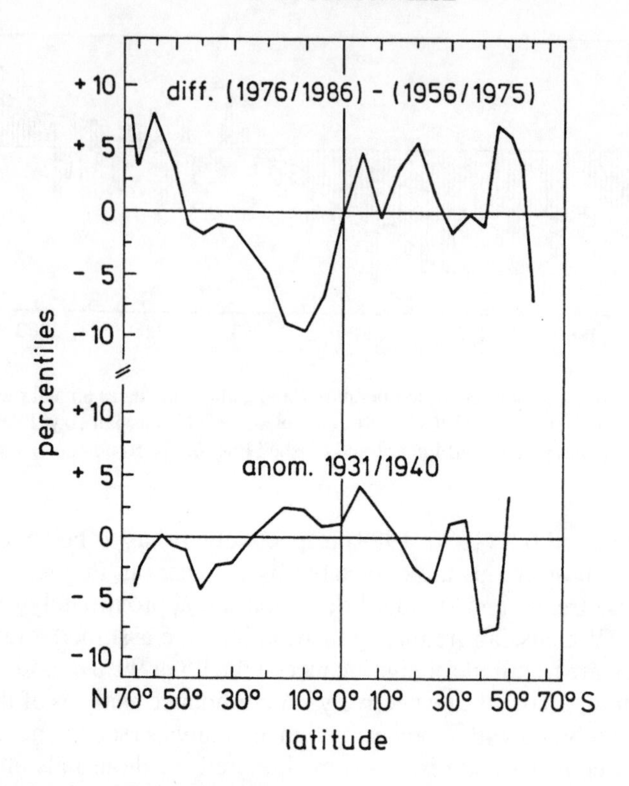

Figure 4. Annual precipitation trends 60 °S–80 °N, percentiles, differences of reference periods (1976–1986)–(1956–1975) and anomalies 1931–1940, in relation to latitude; from Diaz et al. (1989).

Before a somewhat closer discussion of fluctuations and trends is entered, one may reflect on precipitation variations keeping in mind that for this climate element the probability of measurement errors is much larger and the spatial representativeness much smaller than is the case with temperature. On a northern hemisphere scale, Bradley et al. (1987) have analysed monthly, seasonal, and annual precipitation records at 1487 land stations covering mainly the period 1850–1985. The data were transformed to a regular 400-to-400 km grid and percentile deviations from a Gamma distribution were computed. A typical feature of precipitation records is that long-term trends seldom occur. Instead, pronounced interannual variations frequently dominate including fluctuations on a multi-year (a few years to a few decades) scale. Figure 4 (from Bradley et al. 1987) points to the possibility of increasing precipitation rates north of 35 °N and drier conditions in the 5–35 °N zone since about 1950. In a revised global analysis Diaz et al. (1989) see a trend to wetter conditions north of c. 45 °N and drier conditions southward (southern hemisphere is very uncertain); see also trend results presented in

Section 5. Bradley et al. (1987) state that the 1950s were extremely anomalous with four of the wettest years in close succession (1953, 1954, 1956, 1957). Some of the driest years are also clustered: 1912, 1913, 1918, 1919, 1920. Regional and seasonal peculiarities are, however, enormous so that very different precipitation statistics are obtained in different regions (see regional and local contributions in this book). Neither in the case of temperature, nor in precipitation is a systematic long-term tendency towards increasing or decreasing variability statistically confident although when restricted to particular subintervals and particular regions such effects can be seen. There are, however, distinct differences from subinterval to subinterval and from region to region. In total, the picture of precipitation fluctuations is very complicated and nonhomogeneous. Consequently, it is nearly impossible to summarize global precipitation fluctuations in a few sentences.

4. Some aspects of spectral statistics

Climatological records may also be examined with respect to their spectral features. Statistically, this means that a time series is transformed so that the variance is portioned out to various periods. Graphically, spectra are represented with variance as the ordinate versus period as the abscissa. Periods showing up relatively large contributions of variance point to related cycles or quasi-cycles in climate.

The most common methods of time series spectral analysis are: (1) periodogram; (2) variance (power) spectrum analysis, where the Fourier transform of the autocorrelation is computed, called here ASA short for autocorrelation spectral analysis; (3) fast Fourier transform, FFT; (4) maximum entropy spectral analysis, MESA (or maximum entropy method, MEM). For details, see e.g. Blackman and Tukey (1958), Mitchell et al. (1966), Schönwiese (1985, 1987) and Essenwanger (1986).

There are some advantages to the ASA, particularly the availability of comparatively simple significance tests. These tests are usually based on first-order autoregressive (ARI) Markov process and, in consequence, a red noise null hypothesis, where the chi-square distribution is used for determining the confidence limits for the sample spectrum peaks (for definitions and problems, see e.g. Mitchell et al. 1966). Although red noise (or white noise, if the lag 1 autocorrelation coefficient is not statistically significant) is assumed to define the autoregressive structure of the population from which the sample is drawn, a higher order AR process would sometimes be a more appropriate null hypothesis.

The outstanding advantage of the MESA method (periodogram and FFT are not appropriate in this context) is the increased spectral resolution in the low-frequency domain (relatively long periods). When, however, the algorithm parameters (ASA: number of lags; MESA: number of coefficients) are varied

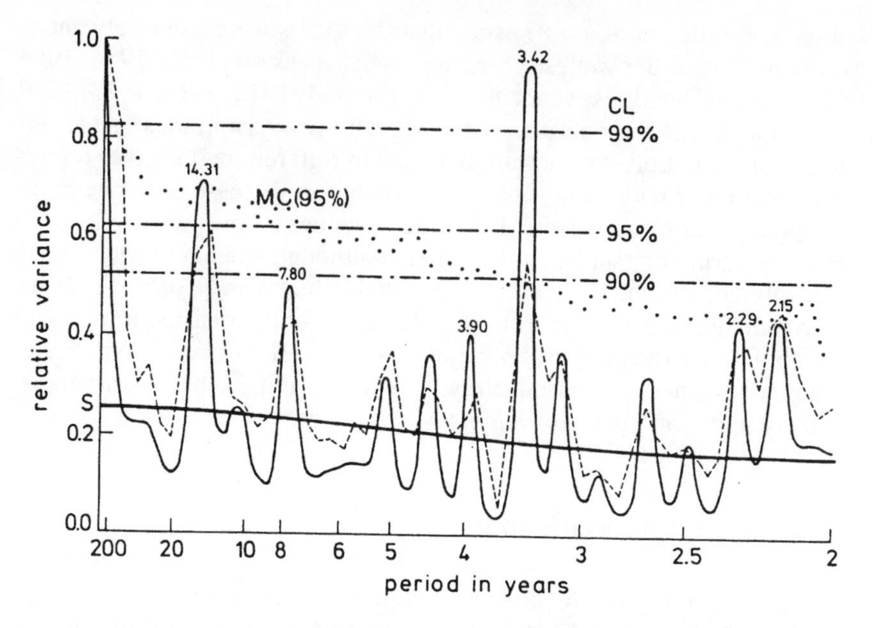

Figure 5. Variance (power) spectrum of Hohenpeissenberg annual surface air temperatures 1781–1980 (compare Figure 2), comparison of two methods: autocorrelation spectral analysis ASA (Blackman and Tukey 1958), dashed line, and maximum entropy spectral analysis MESA (Burg 1972), solid line. CL means confidence levels (related to ASA), MC Monte Carlo testing (related to MESA); from Schönwiese (1987a).

it can be empirically verified that the MESA based variance values vary much more than the corresponding ASA values. This means that the amount of variance estimated by the MESA (for example, algorithm after Burg 1972) at any particular frequency band may not be stable, whereas the frequencies themselves are comparatively stable.

Nevertheless, one should be aware of the fact that every sample variance spectrum may contain peculiar statistical artifacts and, in general, it is not possible to discern these artifacts from real physical effects, particularly if these physical effects are not known or are problematic. Because different algorithms reveal different artifacts (again in general), it is prudent to analyse any given database by means of at least two different and largely independent algorithms. Then, differences in the results may well be due to artifacts, and similarities may be, but not necessarily, more reliable. In comparing ASA and MESA, neither can claim to be the "better" technique. For best results one should try to combine the advantages of both techniques while limiting the disadvantages.

In Figure 5, the ASA (dashed line) and MESA spectra of the Hohenpeissenberg (compare Figure 2) mean annual temperatures 1781–1980 are plotted. Note the similarities in the sample spectrum peaks (signals), although only

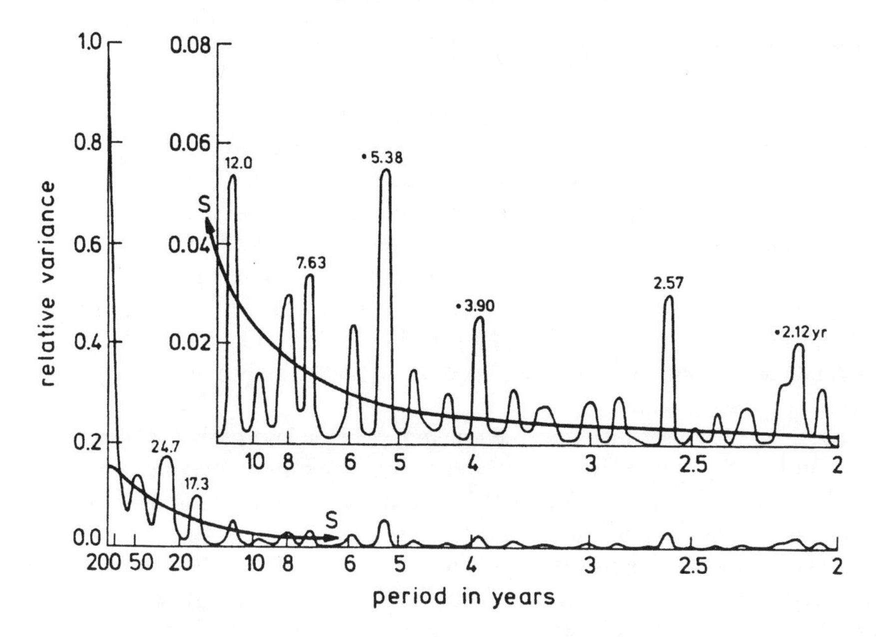

Figure 6. Maximum entropy variance spectrum (MESA method) of mean northern hemisphere annual surface air temperatures 1781–1980 (data source: Groveman and Landsberg 1979; Jones et al. 1982).

the 3.4- and ca. 14-yr cycles (mean periods) exceed the 90% confidence level (or 0.1 error probability), which means relatively weak significance, and only the long-term residuum exceeds the 99% level. The Monte Carlo simulations for the MESA sample spectrum confidence levels reveal similar results where the 3.4- and 14.3-yr peaks and the long-term residuum exceed the 95% confidence level. This is shown in Figure 5 by means of a dotted line. (Details and results of a worldwide spectral analysis of 242 temperature and 164 precipitation time series are given in Schönwiese et al. 1986.)

This spectrum is compared (see Figure 2) with a corresponding MESA spectrum of the reconstructed Northern Hemisphere mean annual temperatures 1781–1980. [Note that continental areas are emphasized; data before 1881 are from Groveman and Landsberg (1979); updates are from Jones et al. (1982).] Here, c. 2.6- and 5.4-yr cycles are much more prominent and instead of a c. 14-yr cycle, a 12-yr cycle is indicated (Figure 6). The latter is , however, not correlated with sunspot variations (see Schönwiese 1983, 1984). A more detailed analysis of the hemispheric record is presented in Schönwiese (1983 and 1984), where it is shown that the quasi-biennial oscillation (QBO) exceeds the 99% confidence level over both the 1881–1980 (mean period 2.05 yr) and 1579–1980 interval (mean period 2.13 yr). It should be noted, however, that the Northern Hemisphere data reconstructions before 1881 and particularly before 1781 are crude because of the use of proxy data (tree ring series) and possible regression instabilities, including coherence problems (see Schönwiese 1987a).

There arise three questions: (1) Are these signals really significant and stable in time? (2) Are they of a global (hemispheric) or a more regional (local) type? (3) What are the physical reasons for them?

No doubt, the last question is the most difficult one. There are many suggestions concerning the QBO. Relatively long-term variations may be due to external forcing of a climate system like volcanism (also affecting the year-to-year variations) and solar irradiation change (see e.g. Gilliland 1982; Hansen et al. 1981; Schönwiese 1984). We are, however, far from understanding many of the other spectral signals which may or may not be statistically significant.

Many of them, for instance, the quasi-5-yr cycle and probably also the ca. 3–4- and ca. 8-yr cycles may be produced by atmospheric-oceanic circulation mechanisms such as the Southern Oscillation/El Niño phenomenon. Most hypotheses, however, are speculative.

Nevertheless, it is still reasonable to persue answers to the first two questions because they, in turn, could enable us to make progress towards an understanding of the physical background to climate variations. The first question can be discussed by means of moving spectral analysis, whereas the second should be approached through an analysis of regional and local spectral variability and its spatial coherence.

In general, all climatic time series statistics vary as time varies. This gives rise to ambivalent consequences. On the one hand, the estimation of the confidence levels of the spectra becomes problematic because time-dependent autocorrelation variations cause time-dependent red noise variations on which these estimations are based. On the other hand, the signals detected in the customary "integrated" spectrum can be studied in a "dynamic" way by considering their variations in time. For adequate tools to perform moving (running, "dynamic") spectral analysis, see Olberg and Schönermark (1981); Schönwiese (1987); Malcher and Schönwiese (1987) who show that in most cases variance spectra are not stable in time. This quality complicates the interpretation considerably but, at the same time, throws some light on the spectral behavior of climate. In a similar manner, correlation analysis can be extended spectrally leading to "integrated" and "moving" coherence spectra of climate elements and particular forcing parameters.

It is not possible to describe all these methods and results in detail here. Instead, some dominant quasi-cyclical components of variance which appear relatively frequently in precipitation time series spectral analysis: c. 2, 3.3, 4, 5, 8 and roughly 15 years will be listed (Schönwiese et al. 1986). Often, however, the confidence of these spectral peaks is questionable, especially with respect to longer periods. Usually, the long-term residuum spectral variance contributions are smaller for precipitation compared to corresponding temperature variance spectra.

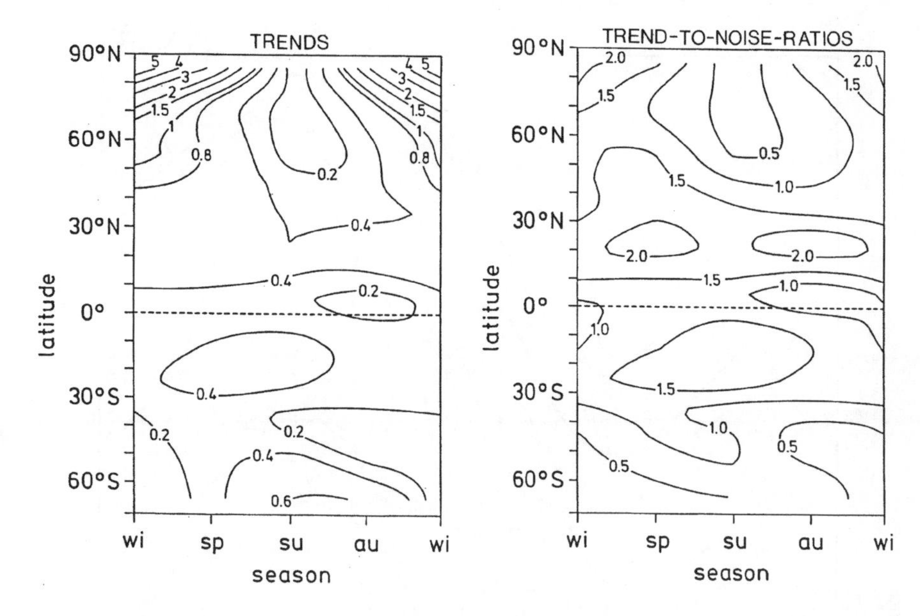

Figure 7. Linear trends in K (left-hand plot) and trend-to-noise ratios (right-hand plot) of global surface air temperatures 1890–1985 (data source: Hansen and Lebedeff 1987), disintegrated in respect to latitudes and seasons; from Schönwiese et al. (1990).

5. Summary of observed long-term trends

All trends detected or assumed in climatological records may be part of long-term, including ultra-long-term cycles or quasi-cycles. (Ultra-long means that the period length is considerably larger than the record length so that this period cannot be identified in the time series data.) Because relatively short-term periods, say < 20 or < 50 yr, can be identified in the long climate records, say > 100 or > 200 yr, it is reasonable to consider real or artificial trends only in the long-term time domain, say at least 100 yr. In the context of the "greenhouse" hypothesis (anthropogenic global climate change, see Chapter 2.6) it is again reasonable to concentrate on this long-term view because the anthropogenic increase of the greenhouse gases, particularly CO_2, has been evident since about 1800 (IPCC 1990).

Serious spatial representativeness problems arise when one examines instrumented data since 1800 for trends in global climates. Some of the problems are highlighted in the following discussion on global linear 100 year trend analysis. First, an examination of the trends and the trend-to-noise ratios evaluated from Hansen and Lebedeff (1987) global box time series data set for surface air temperature 1890–1985 (see Figure 7, global charts for all seasons are available from Schönwiese et al. 1990) shows that there have been only small long-term trends in all seasons in the tropics. Maximum trend

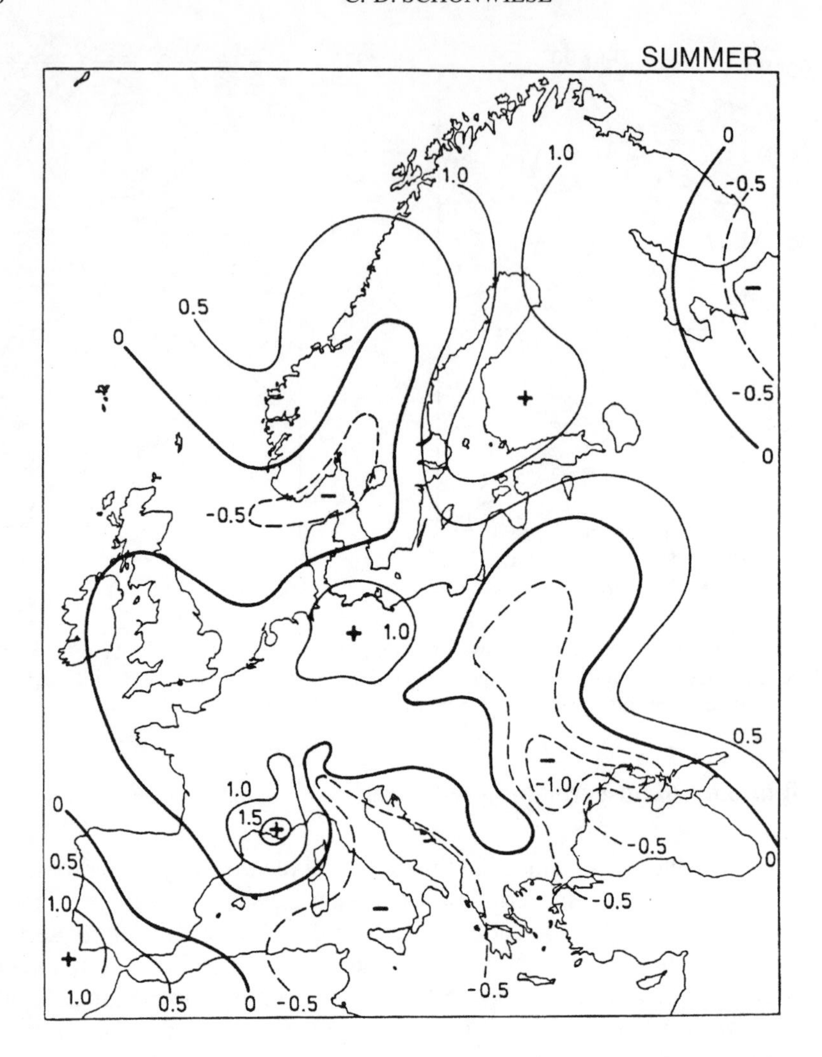

Figure 8a. Linear trends in K of European surface air temperatures 1889–1988 (from Schönwiese et al. unpubl.). (a) Summer season.

values (> 5 K) are found in the arctic winter especially in the Greenland and Northern Canada, and maximum trend-to-noise ratios occur at c. 20 °N in spring and autumn. Note that box data (100 boxes used in Figure 7) involve considerable spatial smoothing. The 100 yr European surface air temperature trend patterns for summer and winter plotted in Figure 8 are rather confusing with warmings and coolings occurring in both seasons.

In the case of precipitation, significant long-term trends are almost non existent. It is typical for precipitation that relatively short-term fluctuations and variability dominate. In addition, we are a long way from having at our

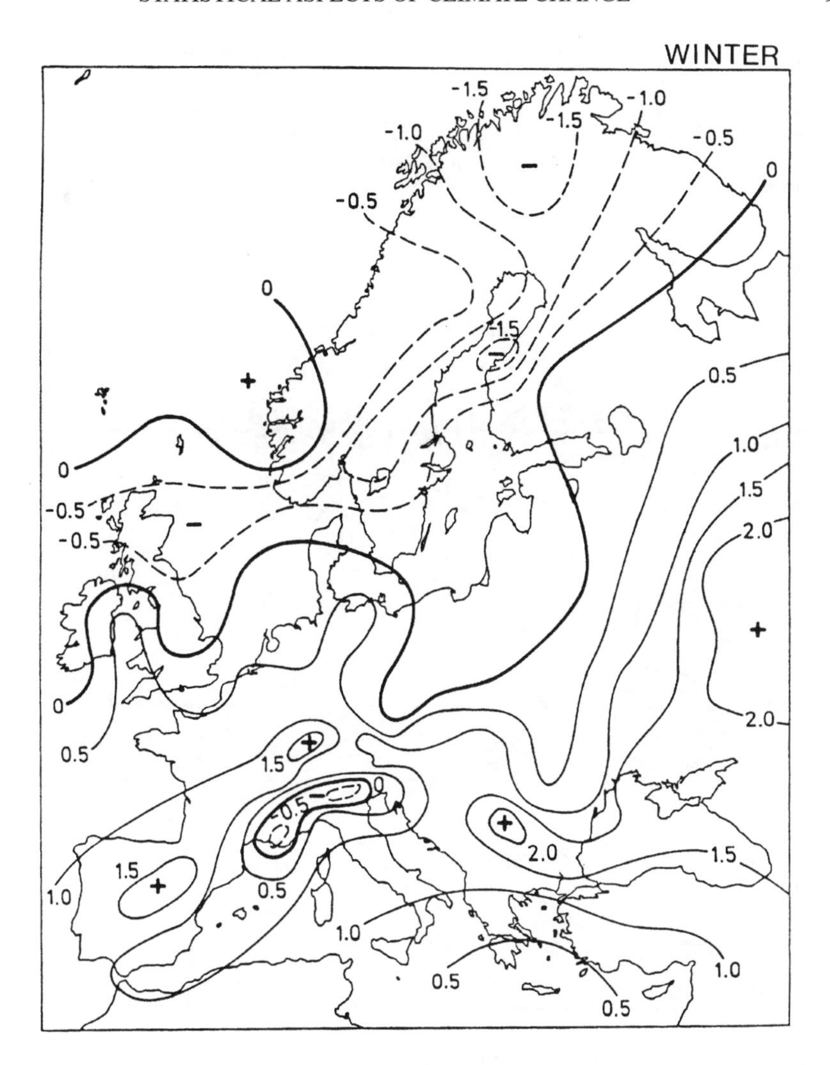

Figure 8b. (b) Winter season.

disposal homogeneous long time series covering the globe in a representative way. However, long-term precipitation trends should not be overlooked even if they are not statistically significant because of the enormous ecological and socioeconomic impact of precipitation trends. Figure 9 (from Ullrich et al. 1991) gives a rough approximation of 100 yr precipitation trends (1889–1988) for some regions of the world where a sufficiently dense network of stations exists. Almost all the trends, however, are not significant. Based on available data, we can speculate that precipitation decreased in some parts of the USA, Southern and Eastern Europe, Northern India, West Sahel (Africa),

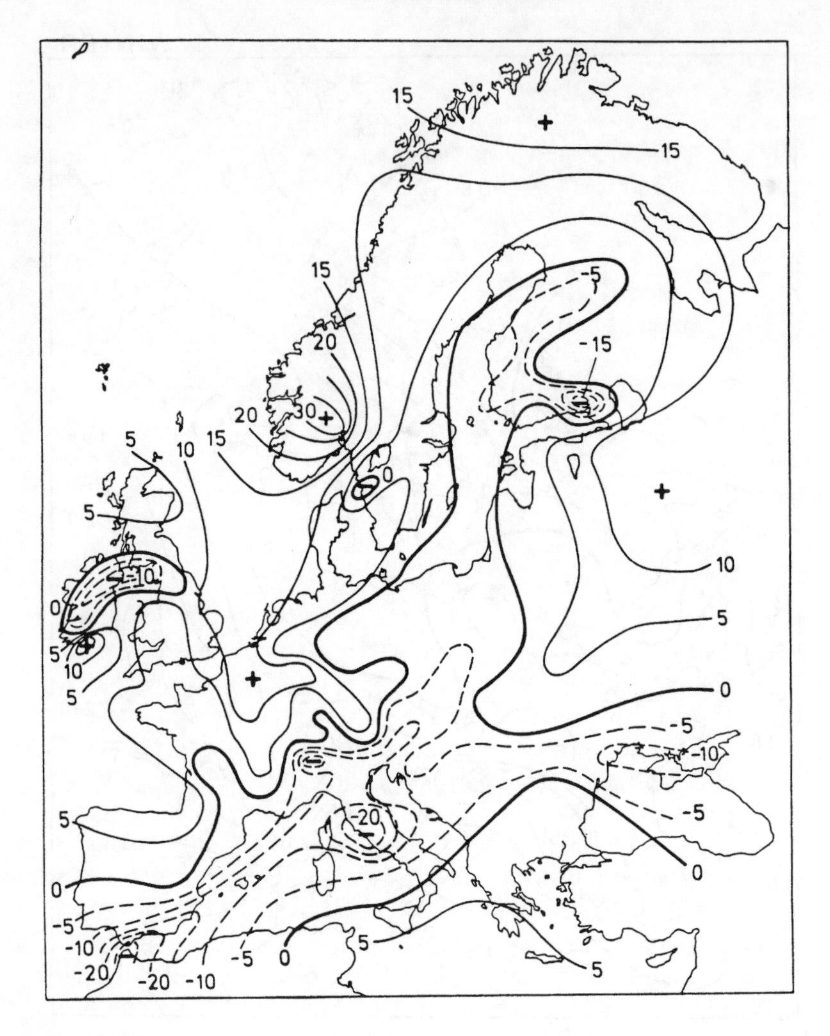

Figure 9. Linear trends in cm of European annual precipitation totals 1889–1988; from Schönwiese et al., unpubl.

South Africa, Japan and Southern Australia; and increased in Eastern USA, North West Europe and Scandinavia, and South West India.

Again, skipping other important climate elements like humidity, air pressure and wind, one feature, however, concerning the atmospheric circulation, shall be mentioned. The IPCC (1990) Section "Observed Climatic Variations and Change" refers to the North Atlantic Oscillation (NAO) in terms of the time series of the mean sea level air pressure difference between Ponta Delgada (Azores) and Stykkisholmur (Iceland). These differences or NAO indices are proportional to the North East Atlantic air flow which influences, in particular, northeastern Western Europe. In the c. 1965–1990 subinterval, there was a pronounced decrease amounting to approximately twice the magnitude of the subsequent increase. Previous to 1905 (since 1870) an increase was observed. In total, a NAO decrease is dominant, although superimposed by pronounced fluctuations. The example shows that trends invariably depend on the time intervals considered and that one should be very careful in description and interpretation, especially if the time intervals involved are significantly smaller than a secular time scale.

6. Possible causes of observed climatic change

All observed climatic change including variations and variability, trends in particular, fluctuations, abrupt change, etc. is caused in a very complex and complicated way by internal and external mechanisms of the climate system comprising the atmosphere, ocean, cryosphere, biosphere and land surfaces. Only a small part of these mechanisms is understood and explained by deterministic laws, theories or hypotheses (including statistical hypotheses on process potentials which may be deterministic). Some parts may be stochastic (random) per se, and others which seem to be stochastic today may be explained by deterministic theories sometime in the future.

In general, mechanisms which are internal to the climate system are interactive. A considerable number of examples can be found in air-sea interaction, e.g. evaporation-precipitation interactions via air and sea temperature, clouds, winds, sea salinity and currents feedbacks (for details, see e.g. Gates 1981; Fisher 1987, 1989; IPCC 1990). Another important example is the El Niño/Southern Oscillation (ENSO) mechanism. Often processes amplify or decrease in intensity (positive and negative feedbacks) and, connections appear in a manner that makes it difficult or even impossible to separate cause and effect. Even in extremely sophisticated climate models (Hansel 1989; IPCC 1990), causes of climatic change are realized incompletely and inadequately such that models are only moderately successful in reproducing the behavior of the climate system and by implication climate variations and variability. The most important external mechanisms which influence the climate system and may force climate variations, when the discussion is restricted to the instrumental period, are as follows:

- volcanism,
- solar effects, and
- anthropogenic forcing.

Note that, within a particular time domain, external mechanisms are defined to be non-interactive, for example, volcanic activity influences climate but not vice versa based on present knowledge. Nevertheless, external mechanisms may be terrestrial or extraterrestrial and more important, may influence and trigger internal mechanisms of the climate system. For instance, variations of solar activity and/or solar diameter may lead to variations in incoming solar radiation and this may change atmospheric circulation patterns. In the end, not only air temperature but also cloud formation and precipitation may be affected.

There are two major aspects of anthropogenic climate forcing: the heat island effect of growing cities (urban effects, similarly industrial regions, power plants) and the "greenhouse" effect. The former effect is a local or regional effect (see Parts 3 and 4 of this book), the latter one global. Both effects, however, address not only temperature but the whole climate.

For details of the "greenhouse" effect which represents radiative forcing of the global climate due to increasing atmospheric concentrations of infra-red-active trace gases like CO_2, CH_4, CFC, N_2O etc., caused by human activities [see Part 4 of this book and IPCC (1990)]. In addition, model projections of anthropogenic global climate change caused by greenhouse gases there are also important observational aspects of this problem, particularly the "detection" of "greenhouse"-induced climatic change in observational data. Because atmospheric CO_2 has increased from the industrial period estimate of c. 280 ppm, roughly in 1800, to 353 ppm in 1990 (IPCC 1990), a related climate change should appear in the climate data time series of the industrial period. This anthropogenic climate change is, however, embedded in natural climate variations of the same or even larger magnitude (IPCC 1990).

Since the climate may be different in a few decades when anthropogenic forcing may become the dominant effect, the present discussions on climate change are relevant to several areas of the environment including ecology, economics and societies. The identification and, even more questionable, the quantification of the "greenhouse"-gas-induced climatic change "signal-s" versus natural variability or "noise" is very complicated and uncertain because methods and hypotheses have to be found which separate these anthropogenic signals from the residuum, i.e. natural climate "noise" and other anthropogenic climate change like the urban effect.

A statistical hypothesis related to the global mean surface air temperature (see also Figure 3) is presented in Figure 10 (see Schönwiese et al. 1990; Schönwiese and Runge 1991; Schönwiese 1991 for details). In this figure, the 10 yr low-pass-filtered observations are compared with the best fit result of a regression model which is driven simultaneously by volcanic, solar, ENSO and CO_2-equivalent greenhouse gas forcing. From this model the temperature component which may be due to greenhouse gases forcing (+0.7 K since 1861, uncertainty +0.2 K) is assessed. This global greenhouse-gas-induced temperature increase accelerates and could produce in the event of a $2 \times CO_2$ a temperature rise of between c. 2.5 and 4.5 K. This estimate is in fair agreement with most of the climate model projections (for details, see Schönwiese and Runge 1991). Seasonal and regional assessments are discussed by Schönwiese and Stähler (1991). In contrast, a simple examination of trends may be misleading. Increasing temperature may be caused by other non-greenhouse gas forcings and coolings may be due to natural (e.g.

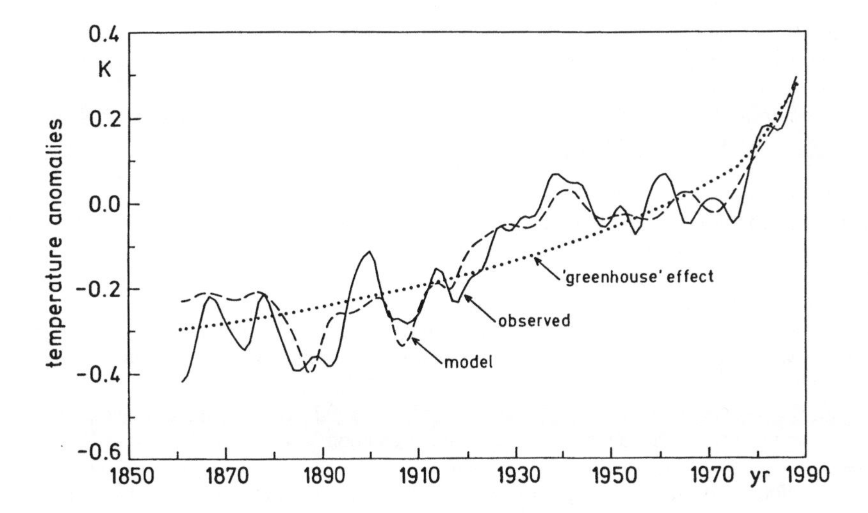

Figure 10. Observed global mean surface air temperature fluctuations, 10 yr smoothed data (solid line compare Figure 3), reproduction by means of a regression model forced by volcanic, solar, and ENSO parameters as well as greenhouse gases, dashed line, and evaluation of the enhanced "greenhouse" effect (temperature signals) based on this model; from Schönwiese et al. (1990).

volcanic) forces which dominate within a particular time subinterval even though greenhouse gas effects forcing may be simultaneously present.

There are some other signs of global climate change which may point to anthropogenic greenhouse forcing but are not discussed here. They include sea level rise, increasing humidity in some tropical areas, decreasing precipitation in the Mediterranean area and parts of USA (see IPCC 1990). Many trends and fluctuations are hard to interpret and are open for discussion with respect to their observational significance and forcing potential. There are many general aspects and special results from analyses of observational data from the instrumental period, but only a selected few have been discussed in this Section. Climate change is not a subject of interest only for scientists, it is also of interest to societies who benefit or suffer from climate conditions and climate change. The risk of an anthropogenically driven climate change is serious and demands improved analysis and interpretation of observed climate variations and variability.

References

Blackman, R.B. and Tukey, J.W. (1958): *The measurement of power spectra*. Dover, New York.

Bradley, R.S., Diaz, H.F., Eischeid, J.K., Jones, P.D., Kelly, P.M. and Goodess, C.M. (1987): Precipitation fluctuations over northern hemisphere land areas since the mid-19th century. Science, 237, 171–175.

Burg, J.P. (1972): The relationship between maximum entropy and maximum likelihood spectra. *Geophysics*, 37, 375–276.

CDIAC (Carbon Dioxide Information Analysis Center, formerly CDIC) (1985): *Climatic data for Northern Hemisphere Land Areas 1851–1980* (contributed by Bradley, R.S. et al.). Oak Ridge National Laboratory, Oak Ridge.

Conrad, V. and Pollack, L.D. (1962): *Methods in climatology*. Harvard University Press, Cambridge (USA).

Diaz, H.F., Bradley, R.S. and Eischeid, J.K. (1989): Precipitation fluctuations over global land areas since the late 1800's. *J. Geophys. Res.*, 94, 1195–1210.

Essenwanger, O.M. (1986): *Elements of statistical analysis*. World Survey of Climatology, Vol. 1B. Elsevier, Amsterdam.

Fischer, G. (ed.) (1987, 1989): *Landolt–Börnstein numerical data and functional relationships in science and technology*, Vol. V/4c, Climatology, Part 1 (1987), Part 2 (1989). Springer, Berlin.

Flohn, H., Kapala, A., Knoche, H.R. and Mächel, H. (1990): Recent changes of the tropical water and energy budget and of midlatitude circulations. *Clim. Dyn.*, 4, 237–252.

Gates, W.K. (1981): The climate system and its portrayal by climate models: A review of basic principles. In: *Climatic variations and variability: Facts and theories*, ed. by A. Berger. Reidel, Dordrecht, 3–19.

Gilliland, R.L. (1982): Solar, volcanic and CO_2 forcing of recent climatic changes. *Clim. Change*, 4, 111–131.

Groveman, D.S. and Landsberg, H.E. (1979): Reconstruction of northern hemisphere temperature: 1579–1880. Univ. of Maryland, Dept. of Meteorology, Publ. No. 79–181, College Park.

Hansen, J. et al. (1981): Climate impact of increasing atmospheric carbon dioxide. *Science*, 213, 957–966.

Hansen, J. and Lebedeff, S. (1987): Global trends of measured air temperature. *J. Geophys. Res.*, 92, 13345–13372. (1988: Update *Geophys. Res. Letters*, 15, 323–326; 1989: unpubl. data on magnetic tape).

Hantel, M. (1989): Climate modeling. The present global surface climate. In: *Landolt-Börnstein*, Vol. V/4c2 (Climatology, Part 2), ed. by G. Fisher. Springer, Berlin, 1–474.

Houghton, J.T., Jenkins, G.J. and Ephraums, J.J. (1990): *Climate change*. The IPCC Scientific Assessment (World Meteorological Organization, United Nations Environmental Programme, Intergovernmental Panel on Climate Change). Cambridge University Press, Cambridge (UK). IPCC, 1990. See Houghton et al.

Jones, P.D., Wigley, T.M.L. and Kelly, P.M. (1982): Variations in surface air temperatures: Part I. Northern Hemisphere, 1881–1980. *Monthly Weather Rev.*, 110, 59–70.

Jones, P.D., Wigley, T.M.L. and Farmer, G. (1991): Marine and land temperature data sets: A comparison and a look at recent trends. In: *Greenhouse-gas-induced climatic change*, ed. by M.E. Schlesinger. Elsevier, Amsterdam.

Lamb, H.H. (1969): Climatic fluctuations. In: *World Survey of Climatology. Vol. 2, General climatology 2*, ed. by H.E. Landsberg and H. Flohn. Elsevier, Amsterdam.

Lamb, H.H. (1972): *Climate: present, past, and future*. Vol. 1, Methuen, London.

Malcher, J. and Schönwiese, C.-D. (1987): Homogeneity, spatial correlation and spectral variance analysis of long European and North American air temperature records. *Theor. Appl. Climatol.*, 38, 157–166.

Manley, G. (1974): Central England temperatures: monthly means 1659 to 1973. *Quart. J. Roy. Meteorol. Soc.*, 100, 389–405.

Mitchell, J.M., Dzerdzeevskij, B., Flohn, H., Hofmeyer, W.L., Lamb, H.H., Rao, K.N. and Wallén, C.C. (1966): Climatic change. WMO Publ. No. 195 (Tech. Note No. 79), Geneva.

Olberg, M.V. and Schönmark, M. (1981): Zur statistischen Struktur von Klimaschwankungen im mitteleuropäischen Raum. *Z. Met.*, 31, 370–374.

Panofsky, H.A. and Brier, G.W. (1958): *Some applications of statistics to meteorology*. Pennsylvania State University, University Park.
von Rudloff, H. (1967): *Die Schwankungen und Pendelungen des Klimas in Europa seit Beginn der regelmäßigen Instrumentenbeobachtungen*. Vieweg, Braunschweig.
Schönwiese, C.-D. (1967): Northern hemisphere temperature statistics and forcing. Part A: 1881–1980. *Arch. Met. Geoph. Biocl. Ser. B*, 32, 337–360; 1984; Part B: 1579–1980. *Arch. Met. Geoph. Biocl. Ser. B*, 35, 155–178.
Schönwiese, C.-D. (1985): *Praktische Statistik für Meteorologen und Geowissenschaftler*. Bornträger, Stuttgart.
Schönwiese, C.-D., Malcher, J. and Hartmann, C. (1986): Globale Statistik langer Temperatur- und Niederschlagsreihen. Report No. 65, Inst. Meteorol. Geophys., Univ. Frankfurt/M.
Schönwiese, C.-D. (1987): Climate variations. In: Fisher, G. (ed.): *Landolt–Börnstein*, Vol. V/4c1 (Climatology, Part 1), Springer, Berlin, 93–150.
Schönwiese, C.-D. (1987a): Moving spectral variance and coherence analysis and some applications on long air temperature series. *J. Climate Appl. Meteorol.*, 26, 1723–1730.
Schönwiese, C.-D., Birrong, W., Schneider, U., Stähler, U. and Ullrich, R. (1990): Statistische Analyse des Zusammenhangs säkularer Klimaschwankungen mit externen Einflußgrößen und Zirkulationsparametern unter besonderer Berücksichtigung des Treibhausproblems. Report No. 84, Inst. Meteorol. Geophys., Univ. Frankfurt/M.
Schönwiese, C.-D. and Runge, K. (1991): Some updated statistical assessments of the surface temperature response to increased greenhouse gases. *Internat. J. Clim.*
Schönwiese, C.-D. (1991): *Das Problem menschlicher Eingriffe in das Globalklima ("Treibhauseffekt") in aktueller Übersicht*. Frankf. Geowiss. Arb. Vol. B3, Frankfurt/M.
Sevruk, B. (ed.): *Precipitation measurement*. WMO/IAHS/ETH Workshop, Swiss Federal Inst. Technology, Zürich.
Sneyers, P. (1975): Sur l'analyse statistique des séries d'observations. WMO Publ. No. 415, Geneva.
Ullrich, R., Schönwiese, C.-D. and Birrong, W. (1991): Recent long-term precipitation fluctuations and trends in Europe and other regions of the world. Poster presented at the XVI General Assembly of the European Geophysical Society, summary. *Ann. Geophys.* Supplement to Vol. 9, c 135.
Vinnikov, K.Y., Groisman, P.Y. and Lugina, K.M. (1990): The empirical data on modern global climate changes (temperature and precipitation). *J. Climate*, 3, 662–677.
WMO (1966): see Mitchell et al.

C.-D. Schönwiese
Institut für Meteorologie und Geophysik
J.W. Goethe Universität
Postfach 11 19 32
60054 Frankfurt a. M.
Germany

2.5. CLIMATE OF THE FUTURE: AN EVALUATION OF THE CURRENT UNCERTAINTIES

H. LE TREUT

1. Introduction: The increasing greenhouse effect

The possible climatic impact of the increasing atmospheric greenhouse effect has become a source of international concern. Numerical models have determined that the continuing use of fossil fuels as our first energy source should elevate the temperature of the Earth by a few degrees within a few decades. The analysis of the Antarctic Vostok ice core has also revealed that the CO_2 and the mean temperature have evolved quasi-simultaneously throughout the ice ages (Lorius et al. 1990). As a global warming of a few degrees would have important social, economical and political implications, there is strong pressure to obtain from scientists a more precise assessment of the expected future changes. But the climatologists are discovering the huge complexity of the problem, which was not fully anticipated from the beginning. In this text we review the strengths and weaknesses of the models used for those climate sensitivity studies, and the variety of mechanisms that must be better understood before any accurate prediction can be made.

We first begin by recalling what the greenhouse effect is and the origin of its increase. This introductory part uses many results from the recent report of the Intergovernmental Panel on Climate Change (henceforth referred to as the IPCC report), a body created by the World Meteorological Organization and the United Nations Environmental Program to assess global climatic change.

1.1. *The increase of atmospheric trace gases*

Direct measurements of air composition during the last decades, supplemented by measurements of the air trapped in bubbles of polar ice cores,

M. Yoshino et al. (eds.), Climates and Societies – A Climatological Perspective, 99–117.

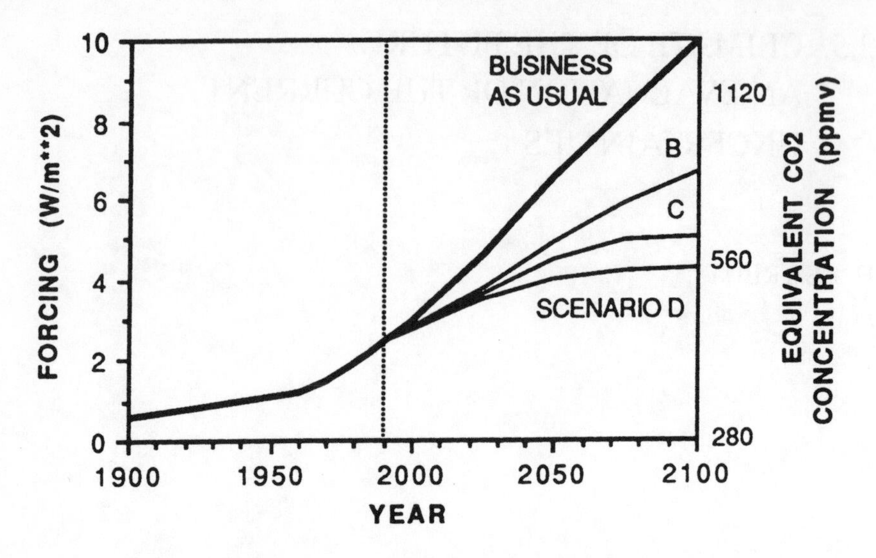

Figure 1. The predicted evolution of the radiative forcing due to the anthropogenic increase of greenhouse gases in the atmosphere (source: 1990 IPCC Report).

which provide detailed information over the last centuries, show a continuous increase in the concentration of minor atmospheric constituents. The carbon dioxide (CO_2) concentration rose from 280 parts per millions in volume (ppmv) at the beginning of the last century to the present 354 ppmv, the current rate of increase being 1.8 ppmv/year. Similarly, methane (CH_4) concentration rose from 0.8 to 1.7 ppmv. As indicated in Table 1 and in Figure 1, the atmospheric composition of many other trace gases also increased significantly throughout this period: this is the case for chlorofluorocarbons (CFCs), nitrous oxides, and tropospheric ozone (O_3), produced by the decomposition of methane. This increase in ozone should not be confused with the depletion of the stratospheric ozone layer.

The increase of these concentrations is almost certainly the result of human activities. But a quantitative estimate of the sources of greenhouse gases is not always possible and the residence time of these gases in the atmosphere also depends upon complex geochemical cycles. This constitutes an important source of uncertainty for the future evolution of climate.

The 25% CO_2 increase since the beginning of the industrial revolution is due to the burning of coal, oil, gas and wood, and to deforestation. In 1987, the respective shares of these two sources were 75 and 25%. But part of these emissions do not stay in the atmosphere, because there are permanent exchanges of CO_2 between the atmosphere, the ocean and the biomass. The ocean constitutes, by far, the largest CO_2 reservoir, equivalent to 50 times the atmospheric reservoir and 18 times the terrestrial biomass including quantities stored in the upper ground layers. As CO_2 solubility increases with decreas-

Table 1. The increase of the atmospheric concentration in greenhouse gases since the beginning of the industrial era (source: 1990 IPCC Report). Summary of key greenhouse gases influenced by human activities.[a]

Parameter	CO_2	CH_4	CFC-11	CFC-12	N_2O
Pre-industrial atmospheric concentration (1750–1800)	280 ppmv[b]	0.8 ppmv	0	0	288 ppbv[b]
Current atmospheric concentration (1990)[c]	353 ppmv	1.72 ppmv	280 pptv[b]	484 pptv	310 ppbv
Current rate of annual atmospheric accumulation	1.8 ppmv (0.5%)	0.015 ppmv (0.9%)	9.5 pptv (4%)	17 pptv (4%)	0.8 ppbv (0.25%)
Atmospheric lifetime[d] (years)	(50–200)	10	65	130	150

[a] Ozone has not been included in the table because of lack of precise data.
[b] ppmv = parts per million by volume; ppbv = parts per billion by volume; pptv = parts per trillion by volume.
[c] The current (1990) concentrations have been estimated based upon an extrapolation of measurements reported for earlier years, massuming that the recent trends remained approximately constant.
[d] For each gas in the table, except CO_2, the "lifetime" is defined here as the ratio of the atmospheric content to the total rate of removal. This time scale also characterizes the rate of adjustment of the atmospheric concentrations if the emission rates are changed abruptly. CO_2 is a special case since it has no real sinks, but is merely circulated between various reservoir (atmosphere, ocean, biota). The "lifetime" of CO_2 given in the table is a rough indication of the time it would take for the CO_2 concentration to adjust to changes in the emissions (see Section 3 for further details).

ing temperature, the ocean releases CO_2 to the atmosphere in the intertropical region oceans which are also affected by biological processes. These biological processes are also present over the continents where they constitute the dominant exchange mechanism. It must be stressed that the anthropogenic fluxes (roughly 6 Gigatons of Carbon per year [Gt. C/year] from burning of fossil fuels and 2 Gt. C/year for deforestation) are small compared to the natural fluxes (roughly 93 Gt. C/year for CO_2 to the ocean and 100 Gt. C/year for the continental biomass). The atmosphere stores 3.4 Gt. C each year, which is less than the anthropogenic emission; the difference then corresponds to a storage by the ocean or the terrestrial biosphere. Estimates of the CO_2 gain by these two components are very uncertain: they lie in the range of 1.0–2.5 Gt/year for oceans, and 1.7–3.2 Gt/year for biomass.

The methane concentration has more than doubled since the preindustrial era, for reasons which are not well understood, at least quantitatively. Methane results from anaerobic processes: fermentation of organic material under water in the absence of air, in rice paddies, enteric fermentation of cattle through rumination, termite activity, and biomass burning. It is also present in underground mines, and in natural gas. The evolution of its concentration seems tightly correlated to the evolution of human population, which seems to indicate that the main anthropogenic activities which influence its production are agricultural. Methane disappears through a complex chemistry, that creates other greenhouse gases such as the tropospheric ozone.

The origin of nitrous oxide seems to be related to the action of microorganisms present in the soils or in water, ground bacterial denitrification, for example.

The chlorofluorocarbons (CFC) are currently used in industries as refrigerating fluids, spray propellers, foaming agents. Their chemical stability, which accounts for their success is also the cause of their very long atmospheric residence time. Their impact on the stratospheric ozone layer, through photochemical processes, has become a major source of concern and has resulted in international protocols to reduce their emissions. Such reduction will also be beneficial for the greenhouse problem. The CFC molecule is infra radiatively very absorbant (around 15000 times as much as CO_2) and may have a strong temperature impact in spite of its low concentration. The proposed substitute products have no known effects on the ozone layer and also have a shorter atmospheric residence time.

Finally, it is important to note that, taking into account the very long life times of carbon dioxide, the CFC and the nitrous oxide, their relative concentrations will remain high for a very long period, at least a century, whatever measures are taken to control emission levels today.

1.2. *The radiative impact of the trace gases*

The climatic impact of these gases is the result of their radiative properties. It is important to recall first the basic principles of the energetics of the atmospheric system. The surface of the earth, heated by the solar shortwave radiation, emits in return longwave radiation to space with a spectrum spread between 4 and 8 microns approximately, and centered around 15 microns. This is necessary to balance Earth's energy budget. A very substantial part of this longwave radiation is absorbed by the molecules of some atmospheric gases: these molecules are not those of the major atmospheric constituents, whose molecules are diatomic like N_2, O_2, H_2, but those of trace gases which have a polyatomic molecule. Note that the longwave emission is not the only manner by which the surface of the Earth returns energy to the atmosphere or space: there is a transport of energy through convective fluxes, and most notably through the latent energy flux, which corresponds to the evaporation and condensation of the water. Under present conditions, this greenhouse effect increases Earth's surface temperature by about 33 K, raising it from 255 to 288 K, or 15°C, which is the present mean Earth surface temperature. Water vapour is a major greenhouse gas. Human activities do not modify its concentration directly. But water vapour loading may still change in the future, as an indirect consequence of global warming produced by anthropogenic trace gases.

To estimate the contribution of a given gas to the greenhouse effect, we need to know not only its concentration, but also its absorptivity for each wavelength of the thermal infrared spectrum. The absorbed energy depends on the position of each absorption line within the whole spectrum of the infrared terrestrial irradiance. It is also possible, for the sake of simplicity, to characterize the radiative importance of each gas for future climate changes, per unit volume, or per unit mass, by a "Global Warming Potential" Index (GWP), which combines its radiative effect and its atmospheric time life. This index depends on the future date at which the radiative importance of the gas is estimated. The radiative importance of the various gases is estimated in Table 2.

It is important to state at this stage that the perturbation of the radiative budget of the Earth due to the anthropogenic increase of greenhouse gases, in spite of its potentially very significant consequences, is relatively modest. The 4 Wm^{-2} associated with a doubling of the CO_2 represents roughly 1% of the mean incoming solar radiation. But a 1% modification of the mean surface temperature, which must then be expressed in absolute temperatures, i.e. in degrees Kelvin, is roughly 3 degrees, which is by no means negligible. This is one of the reasons why the scientific problems involved are so difficult. With this in mind, the IPCC report established four scenarios A, B, C and D, each representing a different assumption concerning the level of trace gas emissions, A being a "Business as Usual" Hypothesis. The contribution of each gas under this scenario is detailed in Figure 2.

2. The physical models used to study the climate sensitivity

The models used to study the potential impact of the greenhouse effect are very complex three-dimensional models which simulate explicitly the atmospheric flow from the Navier–Stokes equations, discretized over the Earth's sphere. A representation of the various sinks and sources of energy, water and momentum is taken into account. It includes a large variety of processes such as

Table 2. Global warming potentials of the main atmospheric gases (source: 1990 IPCC Report). Values are calculated relative to carbon dioxide, and following the instantaneous injection of 1 kg of each trace gas.

Trace Gas	Estimated Lifetime (years)	Global Warming Potential Integration Time horizon Years 20	100	500
Carbon Dioxide	*	1	1	1
Methane-inc indirect	10	63	21	9
Nitrous Oxide	150	270	290	190
CFC-11	60	4500	3500	1500
CFC-12	130	7100	7300	4500
HCFC-22	15	4100	1500	510
CFC-113	90	4500	4200	2100
CFC-114	200	6000	6900	5500
CFC-115	400	5500	6900	7400
HCFC-123	1.6	310	85	29
HCFC-124	6.6	1500	430	150
HFC-125	28	4700	2500	860
HFC-134a	16	3200	1200	420
HCFC-141b	8	1500	440	150
HCFC-142b	19	3700	1600	540
HFC-143a	41	4500	2900	1000
HFC-152a	1.7	510	140	47
CCl4	50	1900	1300	460
CH3CCl3	6	350	100	34
CF3Br	110	5800	5800	3200

exchanges with the surface, unresolved scales, and physics of clouds, which are parameteried. The exception is some radiative processes which are computed explicitly from a radiative transfer equation. These three-dimensional climate models, refered to henceforth as AGCMs (Atmospheric General Circulation Models), require a great deal of computer time. Depending on the spatial resolution, the computer time needed to simulate one seasonal cycle varies from 5 hours of one Cray 2 processor (resolution of about 6° in longitude and latitude, very often used for climate purposes) to 100 hours (resolution of about 2°, much more adequate to resolve the atmospheric general circulation). These figures are only rough approximations and vary widely

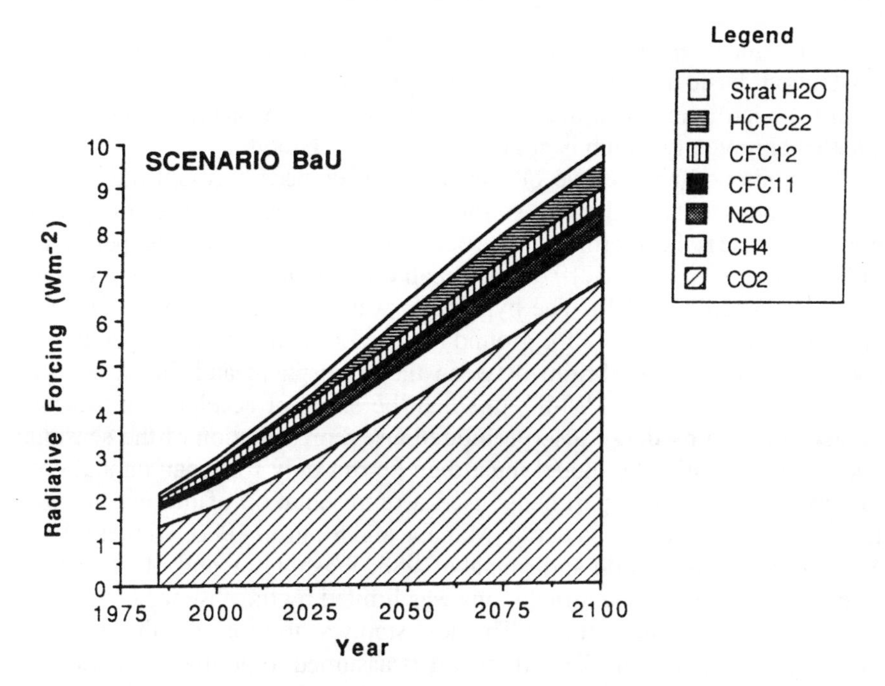

Figure 2. Contribution of each trace gas to the radiative forcing increase in scenario A ("Business as Usual") (source: 1990 IPCC Report).

from model to model, depending on the number of vertical levels used, usually from 10 to 20. For the study of long term climatic trends, and therefore for the study of the increasing greenhouse effect, the models must include some representation of the ocean which can be very crude. The better efforts require a representation of the ocean via a full Ocean General Circulation Model (OGCM).

It is possible to consider simplified versions of these models to explore qualitatively some problems. Indeed, the potential impact of the greenhouse effect was first revealed by one-dimensional vertical models, which simulate the equilibrium vertical temperature profile of an air column in response to solar heating, infrared absorption and emission, exchange with the surface, and vertical redistribution of heat due to convective and radiative processes. These models gave a first approximation of the consequences of a CO_2 increase that was subsequently confirmed, namely, an increase of temperature in the toposphere and a decrease in the stratosphere. At the same time, the height of the tropopause rose. This can be qualitatively explained by saying that the infrared absorption and emission of the CO_2 now occurs higher in the atmosphere. It is also from these simpler models that the first evaluation of the feedback processes that can amplify or dampen the greenhouse warming were made (Schlesinger 1988). As these models are not applicable to realistic scenarios, their results will however not be discussed here. But it is fair to say that our basic understanding of the consequences of greenhouse increase is still a one-dimensional one, and that full three-dimensional models have added a complexity which is not completely understood.

As mentioned above, one of the major differences between the various model experiments carried out so far to determine the impact of the greenhouse increase is the treatment of the ocean. In the first sensitivity experiments of Manabe and Wetherald (1975) the ocean was simulated as if it was a swamp ("swamp ocean models"). The hypothesis was made that there was always enough moisture available for ground evaporation, but no heat capacity was specified, and an equilibrium between the atmosphere and the ocean was reached instantaneously. While, this had the effect of accelerating the convergence of the model, it also prohibited any representation of the seasonal cycle consequently, the model must then be run only for mean annual conditions. More recent experiments used a representation of the ocean as an isotherm slab of prescribed depth (Manabe and Stouffer 1980; Wetherald and Manabe 1986; Wetherald and Manabe 1988; Washington and Meehl 1984). The significance of these simulations was limited by the absence of any representation of oceanic currents. The next step was to represent the transport of energy by the oceans. This transport is assumed to be invariant during a climate modification, which is unrealistic and therefore a serious limitation. The most usual way is to run the model first for the present climate conditions with observed sea surface temperatures (SSTs). The oceanic transport is then determined as the mechanism through which atmospheric fluxes are balanced at the air/ocean interface over an annual cycle. This procedure has been used to complement slab ocean models in Wilson and Mitchell (1987); Hansen et al. (1984) and in other experiments at Goddard Institute for Space Studies (GISS). Until now, there have been very few experiments coupling AGCMs to true representations of the ocean, i.e. OGCMs. Spelman and Manabe (1984) and Manabe and Bryan (1985) have used a simple model with idealized geography, Schlesinger et al. (1985), and Schlesinger and Jiang (1988) have coupled an OGCM to a simple atmospheric GCM with only 2 vertical layers. More recently, Washington and Meehl (1989) and Manabe et al. (1990) have produced the first experiments using truly coupled state-of-the-art AGCMs and OGCMs. Similar, but still unpublished experiments were also carried out at the United Kingdom Meteorological Office and the Max Planck Institute of Hamburg. Using such sophisticated coupled models does not solve all the problems. The representation of the physics of the oceanic upper layers, which is crucial for the exchanges with the atmosphere at the seasonal time scale, is always treated in a simplified manner, as is the physics of the sea-ice. Finally the models, so far, tend to drift away from the observed climate, and need to be stabilized through a "flux correction method".

3. The sensitivity of the climate to a doubling of the atmospheric CO_2 content: An indicative experiment and a tool for the comparision of the models

Many experiments designed to obtain an indication of climate sensitivity to increasing greenhouse effect have considered a somewhat academic case which is to study the "equilibrium response" of the models to a doubling of atmospheric CO_2. It is important to note that this equilibrium is limited to the atmosphere and the ocean surface. The deep oceans would require thousand of years to reach an equilibrium and therefore are not relevant to our concern over the climate of the next century.

There are several advantages to such an approach. First, this equilibrium is reached in 10 to 20 years of simulated time, after the CO_2 level has been switched instantaneously to twice its present value, a much faster process than the slow increase from the preindustrial era. It is, therefore, an experiment which corresponds to the availability of computer time of many groups, and a simple way to compress model results. Also, the study of the equilibrium response of the climatic system diminishes the role of the ocean, which to a large extent, guide the transient behaviour of the climate system. It represents as well a well-defined experiment for models where the ocean is treated in a simple manner. But in interpreting the results, one must keep in mind that they are only indicative. Although they do not correspond to the climate which we will experience when CO_2 has doubled, at the very best, they provide an indication of climate at a slightly later stage when corresponding greenhouse increase has had time to heat the ocean. The time scale of this warming is not uniform. It takes longer in regions where the heat can penetrate deep into the oceans, as in the Antarctic Ocean, and smaller over the continents or in regions where a rather thin and stable ocean upper layer can respond quicker to the atmospheric forcing.

We first review the facts over which most models agree. Some comprehensive reviews of the model results were made by Schlesinger and Mitchell (1987), and by Mitchell (1989). The IPCC report also presents more recent unpublished results from the United Kingdom Meteorological Office (UKMO), the Canadian Climate Center (CCC) and the Geophysical Fluid Dynamics (GFDL), accomplished with models which use slightly higher horizontal resolution.

All models agree on a substantial increase of the mean surface temperature, with estimations ranging from 1.9°C (Mitchell et al. 1989) to 5.2°C (Wilson and Mitchell 1987). Therefore the two extreme predictions were made with two versions of the same model, which differ through their parameterization of cloudiness only. At high latitudes the surface warming is much larger than the mean warming in the winter hemisphere, and, conversely, smaller in the summer hemisphere. The warming is also smaller at low latitudes, where it carries little seasonal cycle. The results of the surface warming simulated

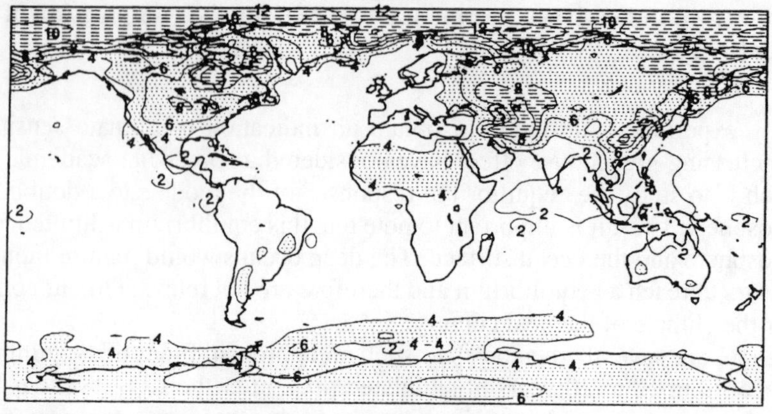

(b) DJF 2 X CO2 - 1 X CO2 SURFACE AIR TEMPERATURE: GFHI

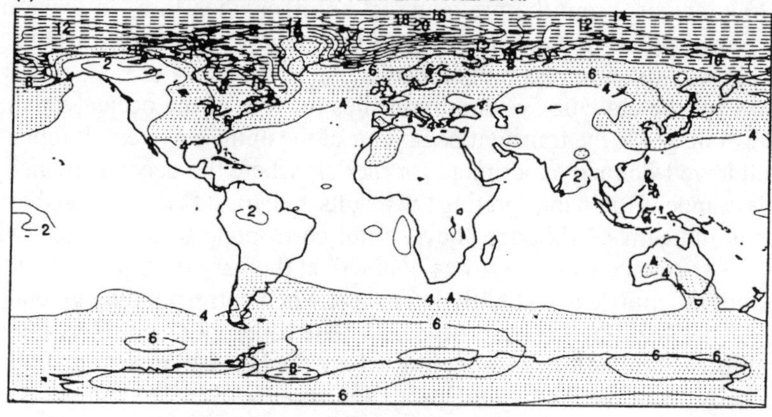

(c) DJF 2 X CO2 - 1 X CO2 SURFACE AIR TEMPERATURE: UKHI

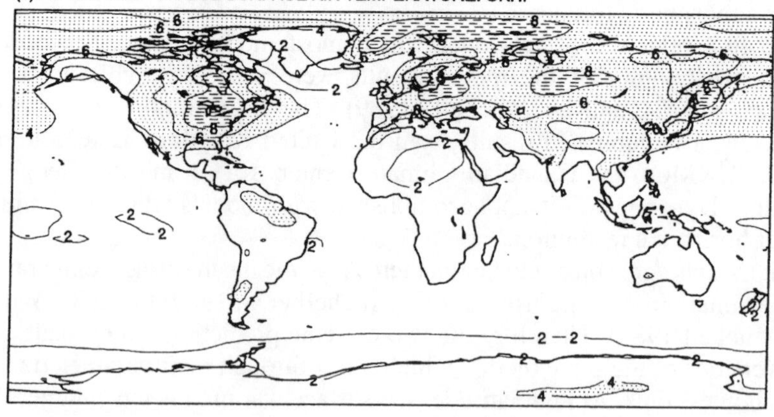

Figure 3. Projected temperature increase for a doubling of the CO_2. Comparison of three model results (see text) for the Northern Hemisphere winter (a–c) and for the Northern Hemisphere summer (d–f) (source: 1990 IPCC Report).

(d) JJA 2 X CO_2 - 1 X CO_2 SURFACE AIR TEMPERATURE: CCC

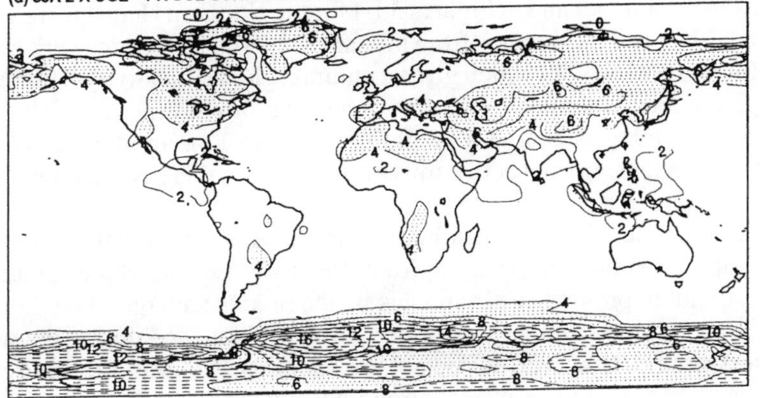

(e) JJA 2 X CO_2 - 1 X CO_2 SURFACE AIR TEMPERATURE: GFHI

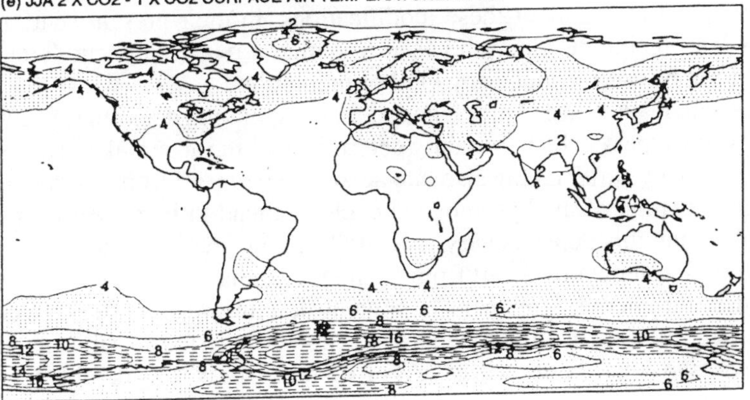

(f) JJA 2 X CO_2 - 1 X CO_2 SURFACE AIR TEMPERATURE: UKHI

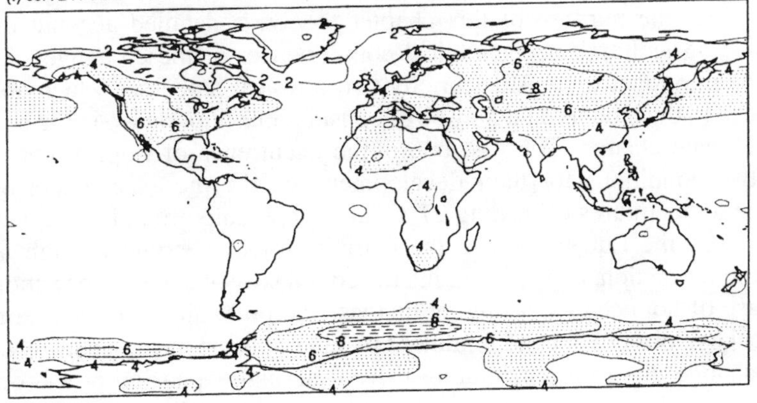

Figure 3. (Continued)

by the UKMO, CCC, GFDL models are in Figures 3a–c for the Northern Hemisphere summer and in Figures 3d–f for the Northern Hemisphere winter. The UKMO result is obtained from the same experimental framework as used by Mitchell et al. (1989) which gives a minimal global sensitivity. The heating at high latitudes is limited to the atmospheric surface layers. At low latitudes a strong warming occurs at a height of about 15 km, an altitude which is reached by the deep convective towers. In all cases there is a cooling in the stratosphere.

The same models produce a marked increase in the global precipitation and evaporation patterns, ranging from 3 to 15%. Enhanced precipitation is generally more pronounced in models which produce a larger warming. It is also larger at high latitudes irrespective of the season, and during the winter in mid-latitudes. However, the sign of precipitation change is uncertain at low latitudes where there is a mix of strong negative and positive modifications in projected precipitation profiles. The models do not agree on the precise geographical locations of these modifications. In their present state, model output on precipitation may be used to define a risk factor, but certainly it should not be used to make any kind of prediction.

The general increase of precipitation and evaporation does not say anything about soil wetness, which is mainly determined as the accumulated difference between the two after taking also into account the run-off from saturated soils. The acceleration of the hydrological cycle can lead to increased wetness or droughts, and this depends very critically on how precisely the models treat the ground water cycle. Apart from a tendency to have wetter winters at high latitudes, the model agreement is poor in defining regions of increased or decreased soil wetness. The models agree, however, on the possible magnitude of these changes, very often in excess of 10 mm. This is considerable for semi-arid regions, which will be particularly vulnerable if a future climate change should occur.

It is not the purpose of this chapter to give a detailed account of the possible agricultural or economical consequences of these changes. But it may be important to consider briefly two physical processes which are not included in the models, and may be of considerable importance in the context of a climate change. The first one is the occurrence of tropical cyclones. The two conditions for their development are the presence of hot ocean surface temperatures exceeding 27°C, and a breeding ground some degrees away from the Equator where the Coriolis force is strong enough. As a first approximation, and recognizing the complexity of cyclone formation, an increase of hot ocean areas might increase the probability of occurrence of tropical cyclones. A second important parameter is the mean sea level which may change due to a temperature driven thermal expansion of the ocean. This could raise sea level by a few tens of centimeters. A change in the mass balance of continental snow and ice should also contribute to a sea level increase. Estimation of the magnitude of this latter term is faced with the same problems as mentioned for soil moisture. Predictions by present models of modifications in the hydrological balance is very imprecise; even its sign cannot be predicted and only a risk factor can be assessed.

4. Climate feedbacks and climate sensitivity

As mentioned already, the problem posed by the increase in atmospheric greenhouse gases is quite new. In fact, we want to determine the sensitivity of the climate to a small radiative forcing at the tropopause of 4 Wm^{-2} due to a doubling of the CO_2. This problem is very different from the one encountered when using a model to reconstruct the mean climate. We need to compute the derivative of the climate and no longer the climate itself. 4 Wm^2 is substantially smaller than the accuracy of present models and observations. The most appropriate models for determining the climate sensitivity are not those which give the "best" account of the present climate, but those which do it for the right reasons and include all the important physical processes. The work of the climatology community during the last years has been to identify more of those feedback mechanisms which are most important to the response of the climate system and must therefore be represented accurately within models.

To study these mechanisms we can first define an "unperturbed" warming corresponding to a doubling of the atmospheric CO_2 as the warming which would occur if the only possible response of the atmosphere was to change its temperature. This value is roughly equal to 1.2°C. The higher values predicted by the models can be explained only through a number of positive feedbacks.

4.1. *Water vapour feedback*

The feedback effect for which there is the greatest agreement among models is the water vapour feedback. When the atmospheric temperature increases, so does the water vapour saturation point. The atmosphere can therefore store more water vapour, which in turn increases the atmospheric greenhouse effect. The response of the atmospheric models to a prescribed modification of the sea surface temperature was used by Cess et al. (1984) as a simple way to compare the sensitivity of 14 models. It is uncertain that the water vapour feedback would operate in a manner equally successful for a doubling of the CO_2. It is quite remarkable, however, that 18 models converge in giving an estimate of the sensitivity associated with water vapour changes around the same value of 0.45 $Wm^{-2}\ K^{-1}$. This is confirmed independently from satellite measurements by Raval and Ramanathan (1980) and Ramanathan et al. (1991), although these estimates, again, are based on the atmospheric response to different types of forcing such as season and latitudinal variations in insolation, and El Niño anomalies. In spite of some ongoing debate on this subject it is more and more likely that the positive water vapour feedback is not some systematic artifact of the models. In most cases, it raises the warming from 1.2°C without feedbacks, to 1.9°C with water vapour feedbacks.

4.2. *Snow and ice feedbacks*

When the temperature of the Earth rises we can expect that the surface covered by snow and sea-ice will diminish. The reflectivity or albedo of the planet is therefore decreased, which means that Earth can absorb more solar radiation. This constitutes a second important feedback process, which explains that colder climates, where the snow or sea-ice areas are more extensive, also responds more strongly to a given forcing (Spelman and Manabe 1984). The sign of this feedback is unambigously positive. It will increase warming from 1.9 to 2.4°C. But this projection is valid only when associated changes in cloud cover are ignored. Clouds have the potential of reversing these results.

It is, however, quite doubtful that present models are able to estimate snow and sea-ice effects quantitatively. The sea-ice and upper ocean physics is generally treated in a very simplified way and the seasonality of any sea-ice response to CO_2 increase is not simulated correctly by most models. The albedo of natural surfaces covered by snow or sea-ice is not clearly known from an observational point of view. Simulations by present day models also ignore many of its dependences such as those on the wavelength of the incident solar visible or near infrared radiation on the state of the snow (wet or cold, young or old), the underlying vegetation cover, and the solar zenith angle.

4.3. *Cloud feedback processes*

The above mentioned comparison of 14 model response to a simple prescribed change in the SST (Cess et al. 1989) has shown that clouds and their representation within models could be held responsible for a large fraction of the uncertainty and disagreement in present models' results.

The very complex role of clouds is due to the fact that their interactions with solar and infrared radiation have the opposite sign. They compensate in a different manner depending on various cloud parameters. For simplicity we may consider three main independent parameters:

- the geographical distribution of the clouds represented by horizontal cover and vertical extension;
- the liquid or ice amount contained in the clouds;
- the size of the particles.

Low clouds are generally thicker and warmer. They reflect efficiently the solar radiation incident on them and have a small greenhouse effect. For these reasons, they tend to cool the surface of the Earth. Conversely, for the colder and thinner high clouds greenhouse effect tends to dominate. Hence, the prediction by most models of a decrease of low clouds and increase of high clouds is a potentially very important feedback. But this effect may be disturbed by a thickening of the low clouds, which may act as a strong negative feedback because such thickening increases solar radiation albedo of the cloud while infrared emissivity is often maximized in such clouds (Roeckener et al. 1987). Mitchell et al. (1989) have also shown that the replacement of ice clouds by water clouds in a warmer climate, cloud lead to a thickening of the clouds and therefore to a negative feedback. Water clouds need to reach a certain density before the autoconversion of cloud water into rain water can take place, whereas ice crystals grow quicker to a size where they can begin to fall. These effects, however, are only known qualitatively (Platt 1989), and are very sensitive to any change in cloud heights. The sign of the feedbacks may therefore very well be reversed (Le Treut and Li 1990).

The size of the cloud particles is also a very important factor because it regulates the cloud albedo, i.e. its interaction with the solar radiation, but has little impact on its interaction with infrared radiation. The differential impact of a change in the droplet or crystal size may be quite important also. An important factor controlling this latter parameter is the number of Cloud Concentration Nuclei. It has been argued (Charlson et al. 1987) that the production of dimethylsulphate (DMS) aerosols by the marine biological activity, which seems very important in the formation of low marine stratocumulus, could be modified under climate change. Indeed such changes have been observed during glacial oscillations (Legrand et al. 1988).

The present models are unable to handle correctly the complexity of mechanisms found in cloud feedback processes. In many cases, the cloud cover is diagnosed on a statistical or empirical basis as a function of some large scale variables including relative humidity, vertical velocity, and vertical stability. Only recently have a few models attempted to simulate explicitly the cloud water cycle, and therefore the cloud water content. This implies a simplified treatment of cloud microphysics. As for the prediction of the cloud droplet size, it is beyond the reach of the present crop of models.

A strong release of heat is associated with the condensation of water vapour within clouds. For convective scales, which are not resolved explicitly, this effect needs to be represented parametrically. The vertical distribution of the heating in the Tropics also turns out to be very sensitive to the details of those parameterizations (Schlesinger and Mitchell 1987).

4.4. *Interactions with the surface*

There are a large number of feedback processes which may involve exchanges between the atmosphere and the ground. The first factor is surface hydrology. A comparison between the models of the National Center for Atmospheric Research (NCAR) and the Geophysical Fluid Dynamics Laboratory (GFDL) has shown the impact of ground water reserve on climatic sensitivity (Manabe and Wetherald 1987). In the mid latitudes in particular, if the soil is saturated in winter, any excess precipitation at that time of the year will disappear as run off, and the corresponding water will not be available for evaporation in the summer; but in a simulation where the soil is unsaturated in winter, an increase in precipitation may have an impact on the summer climate.

The exchanges of water, heat of momentum between the ground and the atmosphere are strongly regulated by vegetation cover. This vegetation cover may evolve with the climate, and reversely affect the climate. One example is the Charney mechanisms, which may have played an important role in the very long Sahel drought (Laval and Picon 1986). It postulates that the surface albedo is increased when vegetations does not occur. As present models do not include vegetation cover as a predicted variable or set of variables, these feedback effects are not found in current simulations of the climate response to a doubling of the CO_2.

Strong biochemical feedbacks may also occur at the surface of the Earth. If CO_2 increases, so would photosynthesis. Reversely, in the case of a warming, the increased transpiration from vegetation, or decomposition of ground organic matter, may release more CO_2 into the atmosphere. Similarly one may expect increased emissions of methane from areas which are presently farmed.

Finally the multiple role of the ocean should be stressed again:

- its strong thermal inertia determines the timing of climate modifications;
- together with the atmosphere, it transports energy from the equator to the poles;
- it determines the sea-ice extent.

This list of feedback processes is certainly not exhaustive. But it is a good indication of the global nature of the problem posed by the greenhouse threat. We cannot consider any part of the climate system separately.

5. Conclusion: Can we predict the climate of the XXIth century?

We have reviewed some of the uncertainties of our present knowledge regarding the physical mechanisms which govern the evolution of the climate system. These uncertainties must be kept in mind when interpreting the results of climate simulations, but should not imped scientists from applying their models to realistic scenarios.

For this purpose, it is necessary to use the climate models in what is generally refereed to as a "transient" mode. This model is forced by a realistic estimate of trace gas concentration since the beginning of the industrial era including a projection of the future based on scenarios defined in Section 1.2. The representation of the ocean then becomes a key feature. Hansen et al. (1988) have used an atmospheric model, coupled to a "slab" ocean with prescribed oceanic heat transport, to estimate the evolution of the climate up to 2020 based on three different scenarios of the future trace gases emissions. The resulting global warming was respectively 0.6, 1.3 and 1.7°C, using a model whose sensitivity for a doubling of the CO_2 is rather high (4.2°C).

The response of fully coupled AGCMs and OGCMs to a 1% yearly increase of the CO_2 was also studied at NCAR (Washington and Meehl 1989) and GFDL (Manabe et al. 1990). Similar experiments are being made at the UKMO and MPI of Hamburg. In general, warming occurs quicker over continents than over oceans. The extreme case is the Antarctic Ocean, where due to a very strong vertical mixing of the water, the warming remains weak. It may be noted that, in those experiments, the variations of the oceanic circulation may induce some local cooling of the surface, for example at high altitudes in the Southern Hemisphere, or in the North Atlantic Ocean. But, generally, these "transient" simulations are a confirmation of the "equilibrium" simulations, with the ocean introducing a mean time lag of about 10 years.

These simulations constitute the best information that will be available for all decision-makers in the near future. The complexity of the problem will require many scientific studies before any significant new step is reached. But, from the best of what we know today, the risk of a global warming of a few degrees by the end of the next century is a serious one that fully deserves the attention it has managed to attract during the last years.

References

Cess, R.D., Potter, J.P., Blanchet, G.J., Boer, S.J., Ghan, J.T., Kiehl, H., Le Treut, Z.X., Li, X.Z., Liang, J.F.B., Mitchell, J.J., Morcrette, D.A., Randall, M.R., Riches, E., Roeckener, U., Schlese, A., Slingo, K.E., Taylor, W.M. Washinton, R.T. Wetherald and I. Yagai (1989): Interpretation of cloud-climate feedback as produced by 14 atmospheric general circulation models. *Science*, 245, 513–516.

Hansen, J., Lacis, A., Rind, D., Russel, G., Stone, P., Fung, I, Ruedy, R. and Lerner, J. (1984): Climate Sensitivity: Analysis of feedback mechanisms. In: *Climate process and climate sensitivity*, Maurice Ewing Series, Vol. 5, ed. by J.E. Hansen and T. Takahashi. American Geophysical Union, 130–163.

Hansen, J., Fung, I., Lacis, A., Rind, D., Lebedeff, S., Ruedy, R. and Russel, G. (1988): Global climate changes as Goddard Institute for Space Studies Three-Dimensional Model, *J. Geophys. Res.*, 93, 9341–9364.

Houghton, J.T., Jenkins, G.J. and Ephraum, J.J. (1990): *Climate Change, the IPCC Scientific Assessment*. Cambridge University Press.

Laval, K. and Picon, L. (1986): Effect of a change of the surface albedo of the Sahel on climate. *J. Atmos. Sci.*, 43, 2418–2429.

Legrand, M., Delmas, R. and Charlson, R.J. (1988): Climate forcing implications from Vostok ice-core sulphate data. *Nature*, 334, 418–420.

Le Treut, H. and Li, Z.X. (1990): The sensitivity of an atmospheric general circulation model to prescribed SST changes: Feedback effects associated with the simulation of cloud optical properties. *Climate Dynamics*, 5, 175–187.

Li, Z.X. (1990): Etude de l'interaction nuage-rayonnement dans le contexte du changement climatique dû à l'augmentation des gaz à effet de serre dans l'atmosphère. Thèse de l'Université de Paris VI.

Lorius, Cl., Jouzel, J., Raynaud, D., Hansen, J. and Le Treut, H. (1990): The ice-core record: Climate sensitivity and future greenhouse warming. *Nature*, 347, 139–145.

Manabe, S. and Bryan, K. (1985): CO_2 induce change in a coupled ocean-atmosphere model and its paleoclimatic implications. *J. Geophys. Res.*, 90, 11689–11707.

Manabe, S. and Stouffer, R.J. (1980): Sensitivity of a global climate model to an increase in the CO_2 concentration in the atmosphere. *J. Geophys. Res.*, 85, 5529–5554.

Manabe, S. and Wetherald, R. (1975): The effect of doubling the CO_2 concentration on a general circulation model. *J. Atmos. Sci.*, 32, 3–15.

Manabe, S. and Wetherald, R.T. (1987): Large scale changes of soil wetness induced by an increase in atmospheric carbon dioxide. *J. Atmos. Sci.*, 21, 580–613.

Manabe, S., Bryan, K. and Spelman, M.J. (1990): Transient response of a global ocean-atmosphere model to a doubling of atmospheric carbon dioxide. *J. Phy. Oceanography* (to appear).

Matthews, E. (1985): Atlas of archived vegetation, land-use, and seasonal albedo data sets. NASA Technical Memorandum 86199, 53 pp.

Mitchell, J.F.B., Senior, C.A. and Ingram, W.J. (1989): CO_2 and climate: a missing feedback? *Nature*, 341, 132–134.

Platt, C.M.R. (1989): The role of cloud microphysics in high-cloud feedback effects on climate change. *Nature*, 341, 428–429.

Posey, J.W. and Clapp, P.F. (1967): Global distribution of normal surface albedo. *Geofis. Int.*, 4, 33–49.

Ramanathan, V. and Collins, W. (1991): Themodynamic regulation of ocean warming by cirrus clouds deduced from observations of the 1987 El Niño. *Nature*, 351, 27–32.

Raval, A. and Ramanathan, V. (1989): Observational determination of the greenhouse effect. *Nature*, 342, 758–761.

Rokeckner, E., Schlese, U., Biercamp, J. and Loewe, P. (1987): Cloud optical depth feedbacks and climate modelling. *Nature*, 329, 138–139.

Sausen, R., Barthel, K. and Hasselmann, K. (1988): Coupled ocean-atmosphere models with flux correction. *Clim. Dynamics*, 2, 145–163.

Schlesinger, M.E. (1988): Quantitative analysis of feedbacks in climate model simulations of CO_2-induced warming. In: *Physically-based modelling and simulation of climate and climate change*, Part 2, ed. by M.E. Schlesinger, NATO ASI Series, Vol. C 243, 653–735.

Schlesinger, M.E. and Jiang, X. (1988): The transport of CO_2-induce warming into the ocean: An analysis of simulations by the OSU coupled atmosphere-ocean general circulation model. *Clim. Dynamics*, 3, 1–17.

Schlesinger, M.E. and Michell, J.F.B. (1987): Climate model simulations of the equilibrium climatic response to increased carbon dioxide. *Review of Geophysics*, 25, 760–798.

Spelman, M.J. and Manabe, S. (1984): Influence of ocean heat transport upon the sensitivity of a climate model. *J. Geophys. Res.*, 89, 571–586.

Washington, W.M. and Meehl, G.A. (1984): Seasonal cycle experiments on the climate sensitivity due to a doubling of CO_2 with an atmospheric circulation model coupled to a simple mixed layer ocean model. *J. Geophys. Res.*, 89, 9475–9503.

Washington, W.M. and Meehl, G.A. (1989): Climate sensitivity due to increased CO_2: Experiments with a coupled atmosphere and ocean general circulation model. *Clim. Dynamics*, 4, 1–38.

Wetherald R.T. and Manabe, S. (1986): An investigation of cloud cover change in response to thermal forcing. *Clim. Change*, 8, 5–23.

Wilson, C.A. and Mitchell, J.F.B. (1987): A doubled CO_2 climate sensitivity experiment with a GCM including a simple ocean. *J. Geophys. Res.*, 92, 13315–13343.
Wigley, T.M.L. (1989): Possible climate change due to SO_2-derived cloud condensation nuclei. *Nature*, 339, 365–367.

H. Le Treut
Laboratoire de Météorologie Dynamique de CNRS
Ecole Normale Supérieure
24, rue Lhomond
75231 Paris Cédex 05
France

2.6. IMPACTS OF THE CLIMATE VARIABILITY ON HUMAN ACTIVITIES

ANNICK DOUGUÉDROIT, JEAN-PIERRE MARCHAND,
MARIE-FRANÇOISE DE SAINTIGNON AND ALAIN VIDAL

1. Introduction

Attention has focused more and more on a better understanding of relationships between climate and human activity. Impacts of climate variability on human activities considered a priority by many decision makers, should be estimated even if they require measurements that are not yet available (Maunder, 1995). One of the four components of the Second World Climate Programme established at the conclusion of the Second World Climate Conference held in Geneva in 1990, involves studies aimed at assessing climate impacts. This programme includes both the present state of impacts as well as their future development under a $2XCO_2$ induced global warming (WMO 1979, 1982, 1990).

The first step in developing an understanding of impacts is a comprehensive assessment of those activities that are climate sensitive. Even here, climate is only one of several components of the environments which impact them. These environments interact in systems such that impacts can only be properly investigated if the systems themselves are considered together.

2. Systems approach to climate impacts

2.1. *Main climate sensitive sectors*

A few activity sectors have a high sensitivity to climate (Hobbs 1980; Phillips 1989; Price-Bridgen 1990), even if it not easy to place a monetary value on it. Of particular importance are climate effects in agriculture, energy, water resources and health (Pyle 1979).

M. Yoshino et al. (eds.), Climates and Societies – A Climatological Perspective, 119–150.

Agroclimatology deals with the relationship between climate variability, and the yield and productivity of food crops (Duckham 1974; Nias 1976). The economic importance of large climate fluctuations, droughts for example, on agricultural production was known in the past as it is nowadays (Boch et al. 1981; Wigley 1981). The percentage decrease of some agricultural productions caused by the 1976 drought in England and Wales was as follows; 22% (wheat), 12% (barley), 12% (oats), and 40% (especially potatoes) (Roy et al. 1978). In 1982–1983, wheat production in Australia was only 63% of the previous five-year average due to the drought conditions which prevailed in the eastern states (Gibbs 1984). The political significance of those large climate fluctuations was discussed by Marchand (1985) and Lamb (1988) for the Irish famine in the 1850s.

By understanding the relationship between climate and crop yields, the climatologist attempts to predict crop yield in advance (Hodges 1991). But he/she faces a factor which cannot be controlled, namely, climate variability whose importance increases as agriculture is modernized. This is because selected productions encouraged by present developments in intensive forms of agriculture are more sensitive to short-term climatic fluctuations than traditional forms. So, recent agricultural crisis associated with climate are preconditioned by the economic and political evolution over the previous decades.

Problems of energy supply involve the production of alternative forms of energy mainly through hydroelectricity and solar radiation. They too deal with the economies of energy consumption which are driven by outdoor temperatures (Johnson et al. 1969). In particular, they must also take into account the energy demands for the heating and cooling designed to maintain indoor comfort. Such demands are highly dependent on the microclimatology of building designs (Benjamin and Davis 1971).

Human bioclimatology represents an interdisciplinary field which has tried to explain various climatic influences on biological activity, human reactions and illness. It also includes the study of weather and climate sensations (heat, comfort, . . .), and the pathology connected with those reactions, as well as with diseases and epidemics due to weather events. For example, death rates have been connected with temperatures. In France, between 1949 and 1977, they increased during each winter in relation to the duration of cold weather (Le Berre et al. 1982).

Water used for economic activities, agriculture, and industry, for example, and for every day life in fast developing towns are in greater demand. Their total amount and their spatial distribution is basically determined by climate, but their use depends on the level of development of each country. The importance of recycling of liquid waste becomes more urgent with increasing industrialization and urbanization.

Beside these large sectors commonly considered as climate-sensitive, many activities such as transport, winter and summer holidays, the construction

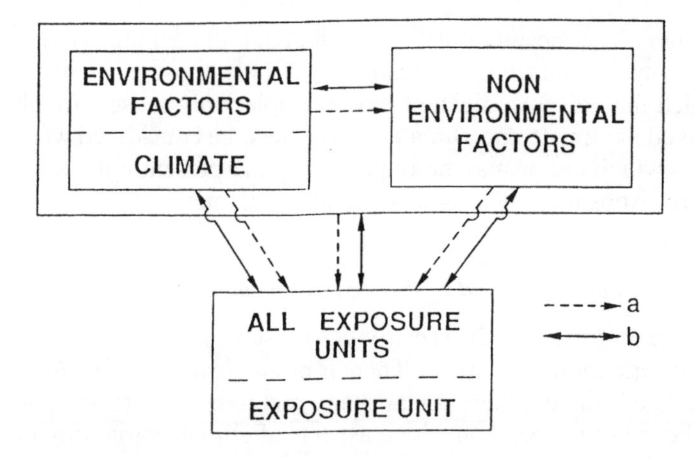

Figure 1. Schema of simple impacts (a) and interactive impacts (b) in relationships between climate and "exposure units" (human activities, health and settlement).

industry, customer behaviour for the sale of products, storage and transport of products etc. depend on weather and climate.

As the global increase of world population takes place mostly in developing countries, the pressure on food and water rises. The need for increasing food with limited funds in most countries makes it necessary to take greater advantage of the climate resources of the agricultural production system and avoid waste of available water.

2.2. *A system approach*

In assessing the impacts of climate on a human activity, an "exposure unit" cannot be separated from the other factors influencing it. The climate is not only one of the environmental factors which impact human activity, it also influences other contributing non-environmental ones. For example, agricultural production depends directly on precipitation but also indirectly on irrigation programmes. Both environmental and non-environmental factors have a combined influence on exposure units (Figure 1a).

But human activities influence the climate too, mainly through air pollution. They create urban heat islands over towns and will enhance (have already begun to enhance?) global warming through the next decades. Such a modified climate will have consequences on both environmental and non environmental factors of the "exposure unit" and on the "exposure unit" itself (Maunder 1995). For example, in the next decades, areas of agricultural production are expected to shift due to global warming. So the connections between climate and human activities do not have a single direction, but interact through several feedback loops (Figure 1b).

Although all "exposure units" are included in the general system, vulnerability in any one unit may be restricted to only those parts of the system with which the unit is associated. For example, only those variables which are involved in climate and climate change need be considered when dealing with units which are or may be impacted by the present climate or a future one. Each "exposure unit" requires a unique treatment.

2.3. *Long-term planning and real-time management*

For each "exposure unit", the sensitivities must be first demonstrated, and the required information determined. There is no single method for characterizing the impacts of climate, there are as many methods as there are sensitivities. The methods used depend on which aspects of climate variability impact the two types of situations faced by decision-makers, namely, long-term planning and real time management.

Many important economic and social decisions on long-term projects are being made today. Analysis of past climate statistical data drawn from records of weather resources and hazards, are useful mainly for strategic planning and long-term investment to which weather sensitive activities should adjust. Its justification lies in the significant benefits made possible by loss prevention and the prospect of optimizing opportunities through suitable planning. For example, information on climate variability is needed to assist planning in irrigation schemes, agro-business enterprises, agriculture, water resources (Duband et al. 1989), and snow-making in ski resorts (see Section 3.1), etc.

Long-term planning is always based on the assumption that the past climate data are reliable indicators of the future. It assumes the stationarity of the climate, an hypothesis suitable only for the next few years, or a little more. It becomes a questionable assumption when planning for the next several decades because of the predicted global warming. We are not able to predict how the climate will change but we must estimate the range of possible climatic changes and their impacts.

But climate impacts can also be used in current operational decision-making and for those planned for the very near future. Real or near-real time data have proved to be a powerful and essential information source for large areas of activities including management decisions in agriculture, building and construction, energy, marketing etc. In agriculture, such data help to estimate the reactions of crop development to specific climatic factors (see Section 3.2), irrigation scheduling, water flow regulation, pesticide applications and so on.

2.4. *Methods used for assessing climate impacts*

Assessing climate impacts has been attempted for decades by many methods (Patil and RAO 1994). They can be summarized by four steps:

– determination of accurate data,
– use of indices,
– use of statistical models,
– use of biological models.

2.4.1. *Accurate data*
Lists of climatic parameters applicable to each climate sensitive sector are long (see Landsberg 1981, for applications in human bioclimatology and agroclimatology). They can be determined in a laboratory, in the field or by statistical relationships (Yarnal 1993). Experiments on the behaviour of human beings or the development of plants grown in laboratories have the advantage of providing accurate measurements. But the drawback of isolation, especially in the case of biological problems, can produce results different from those obtained under open-air conditions (agriculture, human health). Results of such experiments appear reliable in purely technical problems such as transportation and industrial activities. Successful production of snow in ski resorts by snowmakers depends on the temperature and relative humidity of the atmosphere, whose thresholds have to be determined for each machine (see Section 3.1).

At any time, uncertainties in human behavior can disrupt effective field applications of the results obtained by any methods and open a gap between experimental values and real ones. But any attempt to understand how climatic parameters affect a given activity leads to the identification of direct or derived parameters considered most appropriate to the activity, potential evapotranspiration is one example. Also studies of crops showed that productivity varies through the different stages of crop development. This can be presented in a “crop-weather calendar”. Each calendar fits a genotype with its environmental conditions such as climate and soil. The length of the growing period for sugarcane varies from 10 months to two or three years depending on the producing country (Biswas 1988). This type of information could be included as part of a crop-weather calendar.

The derived parameters can be used as raw figures or presented as indices that are simple to calculate and easy to apply. The indices are constructed as single terms to which impact is strongly correlated.

The development of indices is common methodology, particularly in studies of climate impacts on food production and human health. The length of the growing period of a plant is closely related to temperature. It can be estimated by the cumulative daily average temperature above a base temperature (the growing degree days), thereby linking the plant growth to air temperature. The rice yield in North Japan han been correlated with temperature in midsummer, a season that includes the heading period. A yield index was calculated for Hokkaido:

$$I = 1.2035 - 0.04803(T_{7.8} - 22.5)^2$$

with $T_{7.8}$ = average July–August temperature (Uchijima 1981).

The plant stress can be estimated by the R ratios (ET/PE) which is the ratio of actual evapotranspiration to potential evapotranspiration (Yao 1969). Often, indices combine into a single term those climatic parameters that most influence an "exposure unit".

For human health, evaluations of thermal ambient factors were summarized in several indices (Sipple and Passel 1945; Thom 1959).

2.4.2. *Modelling climate impacts*

Two types of models exist, statistical and biophysical. Their common objective is the integration of the complex interactions of natural resources, technology, and management factors (Baier 1977; Duchon 1986; Whisler et al. 1986; Van Keulen and Wolp 1986; Hodges 1991).

Statistical models relate a given impact (for example seasonal crop or forage or animal productivity in agriculture) to climatic data for the same period and area using statistical techniques such as regression analysis or principal component analysis; they oversimplify the complex interactions in a given system. Limitations exist in their development as a result of the lack of data in some areas including climate data, but more importantly, on the other non climatic factors which impact on the system such as the evolution of technology through time (Douguédroit and Bart 1992).

A great number of statistical models have been developed. Because of the impacts of weather and climate on food productions and health, there have been important investigations to establish regression equations linking yield variations and health variations to climate parameters. Many of them divide the impacts into two parts, representing separately, the influence of technology and climate. The equations have a generic form: $I = P_{(t)}/P_{T(t)}$ where $P_{(t)}$ and $P_{T(t)}$ denote, respectively, the present situation and the trend situation in year t. The latter values are constructed by statistical fit (Figures 2a, 2b).

As a whole, empirical-statistical models are considered, after they have undergone a good calibration and validation, as more efficient for predictions in the field than more sophisticated biophysical models. But their coefficents must be adapted to each regional case.

Biophysical models simulate biological development, growth or production by designing mathematical expressions, preferably non linear relationships, to represent the biological response to the environment. Such a system includes many factors other than climate. It has been developed more extensively in agriculture.

A real biological system consists of flow patterns for matter and energy and processes linking the basic components of the system. The contribution of each process is well known but difficult to express mathematically (Thomas and Smith 1989). Such models are very complex and not entirely successful except when they are focused on the growth of a single plant. Landsberg (1981) has produced a scheme for a system analysis of land use in agriculture

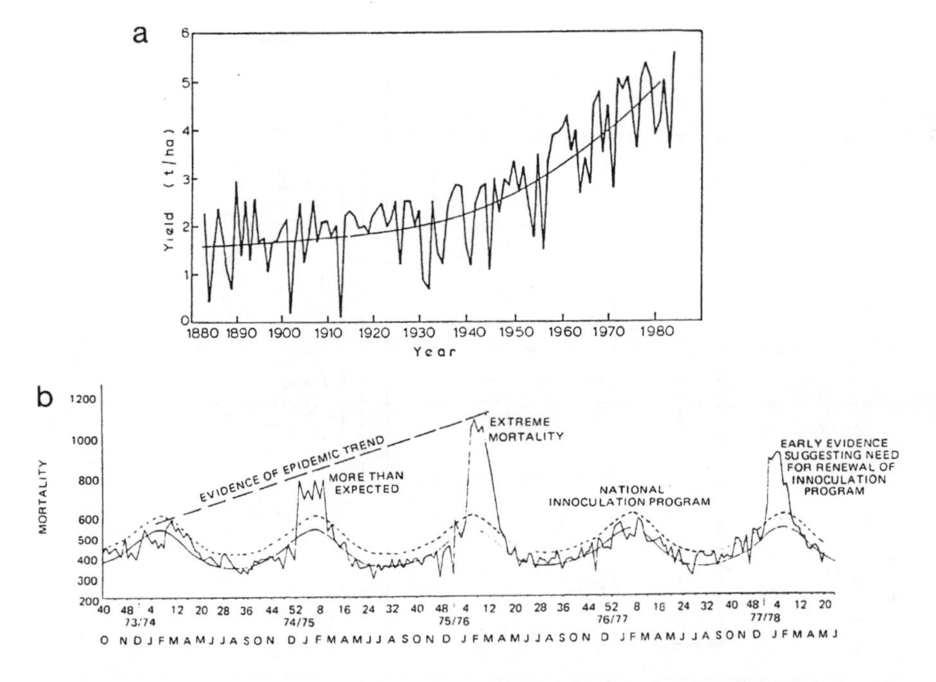

Figure 2. Annual or monthly fluctuations of temperatures and trends. (a) Variations in annual rice yield (tons/hectare) in Hokkaido (1988–1982) (after Uchijima 1983). (b) Cycles of influenza-pneumonia mortality in the United States (1973–1978) (after Pyle 1978).

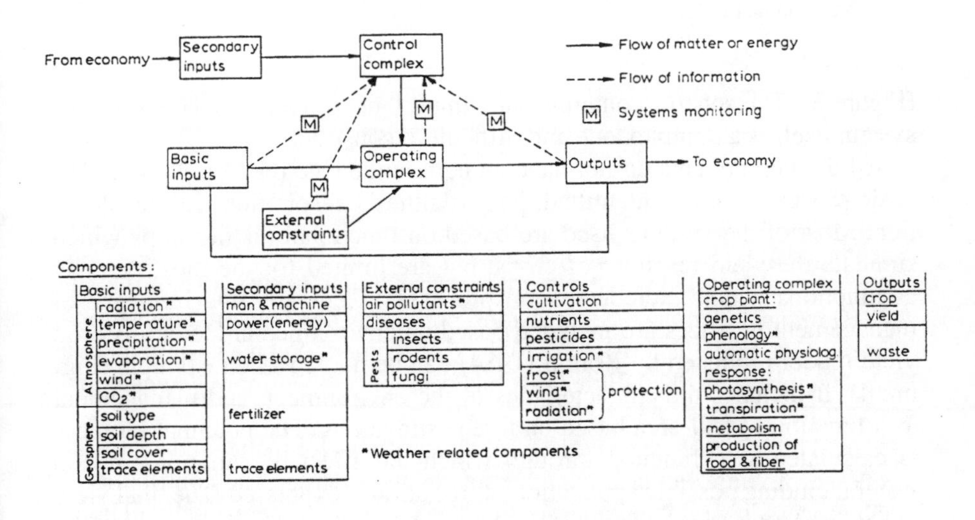

Figure 3. Scheme for a system analysis of land use in agriculture (after Landsberg 1968).

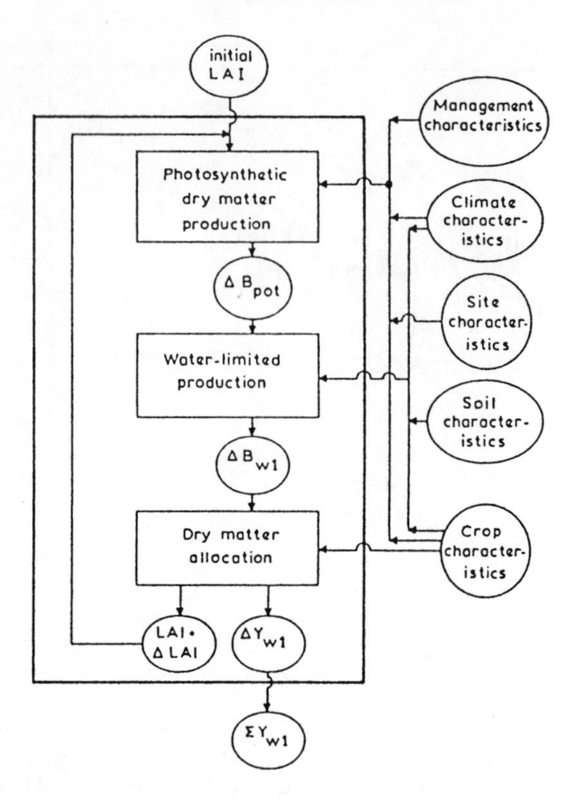

Figure 4. An example of the structure of a crop production model (after Konijn 1984). LAI: Leaf Area Index. B pot: potential biomass production. Bwl: Water-limited biomass production. Ywl: Water-limited yield.

(Figure 3). The scheme contains many important components. However, the system itself is a complex one and difficult to construct.

All the components do not have an equal influence on crop growth. The basic processes can be identified: photosynthesis, respiration and translocation. Most of the systems used are based on functional relationships which simulate the yield response of crops, but are limited for the most part, by assumptions which exclude several processes. Generally, the influence of management practices are included based on environmental factors. A crop yield model produced by Konijn (1984) can be quoted as an example (Figure 4). It includes the characteristics of the environment, and management as a function of leaf area based on three estimations. The potential biomass is calculated as a function through which the effect of variations of soil water including possible application of irrigation is expressed. The final yield is obtained after translocation of the biomass into the different organs of the plant. That system applies particularly to areas with small precipitation where production is based on limited soil water. But it does not consider the possible

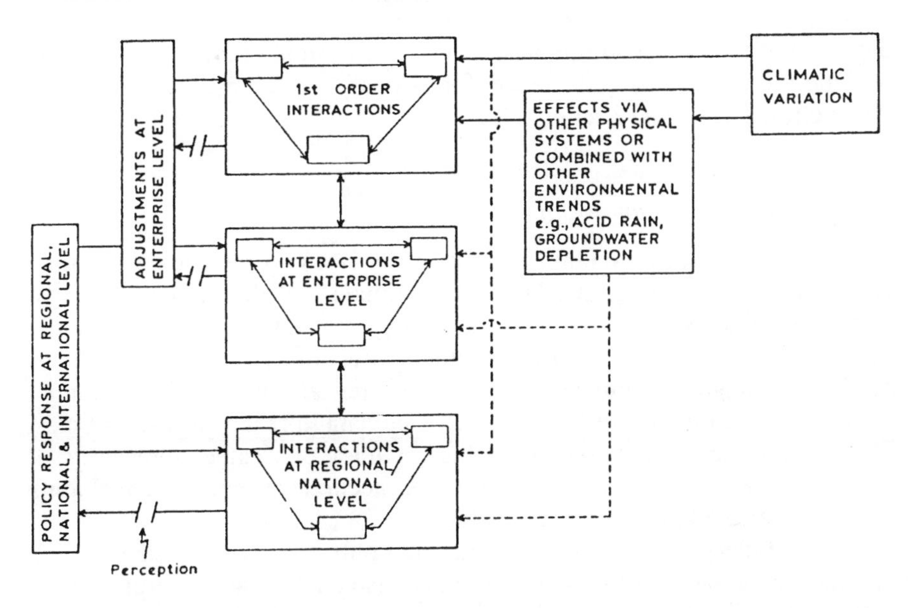

Figure 5. Scheme of the integrated project study method of the IIASA-UNEP. An interactive approach to climate impact assessment with ordered interactions, interactions at each level, and some social and physical feedbacks (after Parry et al. 1988).

development of weeds and pest and the possible lack of nutrients. It has the advantage that it can be adapted for use for different crops and areas after appropriate recalibration. For instance, it has been applied to maize yields in semi-arid Kenya (Parry et al. 1988).

Many systems have been proposed and applied to estimate the response of crop development or animal husbandry to climate (see references in books such as Landsberg 1981; Robertson 1983; Houghton 1985; Kates et al. 1985). The International Institute for Applied Systems Analysis (IIASA) proposed a fully integrated approach to climate impacts in agriculture. The goal was to investigate the effects of climatic change and variability and to evaluate alternative responses to these effects. An interaction approach designed to simulate the relations between climate and human activities considers the cascade of interactions that can appear at several-order levels (first order: the biophysical level, second order: the farm level, and third order: the regional and national levels). Each level is characterized by a series of additional and complex interactions (Figure 5). This form of a fully integrated system has not been successfully implemented because of the lack of ability to connect all models (Parry et al. 1988). Partly integrated systems were used to investigate the effects of climate change in several areas.

Operational systems are different; examples include systems which seek to estimate the date of various phenological stages of crops and monitor their growth in near-real time. As great difficulties are encountered in attempts to build efficient systems, research turned towards another type of modelling helped by remote sensing data (see, for estimation of crops, Delecolle et al. 1991; Groten 1993 and Section 3.2.2).

2.5. *Information on climate data*

Effective use of accurate climate information is a two step process, namely, the acquisition of knowledge of the climate and the dissemination of that knowledge. Strategic planning and long-term investments need statistical climate data drawn from climatological records as well as predictions on future changes in climate if the next decades are included. Operational models depend on the availability of climate data in real or near-real times. Operational decisions are based on the accuracy of weather forecasts. All these involve more and more specific and accurate information including very sophisticated modelling capabilities.

National meteorological services must not only provide basic data (which are inadequate most of the time), but must also analyze and interpret them. In other words, meteorological services should provide information in the most useful form. Two main kinds of users exist, the general public, and the decision-makers. Each user group needs to have specialized information translated into terms that it can understand since knowledge levels vary considerably across groups. As the number of consultants grow, the variety of services supplied by national meteorological bureaus must also increase to match different but specific user needs. National Services may have to expand the range and type of forecasts they now provide and incorporate them into their regular schedule. Information on frost, snow warnings, wind hazard forecasts, fog forecasts and other specialized information for specific industries such as agriculture, are all candidates for the expanded service (Cepremap 1991; Maunder 1995; Starr 1991). National Services will become increasingly service and market driven. These developments parallel the growth of private sector meteorology.

In the near future, climate information will benefit from technological developments in communication as the wider use of computers, the establishment of climate network linked in real-time, the development of compact-disks and automated data collections, etc., transform data management. But the most important point remains the need to provide users with more relevant information and for improved methods of presentation designed to make the information more directly useful.

But better information does not strictly mean better decisions, because a better decision depends also on the way the information is used. A better decision influenced by climate information in a sensitive activity assumes that decision-makers are aware of their potential value and know their current use. Many decisions-makers in the public and private sectors are unaware that climate fluctuations are inherent factors in the management of many activities at all levels, local, national and regional. A great number of them, depending on the country, do not know the value of climate and weather.

Politicians too are not used to looking at the influence of climate on the economy; they become more sensitive to it under public pressure or because of lobbies. But legislation, policies and administration structures and procedures, are not adequate to take into account climate potentials or impacts (see Section 3.3).

Major efforts should be directed towards the development of awareness among decision-makers and politicians of the very considerable potential benefit of wise use of climate information. The ability to manage and properly apply climate information will improve results in many weather sensitive activities. Several theoretical and practical attempts have been conducted to determine how meteorologists and decision-makers could evaluate the forecasts regarding the needs of specific users for real-time management and could choose the best forecast (Murphy and Katz 1985; DeGroot and Fielberg 1986; Krzysztofowicz 1992; Clemens and Winkler 1987).

2.6. *Benefits and costs of the climatic impacts*

The climate has a value which is determined by the benefits and the costs due to the effects of the interannual climatic variability on climate-sensitive economic sectors. Generally, below average production implies a loss of revenue for farmers, enterprises and the national income, even if prices rose because of scarcity.

The cost of extreme events such as droughts in agriculture or frost in the construction industry seems rather easier to calculate than the cost of small climatic fluctuations. In fact, since damages are distributed very irregularly in a country, their estimate are normally left in the hands of a few stake holders. When financial aid is allocated, it is usually designated for a specific area with little attention paid to damages sustained in the wider geographical region. More sophisticated investigations carried out recently with the support of remote sensing demonstrates how technology can help improve estimates of damage to crops (see Section 3.2), roads and other areas of the economy and the social fabric. However, as is often the case, there is a time lag between technological advance and its use in administrative decisions.

Small climatic anomalies such as mild or cold winters or rainfall 25% above or below the average can cause significant losses or benefits when they involve large sectors of the economy. But studies designed to determine the monetary values for those impacts for large sectors of the economy or for a nation as a whole are scarce. Annual benefits of weather forecasts for agriculture in the United Kingdom was estimated to be at least 0.5% of the gross farm income. Annual illness due to weather was estimated to represent a 10% loss in total annual sales (Mason 1966). The annual loss due to weather varies from 3 to 10% of the total potential volume of construction in the United States, and if a relevant weather information was available and appropriately used, a potential saving of about 15% of the estimated loss was possible (Russo 1966). These estimates are in fact very rough figures, not verified ones. Benefits or losses due to climate and weather impacts on human activities, health and settlements are difficult to evaluate in financial terms in technical sectors of the economy and are nearly impossible to assess in social contexts (reduction of risks to lifestyle, maintenance of regular schedules, etc.).

Establishing relations between economic and climate or weather data often requires that both sets be adjusted to bring them, when possible, to the same space and time scales. Most of the economic indicators are usually made for the whole nation and are available too late to be useful in most operational decision-making. A few countries produce and launch commodity-weighted climate information. In New-Zealand, indices are regularly compiled to calculate weights for the climate factor in several economic activities on both regional and national scales (Maunder 1995).

We have witnessed recently the considerable growth in investigations on climate variability. They have increased in number, their scope has widened and their methods have increased in sophistication. Among all the research done on the present state of the art in climate variability, we have chosen four cases which are particularly enlightening.

3. The present state of the art: Three cases

A more accurate awareness of benefits and costs of weather and climate has led to new investigations. To demonstrate the potential power of sophisticated methods, we have selected three cases, the first uses statistical modelling, the second remote sensing, and the third includes the prospect of an enhanced global warming during the next century. But the use of a weather knowledge is limited to a degree by law, and by political and administrative structures and decision-making. Those limitations will be examined in context of French agricultural industry.

3.1. *Assessment of snow-making potential in France based on temperature and hygrometry*

Because of the variability of temperature and precipitation from one winter to another, as well as within each winter, there is a significant risk of insufficient snow cover in winter sports resorts, particularly in lower altitudes in France. The economic consequences of snow shortage are generally very severe, given that skiing is the backbone of the economy in most of these resorts.

When artificial snow-making was introduced in the United States in the early nineteen fifties, the hope was that it will eliminate the risk of shortfalls in natural snow cover in winter resorts. However, because the efficiency of the technology is closely tied to air temperature and hygrometry as well as their variability that promise has never been fully realised. There have been many cases, particularly in French skiing resorts, where snow-makers have proven useless during periods of natural snow shortage precisely because air temperature and humidity conditions were inadequate. Consequently, decision about snow-making should not be taken until after a full evaluation of temperature and humidity profiles of the target location have been made. If long series of records are not available for the area, estimation of these parameters is still possible after a short measuring campaign, using a set of methods developed at the CNRS by Martin et al. (1990–1991; 1991).

3.1.1. *Evaluation of the potential for snow-making based on temperature and hygrometry*

Snow is produced by projecting into the atmosphere very thin droplets of water mixed with a variable quantity of compressed air. If temperature and relative humidity in the atmosphere are below a given level, the droplets freeze into tiny ice-balls which constitute the flakes of the synthetic snow. Each snow-maker has its own temperature and relative humidity threshold specified by a single "wet-bulb temperature". To evaluate the potential for snow-making at a desired location, one should know for how many hours the wet-bulb temperatures remain below each system's threshold value.

Because the duration of period with wet-bulb temperature below a given threshold varies from year to year, decisions on investment should be based on the expected frequency of desired lengths of suitable snow-making periods. With this objective, the authors suggest the production, as a support for decision-making, of a frequency chart that gives the distribution of potential snow-making hours based on varying wet-bulb temperature thresholds (Figure 6a). Potential snow-making hours are often counted only over the months of November and December, because, in most French winter sports resorts, snow-making is essentially viewed as a means to guarantee opening for Christmas holidays.

It is easy to construct a frequency chart for a location where wet-bulb temperature has been recorded over at least 20–30 winters. Such long records are rare, especially along ski-runs. The French climatological network provides only long series of daily extreme dry-bulb temperatures for a limited number of stations. Due to the extremely significant effect of specific local factors on temperature and humidity in mountainous regions, data at a "nearby" station can rarely be used directly to characterize conditions on a particular ski run, nor is it possible to use the vertical lapse-rates that are commonly utilised in studies involving monthly mean temperatures. For these reasons, it is prudent to develop a specific method for constructing long series of hourly wet-bulb temperatures from a limited pool of data obtained from a two-year data acquisition programme in the locations studied.

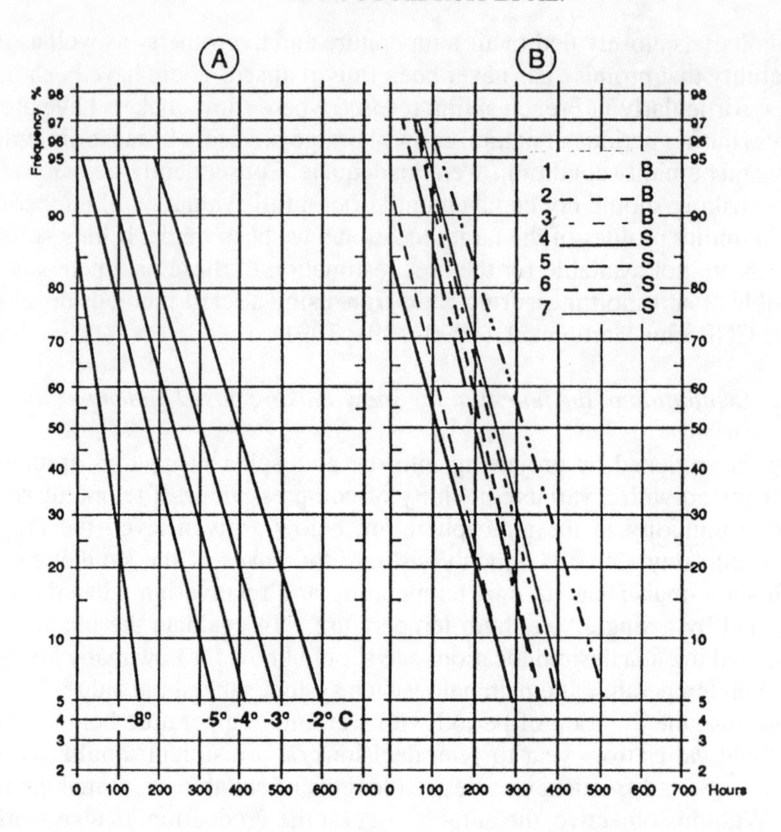

Figure 6. Statistical distributions of potential snow-making hours. (a) For various thresholds of wet-bulb temperature (reconstructed with "regional temperature components") on a slope site at Lans en Vercors (altitude: 1415 m; months of November and December) (in Martin et al. 1991). Note: Lans en Vercors is close to Villard de Lands (Figure 6b). (b) For various climatological stations of METEO-FRANCE's network in the French Northern Alps (wet-bulb temperature threshold: −5°C), (in Martin et al., 1990–1991). 1: Abondance (elevation: 1000 m), 2: Combloux (elevation: 970 m), 3: Villard de Land (elevation: 1050 m), 4: Saint Hilaire du Touvet (elevation: 970 m), 5: Saint Pierre de Chartreuse (elevation: 945 m), 6: La Clusaz (elevation: 1150 m), 7: Col de Porte (elevation: 1325 m). S = slope B = valley bottom, foot of slope or depressed area.

3.1.2. *Assessment of potential*

3.1.2.1. *Method.* The method used in the assessment of viability of snow-making is made up of three major elements.

Reconstruction of a long series of daily extreme (dry bulb) temperatures for a location where they have been recorded during only two winters. For this, three steps are necessary (Martin and Mainguy 1988):

(a) Construction of a non-linear two-way Analysis of Variance (ANOVA) model describing the daily extreme temperatures recorded during the last

20–30 years in climatological stations surrounding the studied location. This procedure condenses the information into a small number (3 or 4) of "regional temperature components" expressed in daily values (°C).

(b) Obtain a relationship between the temperatures recorded in the target location and corresponding daily values derived from the "regional temperature components". This is done through multiple linear regression, whose standard-error indicates the quality of the reconstruction performed in the next step.

(c) Reconstruct all missing data on the target location using the regression equation obtained in step (b). This series will cover the same period as the "regional temperature components". Hourly interpolation of daily extreme temperature series.

The mathematical model used for this interpolation is based on a simplified physical description of the major energy exchanges that take place in the boundary-layer during the daytime and the nighttime (Martin and Mainguy 1988):

- incident solar radiation which can warm the air with varying degrees of efficiency,
- the inertial effect of the environment which tends to pull back temperature toward some "equilibrium state",
- long-wave radiation effects and the warming or cooling caused by incoming air-masses.

In areas with very high moisture conditions, wet-bulb temperature is virtually always very close to the dry-bulb one, consequently snow-making hours may be computed directly from the reconstructed dry-bulb temperature series. In dryer places, the conversion into wet-bulb temperatures is done after a specific statistical procedure has been used to generate a long series of hourly relative humidity data which has the same distribution parameters as the series recorded at the target location during the two-year data acquisition campaign.

3.1.2.2. *Operating mode and results.* A computer software package (CANOVAL) has been developed by the CNRS to operate the three methods.

The locations of instruments for the two-year measuring campaign are chosen jointly by the staff of the winter sports centers and special investigators from the CNRS. Generally, these investigators also supervise the whole measuring campaign, operate the computer software and produce the frequency charts.

The method has already been applied to several stations of the METEO-FRANCE's network in the French Alps and Pyrenees, and to over 31 measuring points distributed in 19 winter sports centers in the French Alps and Pyrenees. The results of these studies confirm several facts:

- Snow-making potential can vary greatly from one location to another, even over very short distances (Figure 6b).

- Because of the very high cost of snow-making systems, the financial implications of an inadequate decision are very high. The French Ministry of Tourism revealed that the average cost of keeping ski-runs snow covered in 1989 was 1 million French francs per hectare. The need for careful assessment of the feasibility of snow-making is economically well justified.
- An assessment for snow-making potential may even, in some cases, influence the development policy of a mountain community. For example, after viewing the results of a study he ordered, the mayor of a skiing center in the Vercors mountains declared that the fairly low snow-making potential of his community's slopes would lead him to focus winter sports development on cross-country skiing rather than downhill.

3.2. *Monitoring drought and its impact over agriculture by using remote sensing: Example of the 1989–1990 droughts in France*

3.2.1. *Introduction*

Faced with an exceptional drought with a strong adverse impact on agriculture, governments may compensate farmers whose crops are damaged. Precise information on stricken areas must be provided before decisions on locations of damage and financial restitution are made.

Climatic information is actually insufficient to specify the spatial variation of droughts given that:

- localized storms may not have been detected,
- soil water capacity which is a key parameter in water stress varies spatially,
- resistance to drought depends on the crop types and variety.

Field collected information may also appear insufficient, as they are not available for all crops (e.g. they do not exist for forage), and, when they are available, they are often late and therefore, incompatible with a timely compensation policy.

The impact of drought on production must thus be measured cartographically, and, if possible, quantitatively. Remote sensing is based on direct and comprehensive observation of vegetation. It can provide a basis for the evaluation of drought impact and water availability at regional and local levels.

3.2.2. *Characterization of drought by remote sensing*

Drought, and more generally stress, changes spectral properties of vegetation. These can be detected from satellite (Lepoutre and Vidal 1989).

Basic properties of vegetation are the following:

- low reflectance in the visible band (maximum 10%), due to a strong absorption of solar radiation by chlorophyll (in red and blue bands);

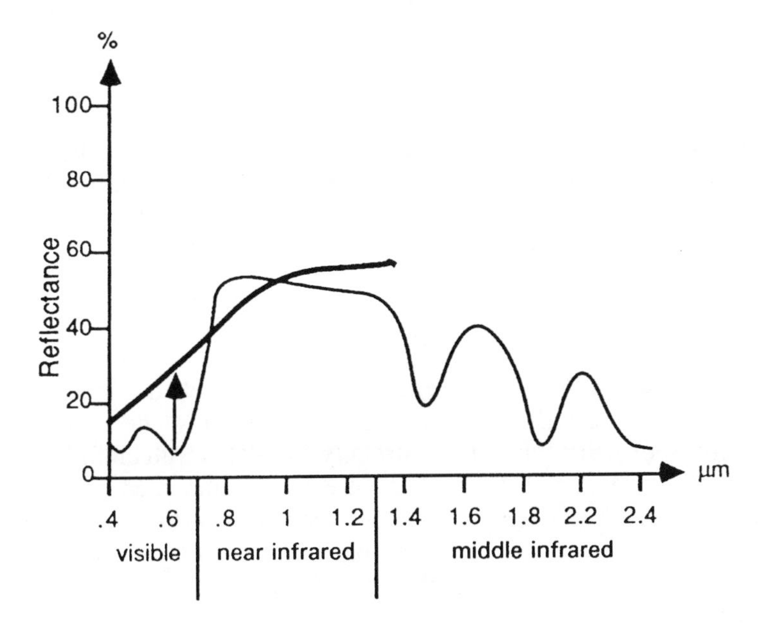

Figure 7. Spectral response of vegetation, and evolution under stress.

– high reflectance in the near infrared band (around 50%), due to the internal structure of the leaves. This feature is strongly correlated with biomass amount;
– thermal emission in the thermal infrared band, depending on the surface temperature of the crop which is a function of the energy budget. This temperature is inversely correlated with crop evaporation, i.e. its water consumption.

Based on these characteristics we can easily describe the effects of drought (Figure 7):

– water stress decreases crop evaporation, which increases the surface temperature;
– this stress affects the crop water content and leaf turgescene thereby decreasing the near infrared reflectance;
– if this stress continues, the chlorophyll activity will be affected, and the visible spectrum reflectance will increase (especially in red and yellow band: the crop "yellows").

These modifications can be detected by the AVHRR (Advanced Very High Resolution Radiometer) sensor of the NOAA (U.S. National Oceanic and Atmospheric Administration) satellite. This meteorological satellite provides information on the same area every 6 hours with a resolution of 1.1 km, in visible (0.58–0.68 μm), near infrared (0.72–1.10 μm) and thermal infrared bands (10.3–12.5 μm). Remote sensing is, therefore, exciting technology for monitoring global vegetation including inter annual changes in the variable.

When considering the spectral bands of most satellites, many authors, following Tucker (1979), have introduced band combinations called vegetation indices. The most frequently used is the normalized difference between near infrared (NIR) and red or visible (R) reflectances, and is called NDVI (Normalized Difference Vegetation Index):

$$\mathrm{NDVI} = \frac{\mathrm{NIR} - \mathrm{R}}{\mathrm{NIR} + \mathrm{R}} .$$

As one can see in Figure 7, a stressed crop will have a decreasing NIR – R difference, as the NIR + R sum increases, its NDVI decreases. Many studies have shown that this index is strongly correlated with biomass amount (for example, Justice et al. 1985). Lepoutre (1987) computed differences in biomass yield between a dry and reference year based on this correlation. For other crops which are more or less sensitive to water stress at various growth stages, a complementary analysis of the impact with field experts is generally necessary.

3.2.3. *Application: Operational monitoring of the 1989–1990 drought in France*

Most of the issues described here have been taken from the E.E.C.-J.R.C. and GEOSS Final Report: *Towards a real-time system for vegetation monitoring*. Their authors are acknowledged here for providing the information used.

3.2.3.1. *Methodology.* The main principle in monitoring drought from remote sensing data is based on the comparative analysis of the evolution of the NDVI of a given area, both for a reference period (or year) and for the "dry" period (or year) (Figure 8). In order to perform this monitoring, 4 successive steps are necessary:

- selection and organization of satellite, meteorological and ground-based data;
- division of the territory into radiometric homogeneous areas;
- definition of warning thresholds on the radiometric evolution of the area;
- drought impact monitoring using these thresholds. The data management consists of:
- archiving historical and other physical data (satellite data of previous years, soil map, . . .);
- collecting as quickly as possible satellite data (1 selection of clear sky NOAA images per month) and agricultural land use data.

A division of the territory into radiometric homogeneous areas is then made from the historical data: all the pixels that have the same radiometric evolution during a whole year are considered similar. The areas obtained are then corrected by taking into account ancillary data (land use, soil), in order

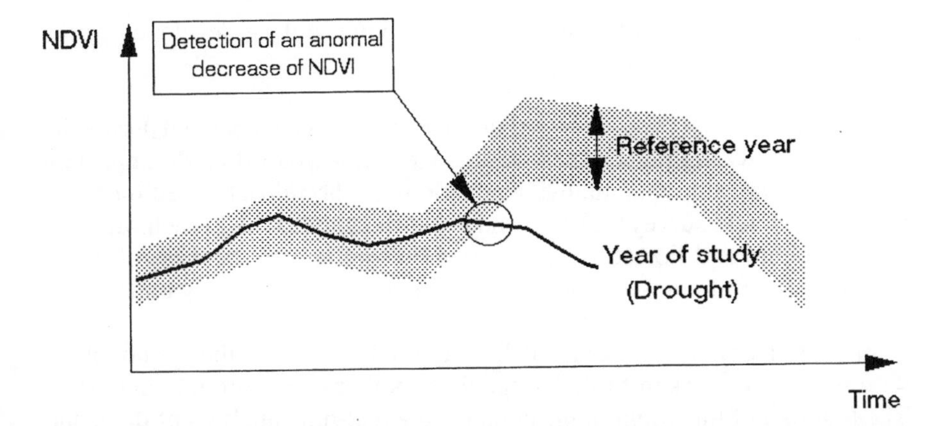

Figure 8. Principles of drought impact detection: the NDVI evolution of the considered year is compared to the one of the reference year.

to eliminate punctual effects appearing on images (e.g. a localized storm may induce an heterogeneity in an otherwise homogeneous area). To get an idea of the size of these areas, the Midi-Pyrénées region covering 1/8th of France, i.e. 45000 km^2, was divided into 21 agricultural areas each with a surface ranging from 637 to 4442 km^2. This division could change between two successive years due to changes in land use if warranted by new satellite and ancillary data.

With each homogeneous area, a warning threshold for drought impact is determined by analyzing the variability of the radiometric qualities in each area. For a given area, the alert threshold is defined, for each time of the year, by the limits $(m - s)$ and $(m + s)$, where:

– m is the spatial mean value of the NDVI for the area at the time,

– s is the standard deviation of the NDVI for the area at the time.

This threshold is simple to use. The main difficulty is the one encountered when comparing the evolution of NDVI during a current year against the one obtained in the reference year (e.g. in France where 1987 was the reference year, the droughts of 1989 and 1990, were analyzed by comparing evolution of their NDVI with 1987). Actually, different climatic conditions produce different crop phenologies, such that the same phenological stages are not attained at the same time in 2 different years, especially if climatic conditions are exceptional in one of those years. For this reason, it is necessary to fit the phenological stages of the year in question with those of the reference year. The weekly fitting of those stages requires the use of global climatic information as well as real time field information. France could be divided into three large areas each with a different fitting. Assuming that these conditions are met, it is then possible to monitor the impact of a drought in homogeneous areas with almost real-time sequence. This point will be developed in the next section.

3.2.3.2. *Results and discussion.* Preliminary results were obtained during the drought of 1989 in France. These results yielded maps of drought impact on vegetation and semi-quantitative evaluations of the impact based on satellite data and field surveys. These products have allowed the French Ministry of Agriculture and Forest to save around 300 million FF because of a better distribution of compensation among damaged and relatively undamaged farmers.

In 1990, a true agricultural watch programme, based on the methodology described above, was instituted to monitor another severe drought in France. The first part of this programme comprised a real-time analysis of the meteorological data necessary to obtain a first approximation of the distribution of drought impact as well as understand the radiometric evolution of the targeted homogeneous areas. NOAA satellite data were then analyzed almost in real-time allowing the final result to be obtained approximately one month after the satellite overpass. After fitting the phenological stage, and removing cloudy areas from the image, the NDVI evolution of each area was compared to the evolution of the reference year (1987). For each area, 3 criteria of drought impact were considered:

- the intensity of the difference between 1990 and 1987 NDVI profiles;
- the surface on which the differences were observed;
- the duration of the observed difference between profiles.

The results of this analysis were then correlated with the crop yield estimated from field survey. From this correlation, quantitative estimate of the impact of the drought was obtained. Working in homogeneous areas as defined in the previous section gives better results because intra-area variations are smoothed. On the other hand, the effect of some very localised phenomena (e.g. hail storms) may be missed.

Figure 9 gives an example of a map of yield losses for corn in the Tarn-et-Garonn region. As can be seen, crop losses are estimated for each district which implies differential compensation rates for farmers across districts.

3.2.3.3. *Conclusion.* A real-time monitoring of drought impact, based on the analysis of NOAA satellite images, meteorological data and field survey, has been carried out in France in 1989 and 1990. This monitoring resulted in a good and locally specific estimation of crop losses and a better distribution of compensation among the farmers affected. But this approach to impact analysis is still limited by many factors including:

- cloudiness in the areas monitored;
- the quality of the preliminary maps of homogeneous areas in regions where only few data sets are available;
- the fitting of phenological stages between the target and reference year;

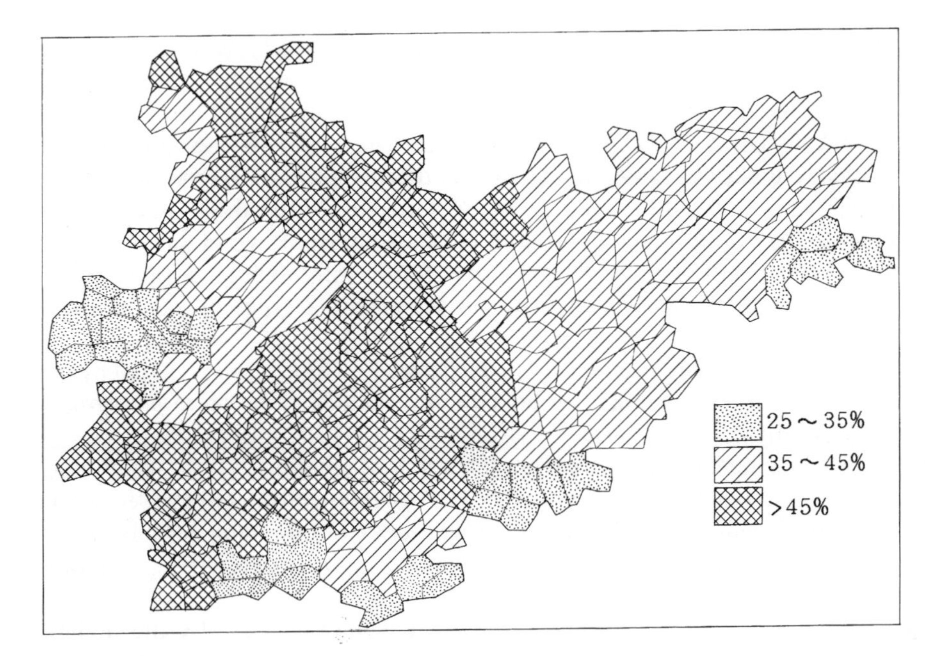

Figure 9. Map of corn yield losses during drought 1990 in France obtained from satellite data and field survey, Tarn-et-Garonne region, Source: GEOSYS and French Ministry Agriculture.

- the lack of historical series of NOAA data that could be used to better define a reference year.

3.3. *Climate, the law and farmers*

Remote sensing and mathematical models help society to obtain a better understanding of the space and time dimensions of the impacts of climatic hazards. But often a conflict arises between results obtained from these instruments and the law. The case of French legislation is a good example of this conflict. Climatic accidents are not yet considered "Acts of God" in France as they are in the United States. In this country, compensation policy is determined by specific laws and administrative regulations.

3.3.1. *What is officially a climatic disaster in France?*

Official texts remain vague as to the definition of climatic disaster. The 1964 Act which created a national guarantee fund for the farm industry made a distinction between disasters and calamities. A disaster must affect at least 25% of the projected yield and 12% of the farm's total production. For calamities, the thresholds are 27 and 14%, respectively.

Thus the law explicitly recognises agricultural disaster or calamity only when the climatic effect is direct and negative in economic terms and does not arise from other physical parameters even if their effect on agriculture was driven by a deterioration of the climate. The respective thresholds of 12 and 14% drop in global farm income which qualifies a farm situated in an area declared as a disaster zone for aid are based on the limits of the "normal climate" as understood by the authorities. It is a margin that society has agreed to grant to the whims of the climate. Within these limits nature remains "neutral or kind" to use Samuelson's phrase. Beyond this limit, one enters the zone of what the French Constitution calls a national catastrophe.

But climatologists make a distinction between 2 types of calamities of atmospheric origin:

- Calamities which can be defined in space and time, easily located by the meteorological office, and linked to the sudden variation of an element of climate.
- Diffused calamities which are the result of a series of days which may not be catastrophic in themselves, but which, if considered as a sequence, cause considerable damage to agriculture. Drought, long periods of persistent rain, and all other factors which limit plant growth or crop projections, come under this category.

Climatic dimensions are not always taken into account by the highest State authorities. Thus in a 1985 report, the Cour des Comptes (Government Accounting Office) allowed compensation only for the "abnormal variability of a natural agent precisely dated and defined". The Office refused to take into account calamities which occurred "during periods which stretch over several months or weeks". This stand aimed at avoiding any abuse and was based on the principle that one must not take as a starting point "registered losses and seek afterwards their meteorological explanation". The Minister of Agriculture refuted these arguments with the contention that "when drought and excessive rainfall become serious, they correspond to the legal definition as there is a variation of a natural agent".

3.3.2. *How is the decision to recognize a disaster taken?*

Climatic calamity of atmospheric origin is recognized as such and compensation is given only after a legal report has been filed and a decision on it taken by the authorities. In the case of a climatic disaster, first, the government representative (Commissaire de la République) is alerted through different channels (Marchand et al. 1986). A preliminary survey is then made by a departmental committee for the evaluation of agricultural disasters. If the decision is positive, i.e. concludes that there has been a disaster, the report is forwarded to the Ministry of Agriculture which takes the final decision, taking into account the opinion of a national committee. Once disaster or calamity is declared, the Agricultural Departmental Boards (DDA) concerned conduct the necessary investigation into claims submitted at the local townhall. At this stage, the smooth functioning of the process and the quality and speed of compensation depend on the mayor's dynamism.

The decision that a disaster has taken place enables the farmer to benefit from improved loan conditions whereas the decision that there has been a calamity carries a right for compensation and tax relief. So, for the farmer, compensation, even if it is only partial, means that a sum of money will be received. On the other hand the granting of loans can be double-edged if disasters occur several years in succession. The weight of debt repayment can drive young farmers into bankruptcy.

3.3.3. *Drop of prices due to an overproduction*

The same climatic event can be construed as both positive and negative. Recent summer droughts in S.W. France in 1989 and 1990 damaged the corn crop but benefited vineyards. When climate factors are favourable, yields improve. However, it is often difficult to measure the economic gains derived in good years against equivalent gains in years with average yields. This difficulty stems from the fact that "good years", can result in over-production which drops prices and lowers farm incomes. When conditions are too favourable, aid to farmers is provided through different economic and political channels including subsidies and bonuses.

For the nation as a whole, compensation is provided through funds guaranteed by the statutes of 4th August 1956 and 10th July 1964. In any case, the funds provided never compensate fully for the cost of the damage as estimated by professional adjusters. The law of 1982 on natural disasters can provide additional relief since it applies to all citizens including farmers. Taken together, the laws add up to significant additional spending by the State. Indeed, the State plays an important role in the way insurance on Acts of God or natural disasters functions. Because it guarantees insurance through a central reinsurance fund, it is an insurer of last resort assuming responsibility for possible deficits. In agriculture as in other impacted areas, public authorities can be viewed as the main guarantee against natural disasters.

It is necessary to separate the economic policy from the social one to ensure that a system designed to protect citizens against natural disasters does not become a cover for guaranteed farm incomes. The authorities cannot relinquish responsibility for protection claimed by citizens who are affected increasingly by the natural environment. Their number continues to grow and it will not be long before farmers are joined by workers in the tourist industry especially in mountain resorts in seeking compensation for adverse weather conditions such as those that cause shortfalls in snow. However, in the national subconscious, the farmer, seen as the provider, still has priority, but for how long faced with an urban society for whom the countryside is becoming a vast budgetary drain.

So the extent of disaster and calamity depends on how society is affected by climate and its ability to face up the hazards and accidents due to climatic variability while under budgetary restraint.

3.4. *Impacts of global warming*

The global mean temperature has been rising for more than a century which is supposed to be due mainly to the large increase of carbon dioxide, methane and other infrared gases in the atmosphere (see Section 2.2). Based on General Circulation Model (GCM) simulation, global warming will be greatly enhanced when CO_2 doubles its pre-industrial concentration sometime in the 21st century (see Section 2.4). Global warming will change climate as we know it and impact a great number of areas of the physical and human environments. As we advance towards the next century, the interactions between human activities and climate will become a dominant global issue (see Figure 1b) (Leggett 1990; Schneider 1990; Correll and Anderson 1991).

3.4.1. *Simulated and arbitrary climate change scenarios*

Estimates of global warming vary considerably among GCMs and depend on several factors as the simulation of the atmosphere processes and ocean influence and projections of fuel consumption i.e. the politics of nations. Lower consumption of fossil fuels will reduce the expected warming (Houghton et al. 1990, 1992, 1995).

The simulated increase of temperature due to CO_2 forcing differs across the globe (see Le Treut, Chapter 2.5, Fig. 3). However, a few features appear consistent at the continental scale. Warming should be 50–100% greater than the global average in high northern latitudes in winter and smaller in the sea-ice areas in summer. As shown by GCMs, warming is also associated with modifications in other climate parameters including precipitation and soil-moisture (Houghton et al. 1992; Wilson and Mitchell 1987).

Precipitation is expected to increase in high latitudes and in the tropics all year round and in the middle latitudes in winter. Elsewhere, changes are either small or uncertain. In general, however, the tropospheric water vapour is expected to increase. As for soil moisture, all models project an increase in the northern high-latitude continents in winter and most indicate a decrease in summer there. All projected changes are tentative with a low confidence. The changes in sea-level pressure associated with the warming are at least as uncertain as those of other projections.

Areas, each a few million square kilometers, have been selected for closer study using very high resolution models or nested regional climate models. The confidence in estimates of change is low, especially for precipitation and soil-moisture and still lower at smaller spatial scales dropping very rapidly as spatial resolution increases. Present models produce realistic but diverse simulations at the space and time scales that are appropriate for impact analyses (Grotch and McCracken 1991; Giorgi et al. 1992; Giorgi and Mearns 1991; Hewitson 1994).

Predicted climate changes are used to determine impacts, but, as their likelihood is not verifiable, scenarios of arbitrary changes in regional climate are utilized. Such scenarios rely on different hypothesis depending on objectives. Total effects of a CO_2 doubling are focused generally on changes in the average parameters as temperature and in the frequencies of extreme climatic events critical to human activities. These extreme events are expected to increase. The most studied risk, however, is drought in agriculture.

3.4.2. *Possible climate impacts in the next century*

Results of simulated enhanced greenhouse effects on the climate of the next century raises considerable concerns over the potential impacts of climate change on society (Glantz et al. 1991; Carter et al. 1994). The ministerial declaration signed at the Second World Climate Conference (1990) recognized future climate change as a common concern of humankind and maintained that there is a need to intensify research on the social and economic implications of climate change and response strategies. Reactions follow two directions, on the one hand is the task of limiting the amount of greenhouse gas of human origin in the atmosphere, based on the knowledge that such reduction will reduce but not completely eliminate the threat of global warming (Houghton et al. 1995). On the other, is the need to determine the likely impacts of projected changes in climate with or without such reduction.

Impacts will affect all climate sensitive resources, economic activities (McIntyre 1978) and lifestyle including stratospheric ozone layer (Megie 1989), natural ecosystems, agriculture (ASA 1995), food supply (Rosenzweig and Maritin 1994), forestry, water, human health (Jones and Wigley 1991) and settlement, etc. Rise in sea level will present major problems as will the submersion of low lying coasts some of which are heavily crowded with people (Titus 1988; Wigley and Raper 1987; Pacpe et al. 1990). Present studies on climate change and impacts mostly deal with terrestrial ecosystems (Bolin et al. 1986), agriculture, crop yields, forage and livestock production (Parry 1990), and less often with areas such as hydrology, water resources (Lettenmaier and Gan 1990; Waggoner 1990; Rampal 1990) and decision-makers (Paoli 1994). General investigation on the impacts of many climate-sensitive physical and economic sectors has been conducted for several countries: for Australia (Pearman 1988), for the U.S.A. (Smith and Tirpak 1989; Adams et al. 1990) and for the United Kingdom (UK Deptartment of the Environment 1991).

3.4.2.1. *Direct effects of the increase of CO_2*. The effects of a doubling of atmospheric CO_2 have been observed in many laboratory experiments in which interactions with other likely participants in a climatic change have been excluded. Due to such exclusions the response of the climate system to factors and interactions encountered in the open field is uncertain.

An enhancement of CO_2 affects the growth rate of plants and their productivity directly (Allen et al. 1990; Strain and Cure 1985; Kimball et al. 1987). CO_2 enrichment increases photosynthesis and biomass accumulation (Lemon

1984) but does so differently in C3 and C4 plants. The allocation of the carbon is modified. Plants develop more quickly but their senescence is accelerated. Growth and yields benefit (0–10% increase of C4 crops such as maize and sugarcane, and 10–50% increase for C3 crops such as wheat and rice based on a CO_2 doubling). But other factors could limit yields and reduce the effects of an enhanced CO_2. Temperature and moisture are important controllers of the carbon budget and development stages. Temperature, moisture and CO_2 interact in a variety of ways that are not well understood. Other environmental conditions such as nutrient availability can also affect plant development.

Climate change is likely to affect plant types within a vegetation community differently, crops and weeds, for example and could introduce changes in competitiveness (Peters and Lovejoy 1992). It could also modify the present spatial distribution of pests and diseases. The entire food chain could be modified because of possible effects of climate change on herbivores and their interactions with their predators and parasites. Finally, potential impacts on terrestrial ecosystems and agricultural production could vary considerably depending on the type of climate change created by an enhanced or a doubled CO_2.

3.4.2.2. *Shifting of thermal limits*. All GCMs agree that more significant temperature increases will occur in mid and high northern latitudes where temperature is an important factor in determining space and time variations of terrestrial ecosystems and agricultural production. Here the growing season in generally defined by temperature. The pattern of forests and agriculture in North American and North Eurasia could change drastically from what it has been since the last glaciation, and advance poleward (Emanuel et al. 1985; Kenny et al. 1993; Chapin III et al. 1992).

Investigations on potential shifts in other areas were also conducted for forests and crops. A few were based on temperature only, growing degree days (GDD in °C) during the frost-free growing season is often used (Newman 1980). Models established for forests have also simulated large scale migrations of biota using several climate components along with temperature, precipitation and evapotranspiration (Emmanuel et al. 1985; Prentice et al. 1989). These studies used statistical relationships which do not infer causal relationship between present day climate and vegetation, yet assume that the same scenario can be applied in the future. Sometimes, scenarios of climatic change are applied to more sophisticated models to simulate the growth and yield of annual agricultural productions (Santibanez 1991; Rosenzweig and Iglesias 1994; Mearns and Rosenzweig 1994).

These scenarios based on a CO_2 doubling do not take into account possible responses of the terrestrial ecosystems at several timescales. In the short term (years to decades), acclimatization and adaptation could limit change.

3.4.2.3. *Problems of adjustments*. Almost all forests are managed to some degree and agricultural productions depend partly on their management. Many investigations on likely climate impacts in the future assume that the relation between climate and production will remain constant. First, they either neglect or underestimate the evolution of technologies. In most northern latitude regions (where concern over a predicted warming is high), increases in total production have resulted mainly from intensification. Adjustments in agriculture will be better managed in industrial countries than in developing ones because such adjustments are functions of economic and political factors. Adjustments can occur at several levels of decision-making including the enterprise and political levels. They can occur hierarchically, at regional, national or international levels depending on policies in place.

Climate change will affect incomes thereby influencing societal adjustment to the new environment. Risk management, the evolution of technology, commodity prices and general policy including the attitude of managers towards climate issues will determine how well humans and the other areas of the physical environment manage change under a future climate (Maunder 1995).

3.4.2.4. *Integrated approach to impacts in agriculture*. A possible approach to the assessment of climate change impacts in agriculture is to use a set of models hierarchically as proposed by the IIASA (Parry et al. 1988). This approach consists of two sequential but linked steps. Firstly, models of climate variations are created. Secondly, the models created in the first step are linked with biophysical models in which first order relationships between climatic variables and the effect of climatic change on farm level production and the regional economy is established. This second step uses economic models in which adjustments in enterprise and policy responses are explicitly expressed models of this type offer a partly integrated approach compared to the general schemes discussed earlier (see Section 2.4.2).

The ultimate objective of climate-impact modelling as demonstrated in these efforts is to determine the effects of short term fluctuations and long-term changes in climate on agricultural output. Case studies were conducted in several regions based on three scenarios of climatic variations involving anomalous weather to which agricultural production was deemed to be particularly sensitive. The results of the studies were used to determine a range of adjustment options that could be used to mitigate adverse climate impacts or exploit beneficial ones. The five cases studied in cool temperature and cold regions examined primarily the effect of CO_2 driven climate change. Six cases in semi arid low latitude regions where change mainly involved precipitation and short term variations due to droughts, yielded inconclusive results. Because crop yields exhibit a non linear response to water stress, changes in the probability of extreme events may be the significant factor which the studies may not have considered.

4. Conclusion

The efficient determination of climate impacts requires a systems approach in which climatic variability is recognized in every facet of the system. This systems approach should apply both to long-term planning and shorter term operational real time management. Awareness of the potential value of climate knowledge range from uneven to low among decision makers in climate sensitive activities and must be remedied.

Greater attention should be paid to the acquisition of accurate data for impact assessment using new and increasingly sophisticated technology. Developments in remote sensing should be fully exploited.

At present, models which do not represent process well remain the most efficient tools for impact prediction in a number of activity sectors such as agriculture. They can be improved with better parameterisation of those physical processes which connect stimulus to response. This applies to both statistical and biophysical models. But while we await these developments, simplified systems which incorporate both the effects of management strategies and the significant impacts of climate on human settlement, health and activities should be adopted and utilized to their fullest potential.

References

Adams, R.M., Rosenzweig, C., Peart, R.M., Ritchie, J.T., McCart, B.A., Glyer, J.D., Curry, R.B., Jones, J.W., Boote, K.J. and Allen, L.H. Jr. (1990): Global climate change and U.S. agriculture. Nature (London), 35, 219–244.

Allen, S.G., Idso, S.B., Kimball, B.A., Baker, J.T., Allen, L.H., Mauney, J.R., Radin, J.W. and Anderson, M.G. (1990): Effects of air temperature on atmospheric CO_2–plant growth relationship. United States Department of Energy, DOE/ER-1450T, TR048, Washington, DC.

ASA (American Society of Agronomy) (1995): *Climate change and agriculture: Analysis of potential international impacts*, Special Publication (59), Madison, Wisconsin, USA, 382 pp.

Atarr, J.R. (1991): Products and services offered to the Agricultural Industry by the UK Meteorological Office. In: *Actes du ler Colloque sur les Applications de la Météorologie et leurs Intérêts économiques*, METEO-FRANCE and Rég. Franche-Comté, Besançon, 171–193.

Bach, W., Pankrath, J. and Schneider, S.H. (eds.) (1981): *Food-climate interactions.* Reidel, Dordrecht, 1981, 504 pp.

Baier, W. (1977): Crop-weather models and their use in yield assessments. WMO Technical Note No. 151, 48 pp.

Benjamin, N.B.H. and Davis, C.C. (1971): Impact weather on construction planning. Paper presented at the National Meeting on Environmental Engineering. American Society of Civil Engineering, St. Louis, U.S.A., 21, October 1971.

Biswas, B.C. (1988): Agroclimatology of the sugar-cane crop. WMO No. 703, Genéve, 90 pp.

Bolin, B., Dooz, B.R., Jager, J. and Warrick, R.A. (1986): *The greenhouse effect, climatic change and ecosystems.* SCOPE 29, Wiley and Sons, Chichester, 541 pp.

Carter, T.R., Parry, M.L., Harasawa, H. and Nishioka, S. (1994): IPCC technical guidelines for assessing climate change impacts and adaptations. IPPC Special Report to Working Group II of IPCC, 59 pp.

Cepremap, M. (1991): Synthése bibliographique des études sur l'évaluation des avantages économiques appottés par l'utilisation des services météorologiques. In: *Actes du Colloque sur les Applications de la Méteeorologie et leurs Intérêts économiques*. METEO-FRANCE et Rég. Franche-Comté, Besançon, 19–55.

Chapin, III, F.S., Jefferies, R.L., Reynolds, J.F, Sahver, G.R. and Svoboda, J. (eds.) (1992): *Artic ecosystems in a changing climate*. Academic Press, New York.

Clemen, R.T. and Winkler, R.L. (1987): Calibrating and combining precipitation probability forecasts. In: *Probability and Bayesian statistics*, ed. by R. Viertl. Plenum, London, 97–110.

Corell, R.W. and Anderson, P.A. (1991): *Global environmental change*. NATO ASI Series, Series 1. Springer-Verlag, Berlin.

Degroot, M.H. and Fielberg, S.E. (1986): Comparing probability forecasters: Basic binary concepts and multivariate extensions. In: *Bayesian inference and decisions techniques*, ed. by P. Goel and A. Zellner. North-Holland, Amsterdam, 247–264.

Delecoller, R., Maas, S.J., Guerif, M. and Baret, F. (1992): Remote sensing and crop production models: present trends. *ISPRS, J. of Photogram. and Rem. Sens.*, 145–161.

Douguédroit, A. and Bart, F. (1992): Précipitations et production de café au Rwanda. *Publ. Assoc. Intern. Climatol.*, 5, 53–63.

Duband, D. et al. (1989): The impact of climate on the operation in the French electric system. In: *Conference on climate and water*, Helsinski, Finland, 11–15 September 1989. Publications of the Academy of Finland, Helsinski, 275–289.

Duchon, C.E. (1986): Corn yield prediction using climatology. *J. of Clim. and Appl. Clim.*, 25(5), 581–590.

Duckham, A.N. (1974): Climate, weather and human food systems: A world review. *Weather*, 29, 242–251.

Emanuel, W.R., Shugart, H.H. and Stenvensohn, M.P. (1985): Climate change and the broad-scale distribution of terrestrial ecosystem complexed. *Climate Change*, 7, 29–42.

Gibbs, W.J. (1984): The great Australian drought: 1982–83. *Disasters*, 8(2), 89–104.

Giorgi, F. and Mearns, L.O. (1991): Approaches to simulation of regional climate chage: A review. *Reviews of Geophysics*, 29, 191–216.

Giorgi, F., Marinucci, R. and Visconti, G. (1992): 2 × CO_2 climate change scenario over Europe generated using a limited area model nested in a general circulation model: II. Climate change scenario. *J. Geophys. Res.*, 97, 10011–10028.

Glantz, M.H., Price, M.F. and Krenz, M.E. (1990): *On assessing winners and losers in the context of global warming*. Environmental and Societal Impacts Group, NCAR, Boulder, CO, United States.

Grotch, S.L. and MacCracken, M.C. (1991): The use of general circulation models to predict regional climate change. *J. Climate*, 4, 286–303.

Groten, S. (1993): From monitoring to management experiences from NOAA-NDVI application research in semiarid Africa. In: *Proceedings of the International Symposium Operationalization of Remote Sensing*, ed. by J.L. Vangenderen, R.A. Vanzeudam and C. Pohl. ITC, Enschede, the Netherlands, 147–158.

Hobbs, J.E. (1980): *Applied climatology*. Butterworths, London, 218 pp.

Hodges, T. (1991): *Predicting crop phenology*. CRC Press, Boca Raton, FL.

Houghton, D.D. (1985): *Handbook of applied meteorology*. John Wiley and Sons, New York, 1461 pp.

Houghton, J.T., Callander, B.A. and Varney, S.K. (1992): Climate change 1992. The Supplementary Report to the IPCC Scientific Assessment, Cambridge University Press, London, 200 pp.

Houghton, J.T., Meirafilho, L.G., Bruce, J., Lee Hoesung, Callander, B.A., Haites, E., Harris, N. and Maskell, K. (1995): *Climate change 1994. Radiative forcing of climate change and an evaluation of the IPCC IS92 emission scenarios*. Cambridge University Press, 339 pp.

Johnson, S.R., Mc Quigg, J.D. and Rohthrock, T.P. (1969): Temperature modification and cost of electric power generation. *Journal of Applied Meteorology*, 8, 919–926.

Jones, R.R. and Wigley, T. (eds.) (1991): *Ozone depletion: Heath and environmental consequences*. Wiley, New York, 302 pp.

Justice, C.O., Townshend, J.R.G., Holben, B.N., Tucker, C.J. (1985): Analysis of the phenology of global vegetation using meteorological satellite data. *Int. J. of Rem. Sens.*, 6, 1271–1318.

Kates, R.W., Ausubel, J.H. and Berberian, M., (eds) (1985): *Climate impacts assessment: Studies of the interaction of climate and society*, SCOPE 27, Wiley and Sons, Chichester, 649 pp.

Kenny, G.J., Harrison, P.A. and Parry, M.L. (eds.) (1993): *The effect of climate changes on agricultural and horticultural potential in Europe*. Environmental Change Unit, University of Oxford, Oxford, 224 pp.

Kimball, B., Mauney, J.R., Akey, D.H., Hendrix, D.L., Allen, S.G., Idso, S.B., Radin, J.W. and Lakatos, E.A. (1987): Response of vegetation to carbon dioxide. Effects of increasing atmospheric CO_2 on the growth, water relations, and physiology of plants grown in under optimal and limiting levels of water and nitrogen. Rep. No. 149. U.S. Dept. of Energy and USDA, Washington, DC.

Konijn, N.T. (1984): A crop production and environment model for long-term consequences of agricultural production. Working Paper, W. 84–51. International Institute for Applied Systems Analysis, Laxenburg, Austria.

Krzysztofowicz, R. (1992): Bayesian correlation score: a utilitarian measure of forecast skill. *Mon. Wea. Rev.*, 120, 208–219.

Lamb, H.H. (1988): *Weather, climate and human affairs*. Routledge, London, 364 pp.

Landsberg, H.E. (1981): *General climatology, 3. World Survey of Climatology*. Elsevier, Amsterdam, 408 pp.

Le Berre, M., Chamussy, H., Charre, J., Lamontagne, F., Martin, S., Peguy, Ch.P., Risser, V. and de Saintignon, M.F. (1982): Climat et mortalité. *L'Espace Géographique*, 3, 176–181.

Leggett, J. (1990): *Global warming. The Greenpeace report*. Oxford University Press, Oxford, 554 pp.

Lemon, E.R., ed. (1984): *CO_2 and plants: The response of plants to rising levels of atmospheric carbon dioxide*. Westview Press, Boulder, CO.

Lepoutre, D. (1987): Monitoring drought impact on forage using remote sensed vegetation indexes. *C.R. Acad. Agric. Fr.*, 73(6), 61–62.

Lepoutre, D. and Vidal, A. (1989): Remote sensing and drought. *C.R. Acad. Agric. Fr.*, 75(8), 95–102.

Lettenmaier, D.P. and Gan, T.Y. (1990): Hydrologic sensitivities of Sacramento-San Joaquim River Basin, California to Global Warming. *Water Resources Research*, Jan., 69–86.

Marchand, J.P. (1985): *Contraintes climatiques et espace géographique. Le cas irlandais*. Paradigme, Caen, 336 pp.

Marchand, J.P., Bauquin, F. and Letondoux, P. (1987): Les calamités agricoles d'origine atmosphérique: Le cas de l'Ille et Vilaine. *Rev. Géog. Lyon*, 3, 319–328.

Martin, S. and Mainguy, J. (1988): Potentialités climatiques de l'enneigement artificiel en moyenne montagne. 20th I.C.A.M. Sestola (Italy). Servizio Meteorologico Italiano and WMO.

Martin, S., Mainguy, J. and de Saintignon, M.F. (1990–1991): Neige à Noël et potentialités de production de neige de culture dans les Préalpes françaises du Nord. *L'Espace Géographique*, 3, 255–272.

Martin, S., Mainguy, J. and de Saintignon, M.F. (1991): L'évaluation des potentialités thermiques pour l'enneigement artificiel: Exemple de valorisation des réseaux climatologiques. In: *Actes du 1er Colloque sur les Applications de la Météorologie et leurs Intérêts économiques*. METEO-FRANCE et Rég. Franche-Comté, Besançon, 420–428.

Mason, I.B. (1982): A model for assessment of Weather forecasts. *Australian Meteorol. Magazine*, 30, 291–303.
Maunder, M.J. (1995): *The human impact of climate uncertainty*, 3rd ed., Routledge, London, 170 pp.
McIntyre, A.J. (1978): Effect of drought on the economy. In: *The environmental Economic and Social Significance of Drought*, ed. by J.V. Lovett. Sydney, 181–192.
Mearns, L.O. and Rosenzweig, C. (1994): Use of a nested regional climate model with changed daily variability of precipitation and temperature to test related sensitivity of dynamic crop models. In: *American Meteorological Society, Symp. on Global Change Studies*, Nashville, TN, 23–28 January 1994, Boston, 142–155.
Megie, G. (1989): *Ozone, l'équilibre rompu*. Presses du CNRS, Paris, 260 pp.
Murphy, A.H. and Katz, R.W. (1985): *Probability, statistics and decision making in the atmospheric sciences*. Westview Press, Boulder, CO.
Newman, J.E. (1980): Climate change impacts on the growing season of the North American Corn Belt. *Biometeorology*, 7 (part 2), 128–142.
NIAS, Division of Agrometeorology (1976): Crop production and unusual weather. Misc. Report, National Institute of Agricultural Sciences, Tokyo, 30 pp.
Pacpe, R., Fairbridge, R.W. and Jelgersma, S. (1990): *Greenhouse effect, sea level and drought*. Kluwer Academic Publishers, Dordrecht.
Paoli, G., ed. (1994): Climate change, uncertainty and decision making. Institute for Risk Research, Waterloo.
Parry, M. (1990): *Climate change and world agriculture*. Earthscan, London, 157 pp.
Parry, M.L., Carter, T.R. and Konijn, N.T. (eds.) (1988): *The impact of climatic variations on agriculture*. Vol. 1: *Assessments in cool temperate and cool regions*, 876 p. and Vol. 2: *Assessments in semi-arid regions*. Kluwer Academic Publishers, Dordrecht, 764 pp.
Patil, G.P. and Rao, C.R. (1994): *Handbook of statistics XII: Environmental statistics*. Elsevier/North Holland, New York.
Pearman, J.I., ed. (1988): *Greenhouse: Planning for climate change*. Victoria, Australia, CSIRO.
Peters. R.L. and Lovejoy, T.E. (eds.) (1992): *Global warming and biological diversity*. Yale University Press, New Haven.
Philipps, M.J. (1989): Applied climatology. WCAP, 6, WMO, No. 281.
Pittock, A.B., Frankes, L.A., Jenssen, D., Peterson, J.A. and Zillman, J.W. (1978): *Climatic change and variability: A southern perspective*. Cambridge University Press, Cambridge.
Prentice, I.C., Webb, R.S., Ter-Mikhaelian, M.T., Solomon, A.M., Smith, T.M., Pitovranov, S.E., Nikolov, N.T., Minin, A.A., Leemans, R., Lavorel, S., Korzukhin, M.D., Hrabovsy, J.P., Helmisaari, H.O., Harrison, S.P., Emanuel, W.R. and Bona, G.B (1989): *Developing a global vegetation dynamics model: Results of an IIASA Summer Workshop*. International Institute for Applied Systems Analysis, Laxenburg, Austria.
Price-Bridgen, A. (1990): *Using meteorological information and products*. Ellis Horwood, NY, 491 pp.
Pyle, G.F. (1979): *Applied medical geography*. Wiley, New York.
Rampal, S. (1991): 2030: Conjoncture sur la ressource en eau en Languedoc karstique. *La Méteorologie*, VII, 38, 2–7.
Robertson, G.W. (1983): Weather-based mathematical models for estimating development and ripening of crops. WMO Technical Note, No. 180, 99 pp.
Rosenzweig, C. and Iglesias, A. (eds.) (1994): *Implications of climate change for international agriculture: Crop modeling study*. EPA, Washington, DC, n.p.
Rosenzweig, L. and Martin, L.P. (1994): Potential impact of climate change on world food supply. *Nature*, 367, 6459, 133–138.
Roy, M.E., Hough, M.N. and Stark, J.R. (1978): Some agricultural effects of the drought of 1975–1976 in the United Kingdom. *Weather*, 33, 64–74.

Russo, J.A. (1966): The economic impact of weather on the construction industry of the United States. *Bull. Am. Met. Soc.*, 47, 967–971.

Santibanes, F. (1991): Possibles variations agroclimatiques en Amérique du Sud, au XXIe siécle. *La Météorologie*, 7, 38, 17–24.

Schneider, S.H. (1989): *Global warming*. Sierra Club Book, San Francisco, U.S.A., 317 pp.

Smith, J.B. and Tirpak, D.A. (eds.) (1989): The potential effects of global climate change on the United States. USEPA-230-05-89-050, USEPA, Washington, DC.

Starr, J.R. (1991): Products and services offered to the Agricultural Industry by the UK Meteorological Office. In: *Actes du 1er Colloque sur les Applications de la Météorological et leurs Inérêts économiques*. METEO-FRANCE et Rég. Franche-Comté, Besançon, 17–193.

Strain, B.R. and Cure, J.D. (eds.) (1985): Direct effects of increasing carbon dioxide on vegetation. DOE/ER-0238, US Dep. of Energy, Carbon Dioxide Res. Div., Washington DC.

Thom, E.C. (1959): The discomfort index. *Weatherwise*, 12, 57–60.

Thomas, T.H. and Smith, A.R. (eds.) (1989): *Mechanisms of plant perception and response to environmental stimuli*. Brit. Plant Growth Reg., Monogr. Vol. 20, Bristol, U.K.

Titus, J.G. (ed.) (1988): *Greenhouse effect, sea level rise and coastal wetlands*. U.S. Environmental Protection Agency, Washington, D.C.

Tucker, C.J. (1979): Red and photographic infrared linear combinations for monitoring vegetation. *Remote Sensing Environ.*, 8 (2), 127–150.

Uchijima, Z. (1981): Yield variability of crops in Japan. *GeoJournal*, 5(2), 151–164.

United Kingdom Department of the Environment (1991): The potential effects of climate change in the United Kingdom. Climate Change Impacts Review Group, Her Majesty's Stationery Office (HMSO), London.

Van Keulen, H. and Wolf, J. (1086): *Modelling of agricultural production: Weather, soils and crops*. Pudoc Wageningen, the Netherlands, 479 pp.

WMO (1979): *Proceedings of the 2nd World Climate Conference*, Geneva, 12–23, February 1979.

WMO (1982): The effects of meteorological factors on crop yields and methods of forecasting the yield. Technical Note No. 174, 54 pp.

WMO (1990): *Economic and social benefits of meteorological and hydrological services*, Proceedings of the Technical Conference, Geneva, 26–30, March 1990, WMO, Geneva No. 733.

Waggoner, P.E. (1990): *Climate change and US water resources*. Wiley and Sons, New York.

Whisler, F.D., Acock, B., Baker, D.N., Fye, R., Hodges, H.F., Lambert, J.R., Lemmon, H.E., McKinion, J.M. and Reddy, V.R. (1986): Crop simulation models in agronomic systems. *Adv. Agron.*, 40, 141–208.

Wigley, T.M.L. and Raper, S.C.B. (1987): Thermal expansion of sea water associated with global warming. *Nature*, 247: 127–131.

Wigley, T.M.L., Ingram, M. and Garmer, G. (eds.) (1981): *Climate and history: Studies in past climate and their impacts on man*. Cambridge University Press, Cambridge.

Yao, A.Y.M. (1969): The R index for plant water requirements. *Agric. Meteorol.*, 6, 259–273.

Yarnal, B. (1993): *Synoptic climatology in environental analyses*. Belhaven Press, London, 195 pp.

(Author for correspondence)
Annick Douguédroit
Institute of Geography, and CNRS/CAGEP URA 903
University of Provence (Aix-Marseille 1)
29, av. Robert Schuman
13621 Aix-en-Provence Cédex
France

PART 3

REGIONAL SCALES CLIMATES

3.1. SOCIETY-CLIMATE SYSTEMS IN TROPICAL AFRICA

O. OJO

1. Introduction

The society-climate systems in Tropical Africa, as in any other part of the world, may be most usefully discussed as components of the global society-climate system which involves atmosphere-ocean couplings, atmosphere-land couplings and ice-ocean couplings. Over Tropical Africa, these couplings, particularly the atmosphere-ocean and atmosphere-land couplings, result in changes in climate on a variety of spatial time scales, whose underlying mechanisms are related to the restless nature of the atmosphere.

In general, three spatial time scales of the atmospheric processes within these couplings may be discerned. These include the local, the meso and the macro or global time scales. While the local time scales focus upon the short-lived activities, particularly near the ground at different places, the global time scales include the long lived general atmospheric circulations. The meso time scales form transitional activities which influence zones and regions in different parts of tropical Africa.

It is these three timescale processes which produce the various changes in weather and climate and which result in the "climatic drama" played out between the atmospheric environment and the other components of the ecosystems in all parts of Tropical Africa. It is also this concept of the "climatic drama" which significantly affects societies in different parts of the world in general and tropical Africa in particular. The "climatic drama" in particular, and society-climate systems in general no doubt form a significant part of human-environmental systems and involves the interactions between humans and the total of their surrounding external conditions.

As in other parts of the world, four categories of the "climatic drama" may also be temporally distinguished for society-climate systems in Tropi-

M. Yoshino et al. (eds.), Climates and Societies – A Climatological Perspective, 153–176.

cal Africa. The first category refers to the "drama" which results in climatic revolutions or climatic changes at scales greater than 1000 years. The second category results in climatic fluctuations at time scales of 10–1000 years while the third category results in erratic pulsations or fluctuations usually less than 10 years duration. The fourth category results in human-made climate fluctuations whose time scales may be 10 years or less (Landsberg 1976). For unravelling the environmental problems which involve society-climate systems, and particularly for practical applications of climate to human society, the last two categories are of particular importance and will be emphasized in the present paper. Particularly during the last three decades, these two categories have created the awareness for the need to respond to the challenges of understanding the "climatic drama" and the resulting impacts on human society.

No matter the category or time scale involved in the discussions of society-climate systems in this chapter, the society-climate systems, which form a major component of the society-environmental systems depend mostly on the interactions between the amount of solar energy received and the amount absorbed by the earth-atmosphere systems. Over most part of Tropical Africa, the amounts of solar energy received are high throughout the year, normally varying between more than about 400 ly/day in the tropical rain forest region to more than 500 ly/day in the arid and semi-arid areas. It is the need to achieve a radiation balance of the global earth-atmosphere systems that results in the various climatic systems and the potentials for climatic variations and changes, and their impacts on human society in different parts of the world.

In the past, such climatic variations were entirely attributed to natural causes; but recent developments have shown that humans are capable of influencing and have indeed considerably influenced the "climatic drama" in different parts of the world, and particularly in Tropical Africa. As Henderson-Sellers (1991) remarked, the influence of humans has shown for the first time that one species, humanity, has the ability to alter the global environment within the lifetime of an individual.

In the following discussions, an analysis will be made of our present knowledge of the climate of Tropical Africa. The characteristics of the recent climatic variations will also be discussed.

2. Our present knowledge of climate of Tropical Africa

As already declared by the first World Climate Conference (1979), global climate has varied slowly in the past millennia, centuries and decades, and will continue to vary in the future. Over Tropical Africa, humankind has taken advantage of favourable climates and has been vulnerable to changes and variations of climates and the occurrence of extreme events, such as droughts and floods. Thus, food, water, energy, shelter and health, all of

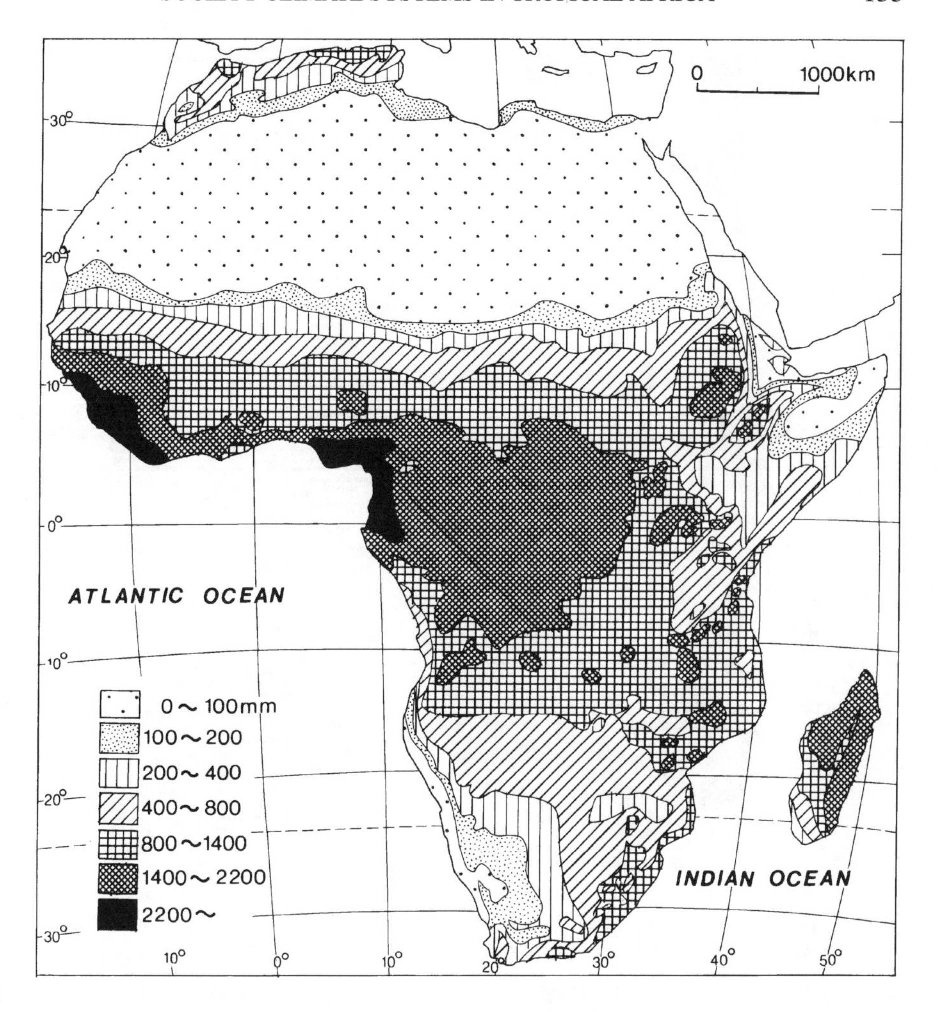

Figure 1. Mean annual rainfall distribution in Africa.

which are aspects of human life that depend critically on climate have been significantly influenced by the various climatic events. No doubt, all countries of Tropical Africa and particularly those in the arid and semi-arid areas have suffered at one time or another under the impacts of climate and climatic variations and changes, and are vulnerable to them.

Figure 1 shows the mean annual distribution of rainfall in Africa. Annual rainfall over Tropical Africa is highest in the low equatorial areas along the Coast of Guinea and in the Zaire regions. Rainfall is also high in the highlands of Ethiopia, the Malagassy Republic, and in parts of the East African Plateau. In these regions, mean annual rainfall generally exceeds 1400 mm, and in some cases even 3000 mm (for example in the Cameroon Highlands with

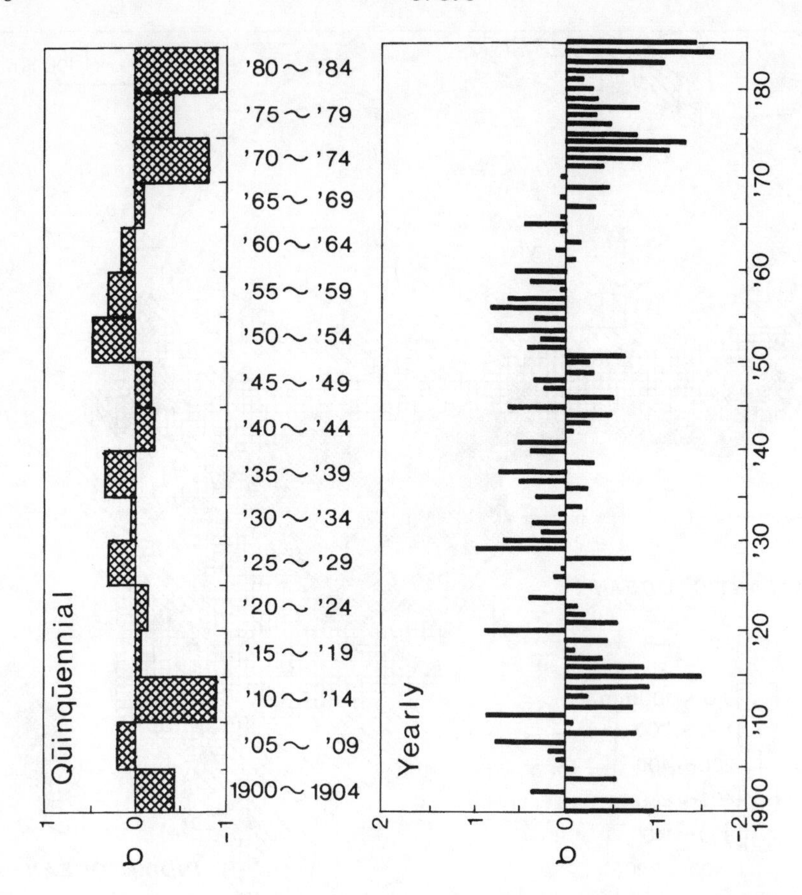

Figure 2. Rainfall variability indices for West Africa.

over 7600 mm). Mean annual rainfall is lowest in the Sahara and Namib deserts and parts of Somalia and Ethiopia. Unfortunately, these deserts also experience the highest variability in rainfall.

Much of Tropical Africa has distinct wet and dry seasons with the rain forest region having a relatively even distribution of rainfall throughout the year, and the savanna areas having a pronounced summer maximum. In January, precipitation is highest to the south of the equator especially in Malagasy, Zaire, and Zambia. In July, rainfall is heaviest north of the equator, especially in Ethiopia and along the Guinea Coast.

Considerable temporal variations also take place in the rainfall of Tropical Africa. Figure 2 shows the rainfall variability indices for West Africa averaged for 60 stations and for 1901–1985. The century began with a relatively long period of drought which lasted until about 1926 with occasional breaks of relatively normal or wet periods which lasted one or two years each. Over the period 1901 to 1926, only three years may actually be regarded as wet with indices greater than $+s/2$. In contrast, there are about 10 years with indices equal to or less than $-s/2$. The second period is a relatively wet period, then two or three wet years were generally followed by two or three years characterized by fairly normal, but slightly drier conditions. About 11 years of the 33-year period have variability indices which are greater than $+s/2$ while only three years have indices which are $-s/2$ or less. The remaining 19 years have deviations which are between $\pm s/2$. Between 1967 and 1985, droughts have been relatively persistent and the indices for most of the years are equal to or less than $-s/2$.

A large number of variations different from the general patterns described above however characterize the climate of Tropical Africa. For example, although droughts have been widespread in most parts of Tropical Africa for most of the period, these droughts vary in intensity, persistence, and spatial coverage. In the forest regions, for instance, the rainfall characteristics show that droughts were less persistent and less severe than in the savanna regions (Figure 3). Within the savanna regions, a comparative analysis also shows that the Sudan and Sahel savannas were characterized by relatively more persistent drought conditions than for the Guinea savannas. In contrast to the Sudan and Sahel savannas, many more years were near normal in the Guinea savanna. Moreover, whenever droughts occur in the forests and the Guinea savannas, they were usually less severe than in either the Sudan and the Sahel regions.

Intra-regional and locational analysis in rainfall variations further illustrates the fact that many variations can occur in the intensity, persistence, severity and spatial patterns of climatic events. In West Africa, climatic conditions were less variable to the western side of the forest region than the eastern side. Similarly, in both the Guinea and Sudan savannas, conditions were more variable to the east than to the west. In the Sahel region, on the other hand, conditions were relatively more humid in the central parts than in the east or west. A comparison of the conditions in both Maiduguri and Sokoto, Nigeria, which are in the same climatic zone, and to the east of the Sahel region, also shows that droughts were more persistent and more severe in Maiduguri than in Sokoto.

These regional, intra-regional and locational characteristics of variation in climatic events can also be illustrated by examining the spatial variations of climatic events in Tropical Africa. Examples are given in Figures 4 and 5 which show percentage frequencies of moderate and severe droughts between 1901–1989 in the Sudan and the Sahelian regions of West Africa. Notice that whereas moderate droughts occur for about 20–80% of the time in the central parts of the region, the frequencies are less to the east and west, and are generally between 10–40% of the period covered. This situation is different for the frequencies of severe drought occurrences. As shown in Figure 5, the

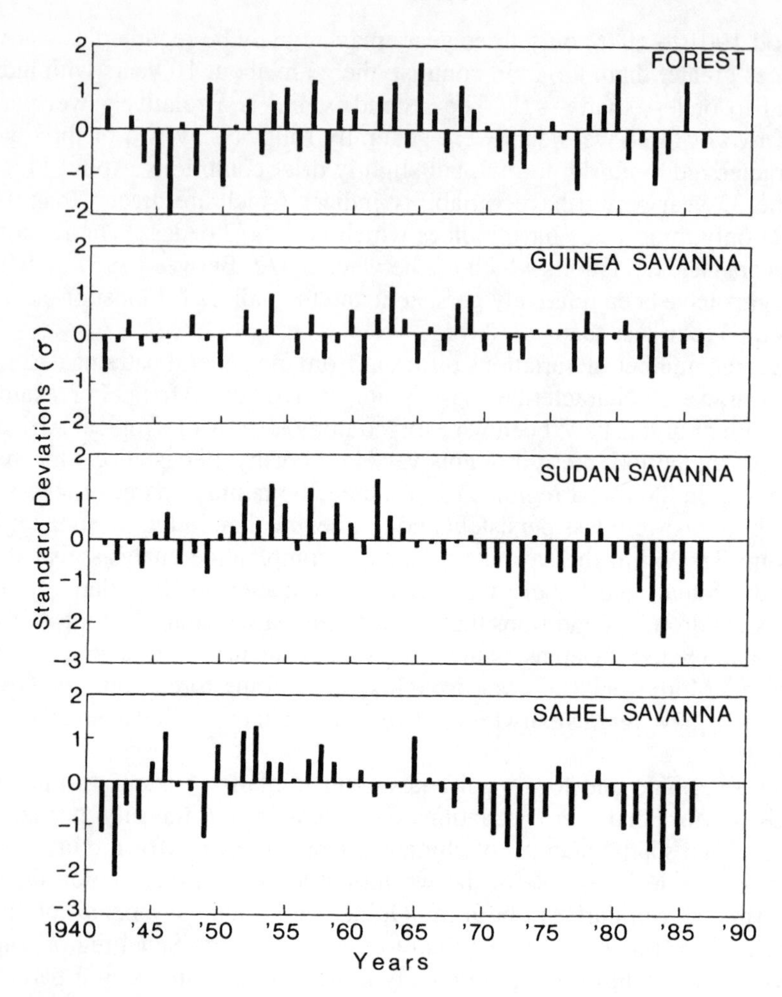

Figure 3. Rainfall anomalies for various zones of Tropical Africa.

frequencies of these droughts are between 20–50% to the west, and 20–30% to the east.

3. Society-climate systems as feedback mechanisms

The characteristics of climate and climatic variability of Tropical Africa discussed above are no doubt a feature of the climatic systems; however, they may also result from human influence which in recent years has become an added factor influencing society-climate systems in Tropical Africa. As already noted above, the rising human population with expanded food needs puts greater pressure on the stabilizing elements of the ecosystem. With

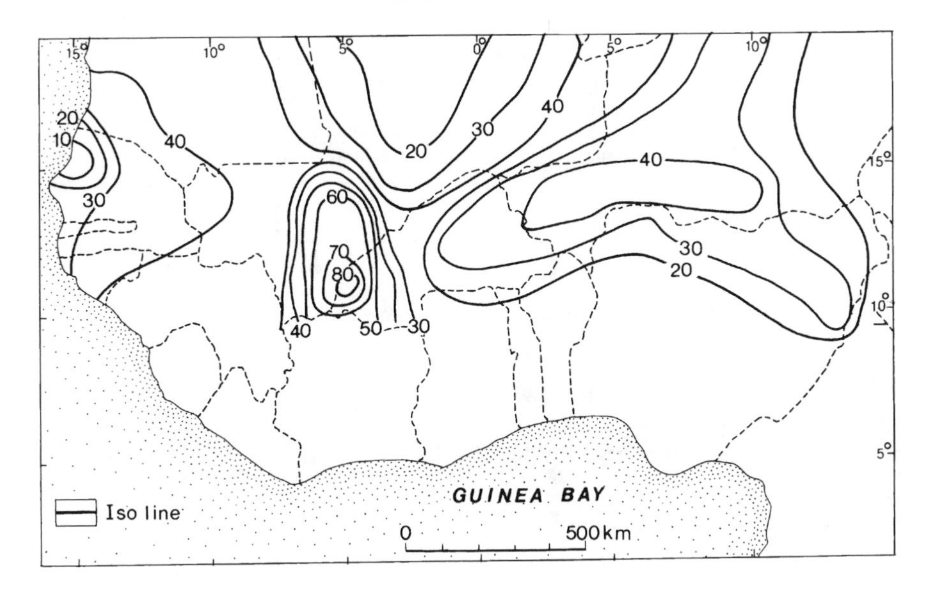

Figure 4. Frequencies of moderate droughts in the Sahel and Sudan, 1901–1989.

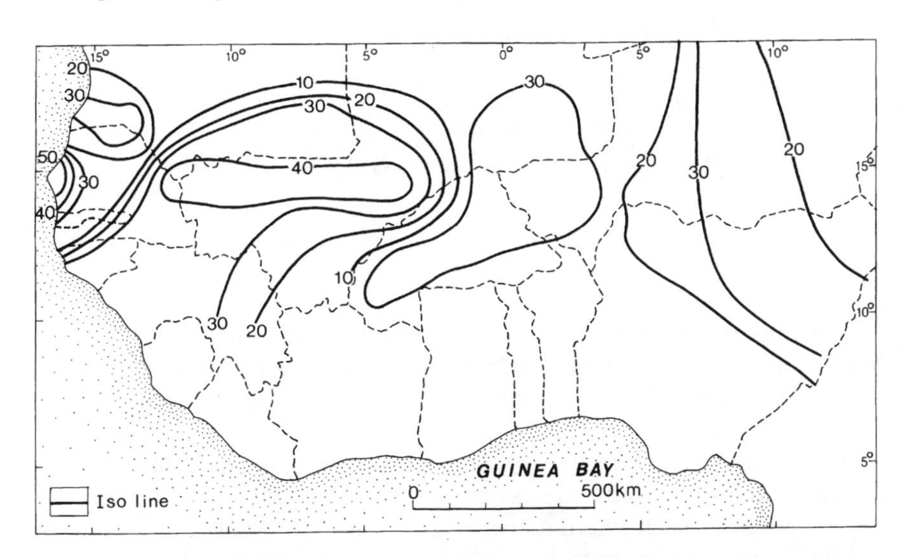

Figure 5. Frequencies of severe droughts in the Sahel and Sudan, 1901–1989.

human interference, particularly in arid and semi-arid and Sudan–Sahelian zones, vegetation is reduced and in some cases steadily disappears and the ground surface is further exposed. Rain, which falls on the ground causes erosion and losses in humus and soil structure, while solar radiation increases the evaporative losses of water from soil and vegetation, and further bakes the

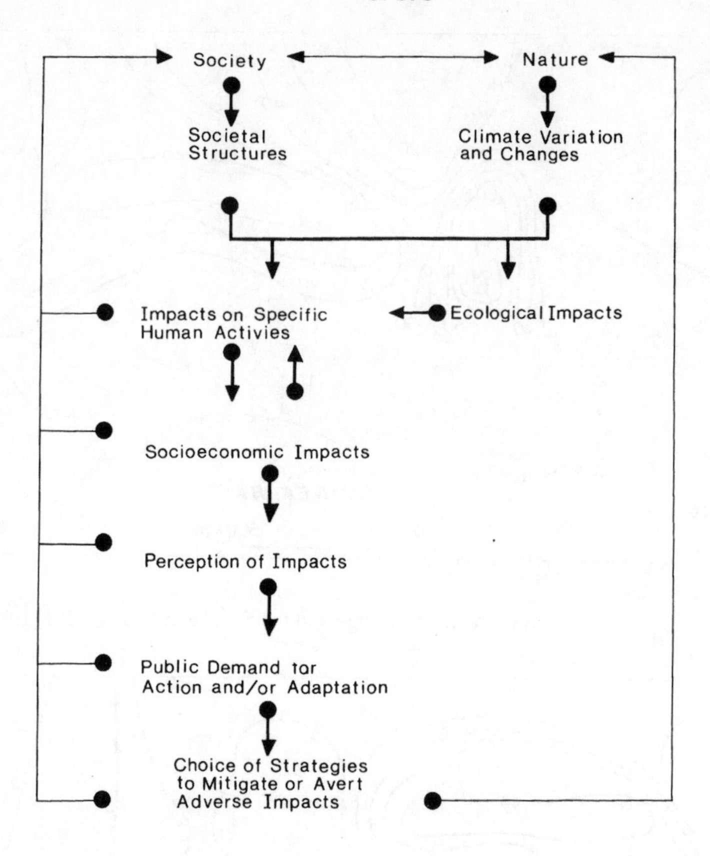

Figure 6. Cascades and feedbacks of causes and effects in climate-society systems.

soil, thereby preventing additional water from sinking into the ground. The groundwater table level drops under the soil beneath, the level of soil water falls and the availability of water from the ground and surface water sources become a problem.

Human interference in the ecosystem usually results in unintended and generally undesirable side effects, causing droughts and desertification. The consequences of such interference, which is usually through crop cultivation and grazing of animals or felling of trees for timber and firewood, are particularly grave with population pressure. Thus, with this human factor, micro-climatic conditions on which the natural vegetation and crops depend, are altered.

In general, therefore, it may be emphasized that if the nature of the problem of society-climate system in Tropical Africa is examined, it involves dealing with a sequence of hierarchy of problem areas beginning with the human and nature (Figure 6). The structure starts with both nature and people interacting with each other. Nature which in this case is climate interacts with the other components of the environmental systems, while human activities (for example, agriculture, lumbering, industrial and transportational activities) create human-made climates on global, regional and local scales. The resulting regional and local climatic changes and variations, determine the effects on environmental processes and specific activities as well as the characteristics of the economic sectors. Such changes and variations have consequences on agriculture, water resources, industry, energy supply and demand, fisheries and marine resources, human health, transportation, diseases, population processes, tourism and recreation. The effects on specific socio-economic activities in turn determine the perception of the climatic impacts, public awareness, and public demand for action or adaptation. They thus determine the probable or desirable responses of the societies. The whole processes result in "cascades" and "feedbacks" of "causes and effects" which close back on themselves in the last step (Figure 6).

All these processes are likely to continue in Tropical Africa but with humans having greater influence in the mechanism by quickening and intensifying the processes of climate and climatic variations and change through their actions on socio-economic activities which in the final analysis would create human-made climates.

4. Climate as a resource in Tropical Africa

The above discussions emphasize climate as a factor within the society-climate system in Tropical Africa. In addition to the influence of climate as a factor, its significance as an environmental resource can also be noted. In general, three aspects of climate as a resource within the society-climate systems may be noted.

First, there are the atmospheric parameters such as rainfall, insolation and wind which can be used to the advantage of the society. Rain water is useful for agriculture while we breathe air. Solar radiation and wind can also be used as sources of energy. The fact that the values of these climatic parameters vary in time and space makes them useful for economic purposes like any other resource. Variations in their abundance and value make the economics of climate as a resource worth considering.

Secondly, there are those climatic conditions which provide humans with satisfaction or social objectives. For example, pleasant atmosphere can provide an environment conducive to tourism, if there is bright sunshine. Along suitable beaches, tourist activities are important.

Thirdly, climatic conditions may create other state of affairs which are useful to humans. As already noted, water, solar radiation and wind can be used for generating power.

It can thus be emphasized that, like any other resource, climate in Tropical Africa can be held, and is indeed sometimes held, as a form of property in some countries. Indeed, climate is sometimes "purchased" through ownership of a particular geographic location, and, as the possibility of this ownership increases, the society alters climate in its environment either inadvertently or intentionally. By choosing one or another policy option, society sometimes eventually creates a certain kind of climate for itself. However, climate is no doubt a common property resource, and must be regarded as such. The fact that climate depends upon the atmospheric and oceanic circulations also suggests that it has the characteristics of a common property. Atmospheric processes occur on global, regional and local basis and the processes do not stop at national borders.

5. Economic impacts of society-climate systems in Tropical Africa

With the variability characteristics of climate over Tropical Africa, and with the recognition of climate as a factor and a resource in the region, there occurs several impacts of weather and climate on practically every aspect of the national economy, and the climatologist has the opportunity to become an applied scientist of almost unlimited scopes and influence. Indeed, in many parts of Tropical Africa, climate directly determines the physical characteristics of the environment which in turn influences the levels of economic productivity. In particular, climate has always been a major factor influencing agricultural productivity, the abundance or lack of fresh water, the comfort and safety of different kinds of transport, the most economic operations of dams and irrigation facilities and the demand and supply of power for both domestic and industrial proposes.

In the following discussions, some of these fields in which the impacts of weather and climate have been noticeably felt will be examined.

6. Agriculture

Agriculture is the most important primary economic activity pursued in Tropical Africa. However, the low level of agricultural productivity in the region has been a basis for world food problem which is rapidly emerging and the wide disparity between food production and food consumption in the region. Weather and climate no doubt form significant environmental factors which are also responsible for the low level of productivity. In this connection, the significance of climate is both direct and indirect.

Probably the most important climatic factor influencing agriculture in Tropical Africa is rainfall. Rainfall determines the characteristics of water available to plants and the farming calendar in different parts of the region. Cropping types and cropping patterns in areas with monomodal regimes, such as the Sudan savanna, are quite different from those in areas with bimodal regimes such as along the coastal areas of West and Central Africa (Figure 7). In addition to rainfall, climatic elements such as solar radiation, temperatures, evaporation and evapotranspiration and winds also directly influence crop productivity. Solar radiation provides energy for photosynthesis while plant growth will cease when temperature falls below a certain minimum or exceeds a certain maximum value. Winds may also do considerable damage to crops and seed distribution processes.

Climate indirectly affects other factors which are significant for agriculture. For example, climate affects soil fertility through leaching of soil nutrients or soil erosion. Climate also influences diseases and pests, being generally an important factor for reproduction, growth and survival of these pests and diseases. Climate also indirectly influences management of farms and farming practices, for example through problems arising from weeding, pests and diseases, storage, handling and marketing.

In recent years, large monetary losses have been caused in agricultural activities because of adverse weather conditions in Tropical Africa. These losses which arise mostly from the impacts of precipitation have caused considerable damages and disrupted lives at one time or another in all countries of this region. Drought impacts on agriculture are probably more widespread and provide more dramatic evidence of the consequences of climatic variations and climatic changes on agriculture. Following the five years (1969–1973) of drought impacts in the Sahelian zones of Africa, about 10 million lives were threatened by hunger, and catastrophe was averted only by massive international food relief efforts. In Nigeria alone, considerable losses were incurred on farm products and most of the victims were forced to migrate to the southern states and in many cases to the urban centers. Indeed, as noted by ECA (1983), Africa's food situation is the single most critical area of concern in the region. For the whole of the decade of the 1970s, total food production was rising by no more than 1.5% compared with an expanding population at an average annual rate of about 2.8% (Table 1). In many countries the average annual changes in food production are negative (Figure 8). Food sufficiency ratio had dropped from 98% in the 1960s to about 86% in the 1980s implying that each African country had about 12% less home grown food in the 1980s than 20 years earlier.

To cover the widening food deficits, Africa has had to import increasing amounts of food or continue to receive aid. Indeed between 1970 and 1980, the volume of total food imports increased at an average annual rate of 8.4%. Food aid to Africa reached 1.5 million tons in 1980. The increasing reliance of the African region on food aid and imports no doubt threatens to create a new and dangerous structural dependence on cereals such as wheat which cannot be easily grown in many parts of the Africa. Although other factors had contributed to the disappointing performance in agriculture and food production, weather and climate have been the most important factor.

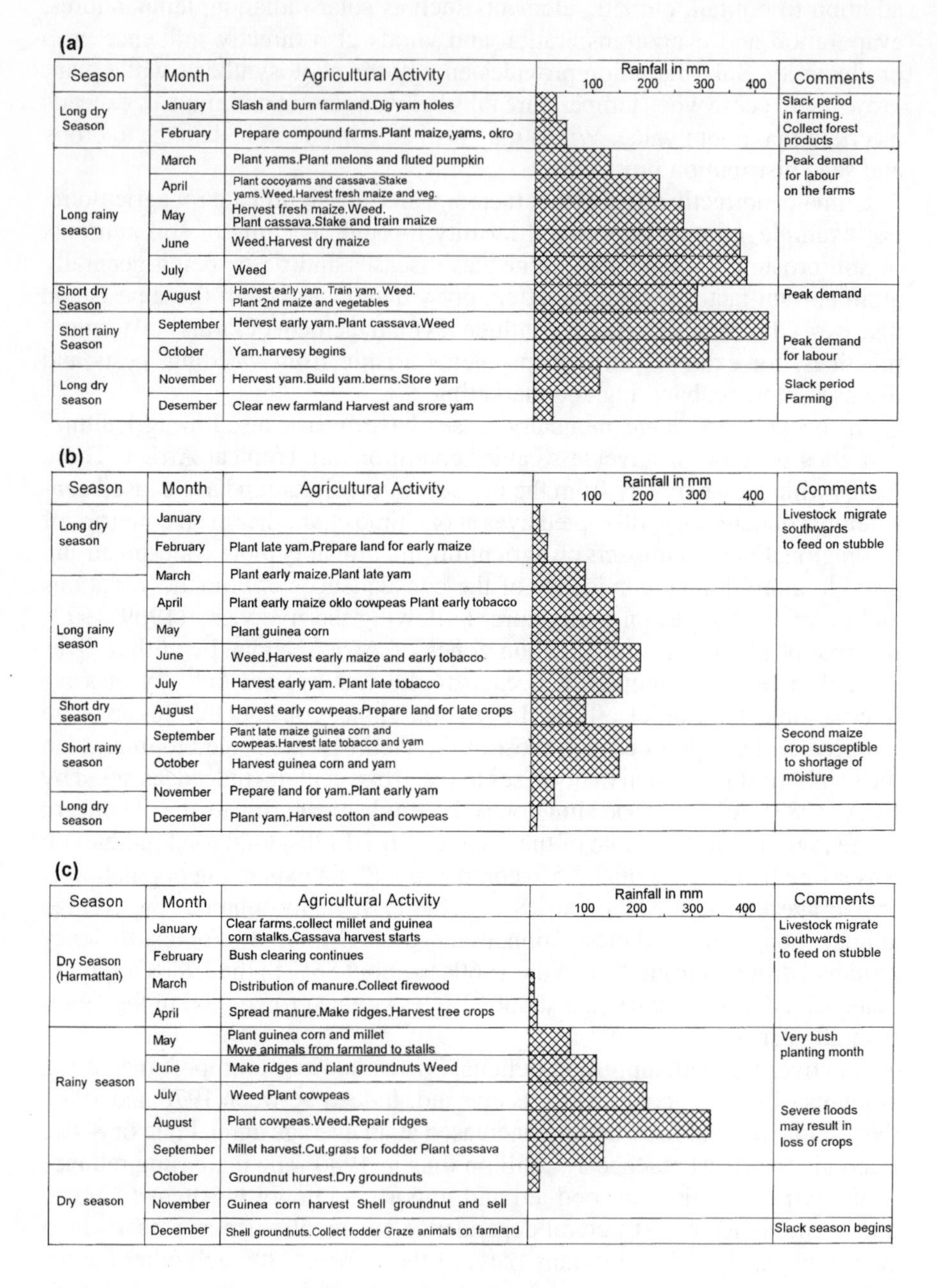

(a)

Season	Month	Agricultural Activity	Rainfall in mm 100 200 300 400	Comments
Long dry Season	January	Slash and burn farmland.Dig yam holes		Slack period in farming. Collect forest products
	February	Prepare compound farms.Plant maize,yams, okro		
Long rainy season	March	Plant yams.Plant melons and fluted pumpkin		Peak demand for labour on the farms
	April	Plant cocoyams and cassava.Stake yams.Weed.Harvest fresh maize and veg.		
	May	Hervest fresh maize.Weed. Plant cassava.Stake and train maize		
	June	Weed.Harvest dry maize		
	July	Weed		
Short dry Season	August	Harvest early yam. Train yam. Weed. Plant 2nd maize and vegetables		Peak demand
Short rainy Season	September	Hervest early yam.Plant cassava.Weed		Peak demand for labour
	October	Yam.harvesy begins		
Long dry season	November	Hervest yam.Build yam.berns.Store yam		Slack period Farming
	Desember	Clear new farmland Harvest and srore yam		

(b)

Season	Month	Agricultural Activity	Rainfall in mm 100 200 300 400	Comments
Long dry season	January			Livestock migrate southwards to feed on stubble
	February	Plant late yam.Prepare land for early maize		
Long rainy season	March	Plant early maize.Plant late yam		
	April	Plant early maize oklo cowpeas Plant early tobacco		
	May	Plant guinea corn		
	June	Weed.Harvest early maize and early tobacco		
	July	Harvest early yam. Plant late tobacco		
Short dry season	August	Harvest early cowpeas.Prepare land for late crops		
Short rainy season	September	Plant late maize guinea corn and cowpeas.Harvest late tobacco and yam		Second maize crop susceptible to shortage of moisture
	October	Harvest guinea corn and yam		
Long dry season	November	Prepare land for yam.Plant early yam		
	December	Plant yam.Harvest cotton and cowpeas		

(c)

Season	Month	Agricultural Activity	Rainfall in mm 100 200 300 400	Comments
Dry Season (Harmattan)	January	Clear farms.collect millet and guinea corn stalks.Cassava harvest starts		Livestock migrate southwards to feed on stubble
	February	Bush clearing continues		
	March	Distribution of manure.Collect firewood		
	April	Spread manure.Make ridges.Harvest tree crops		
Rainy season	May	Plant guinea corn and millet Move animals from farmland to stalls		Very bush planting month
	June	Make ridges and plant groundnuts Weed		
	July	Weed Plant cowpeas		Severe floods may result in loss of crops
	August	Plant cowpeas.Weed.Repair ridges		
	September	Millet harvest.Cut.grass for fodder Plant cassava		
Dry season	October	Groundnut hurvest.Dry groundnuts		
	November	Guinea corn harvest Shell groundnut and sell		
	December	Shell groundnuts.Collect fodder Graze animals on farmland		Slack season begins

Figure 7. Rainfall and agricultural patterns in Tropical Africa.

Table 1. Size and growth of population.

	Population (millions)			Average annual population change (percent)			Average annual increment to the population (thousand)		
	1960	1990	2025	1965–70	1975–80	1985–90	1965–70	1975–80	1985–90
WORLD	3,019.4	5,292.2	8,466.5	2.06	1.74	1.73	72,398	74,061	87,666
Algeria	10.8	25.4	50.6	2.85	3.06	3.12	365	530	733
Angola	4.8	10.0	24.0	1.52	3.39	2.70	82	241	253
Benin	2.3	4.7	13.0	2.06	2.77	3.10	53	90	138
Botswana	0.5	1.3	3.4	2.54	3.54	3.51	15	29	41
Burkina Faso	4.5	9.0	22.7	2.14	2.30	2.67	114	151	225
Burundi	2.9	5.5	13.1	1.45	1.80	2.80	48	71	46
Cameroon	5.5	11.2	26.2	2.11	2.57	2.60	135	208	274
Cape Verde	0.2	0.4	0.9	3.04	0.87	2.81	8	3	10
Central African Rep.	1.6	2.9	6.8	1.63	2.22	2.46	29	48	67
Chad	3.1	5.7	13.2	1.82	2.10	2.40	64	89	132
Comoros	0.2	0.5	1.3	2.43	3.40	3.116	12	15	
Congo	1.0	2.0	5.0	2.18	2.46	2.73	25	35	51
Cote d'Ivoire	3.8	12.6	39.8	4.05	4.19	4.12	202	315	469
Djibouti	0.1	0.4	1.1	7.68	4.48	2.96	11	12	11
Egypt	25.9	54.0	94.0	2.35	2.69	2.55	733	1,046	1,296
Equatorial Guinea	0.3	0.4	1.0	1.55	1.99	2.34	4	7	10
Ethiopia	24.2	46.7	112.3	2.41	2.44	2.0	695	888	894
Gabon	0.5	1.2	2.9	0.36	4.70	3.45	2	34	37
Gambia, The	0.4	0.0	1.9	2.77	3.15	2.83	12	19	23
Ghana	6.8	15.0	37.0	1.91	1.76	3.14	157	181	436
Guinea	3.7	6.9	15.7	1.95	2.17	2.48	81	111	160
Guinea-Bissau	0.5	1.0	2.2	0.06	5.04	2.08	0	36	20
Kenya	6.3	25.1	77.6	3.30	3.82	4.22	350	578	955

Table 1. (continued)

	Population (millions)			Average annual population change (percent)			Average annual increment to the population (thousand)		
	1960	1990	2025	1965–70	1975–80	1985–90	1965–70	1975–80	1985–90
Lesotho	0.9	1.8	4.3	2.01	2.41	2.85	20	30	47
Liberia	1.0	2.6	7.2	2.85	3.09	3.18	36	53	75
Libya	1.3	4.5	12.8	4.04	4.37	3.65	73	119	52
Madagascar	5.3	12.0	33.0	2.28	2.90	3.18	145	237	352
Malawi	3.5	8.4	22.8	2.56	3.00	3.31	109	169	257
Mali	4.6	9.4	24.1	2.15	2.19	2.94	116	146	256
Mauritania	1.0	2.0	5.0	2.17	2.46	2.73	25	36	52
Mauritius	0.7	1.1	1.5	1.83	1.91	1.25	15	17	13
Morocco	11.6	25.1	44.4	2.78	2.27	2.56	397	415	604
Mozambique, P.R.	7.5	15.7	34.4	2.39	2.83	2.65	212	320	389
Niger	3.2	7.1	18.9	2.08	2.59	3.01	82	129	199
Nigeria	42.3	113.0	301.3	3.24	3.49	3.43	1,709	2,577	3,564
Rwanda	2.7	7.0	18.1	3.16	3.27	3.40	109	156	226
Senegal	3.0	7.0	16.4	2.89	3.46	2.69	108	180	185
Sierra Leone	2.2	4.2	9.6	1.79	2.15	2.49	45	66	97
Somalia	2.9	7.6	18.9	2.31	5.06	2.32	80	239	231
South Africa	17.4	35.2	63.2	2.49	2.23	2.19	526	595	731
Sudan	11.2	35.2	63.2	2.49	.223	2.19	526	595	731
Swaziland	0.3	0.8	2.2	2.52	3.12	3.43	10	16	25
Tanzania	10.0	27.3	84.8	3.08	3.42	3.67	385	593	915
Uganda	6.6	18.4	55.2	3.96	3.20	3.49	352	387	590
Zaire	15.9	36.0	99.5	2.11	3.27	3.17	390	796	1,056
Zambia	3.1	8.5	25.5	2.96	3.40	3.76	115	179	290
Zimbabwe	3.8	9.7	22.6	3.28	2.96	3.15	159	197	283

Source: World Resources (1990–1991).

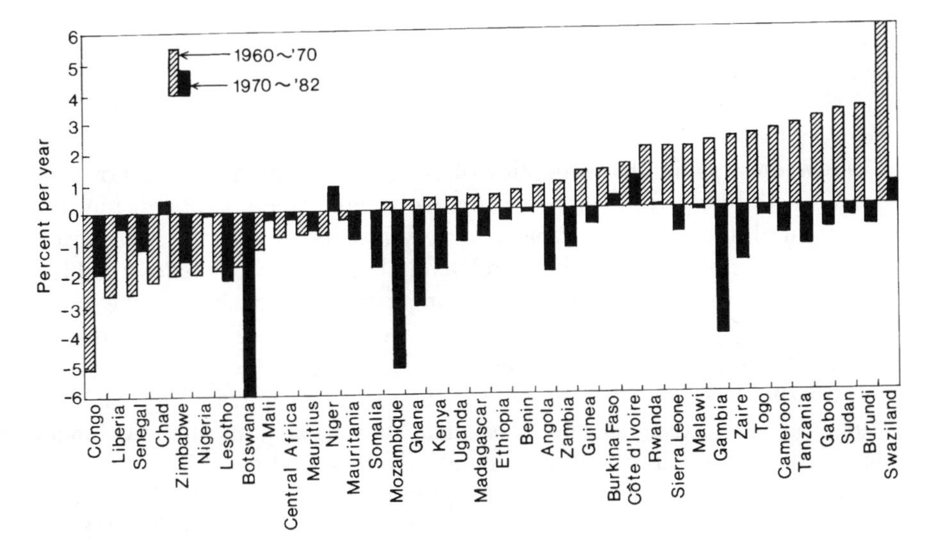

Figure 8. Mean changes in average annual food production in 1960–1970, and 1970–1982.

7. Water resources

The characteristics of fresh water resources in Tropical Africa are largely governed by the characteristics of precipitation, which is the main source of the water supply systems. Similarly, evaporation and evapotranspiration, the main depletion avenue of water are dependent upon climate. In Tropical Africa, both surface and groundwater resources are usually affected by society-climate systems, although the characteristics of water resources are further complicated by relief, vegetation and soil factors. The magnitudes of runoff vary with the climate systems and in general, runoff is highest in areas of highest rainfall and lowest in areas of lowest rainfall. Thus, the percentage of rainfall which becomes runoff tends to increase from the arid and semi-arid areas to the tropical rainforest areas. Over highland areas, such as the Futa Jalon highlands, the Ethiopian highlands, the East African highlands and the Cameroon mountains, rivers have greater mean discharges and higher runoff coefficients than rivers which flow over relatively lower drainage basins. In the absence of effective vegetation cover, runoff coefficients may be as high as 80% for individual storms in these areas.

Perhaps the greatest impacts of society-climate systems on water resources in Tropical Africa, are associated with the marked seasonal alternation of high and low flows of streams and rivers, and the variabilities which have particularly characterized these rivers in Tropical Africa over the past three decades. In general, the flows of the river decline very rapidly after the rainy season although some rivers are capable of sustaining permanent flow throughout the dry season. In equatorial regions with long wet seasons and sometimes rainfall throughout the year, a larger water surplus and relatively lower evaporation rates than in the savanna and semi-arid areas combine to maintain the dry season flow. In the drier areas, permanent flow occurs only occasionally.

The hydrographic characteristics of relatively small rivers are however somewhat different from those of relatively large rivers although the discharges in all the rivers fall rapidly to very low levels once the wet season has ended. The period taken by the relatively large rivers to recede from high wet season to low dry season flow is longer than those taken by the small rivers. This means that the period of really low flows begins later and lasts for a shorter time than in small rivers.

The patterns of the river flows are, however, complicated by, and are a function of the channel and flood-plain morphology and the catchment characteristics particularly of the tributary rivers. Flood-plains such as those of the inland Niger delta and the Sudd region of the Nile river, are able to store much of the river water in the rainy season, and their waters are released back into the river channels in the dry season. Thus they are able to (a) sustain flow for some distance downstream, (b) flatten and elongate the flood wave so that the length of the period of low base flow dimishes with distance downstream, (c) delay the rate at which flood peaks move downstream so that the early part of the dry season is a period of rising, rather than falling discharge, and (d) increase the amount of base flow so that the minimum flows in the rivers exceed the minimum flows into them from headwater areas.

As for the surface water, groundwater resources also always reflect the pattern of rainfall characteristics in Tropical Africa. Over the past three decades the impacts of the relative persistence of droughts in the region have been remarkable. Groundwater tables usually rise during years of relatively high positive rainfall variabilities and fall during years with negative deviations. The areas of permanent groundwater discharge are areas of relatively high rainfall as can be noted for West Africa in Figure 9. It is significant to note that the relatively dry southeast Ghana, southern Togo and southern Benin are not areas of permanent groundwater discharge, in spite of the relatively long wet season, because water surplus during the year is generally insufficient to produce significant groundwater recharge.

8. Socio-economic impacts

Society-climate systems in Tropical Africa also have significant impacts and implications on other socio-economic aspects. Indeed, many of the socio-economic activities have been destabilized and this has considerably affected the already shaky and fragile economy of the region. Among the socio-economic activities affected are transportation, commerce and industry, tourism and recreation.

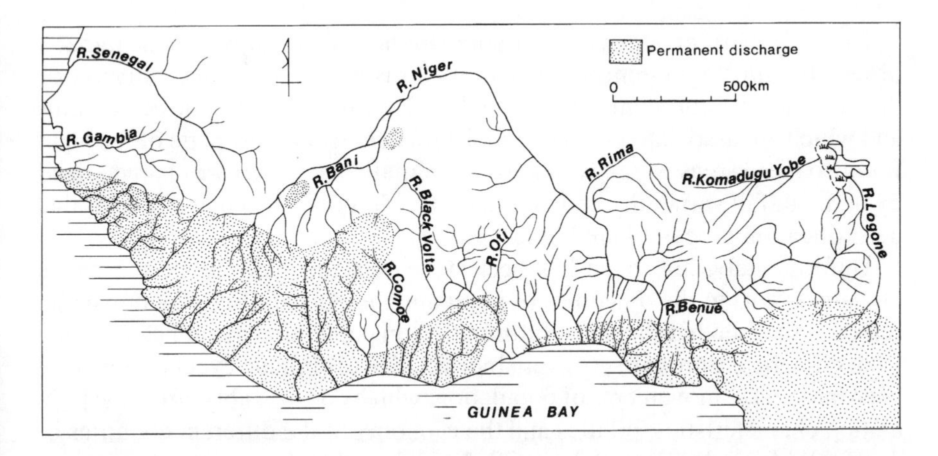

Figure 9. Map of West Africa showing areas of permanent groundwater discharge.

As a case in point, when floods occur during the rainy season, the movements of goods and people are always chaotic and with many more slippery roads, the rates of accidents become higher. Increased incidence of flooding also usually has disruptive impacts on commercial and industrial activities, resulting in economic and human hardship. Both air and water transport (rivers, lakes and canals) are also sometimes adversely affected by poor visibility while water transport may be disrupted by fluctuations in the available water due to floods and droughts. Indirectly, industry and commerce in the region have been influenced by the society-climate systems through the impacts of climate on the availability of raw materials and labour, or through the demand for some particular goods, for example, air conditioners.

9. Population and human settlements

In 1955, sub-Saharan Africa had a population of 190 million, 7% of the world total. By 1985, this population had more than doubled to 432 million. Recent United Nations and World Bank estimates predict another doubling by 2010, raising the population to over 13% of world population. A third doubling to 1,700 million people could occur by 2055.

Both the present population trends and the future perspectives could have serious impacts on society-climate systems. Such impacts are, and would continue to be important in Tropical Africa, particularly in those areas where the rates of population growth are highest. These include eastern and western parts of Africa where more than half of Africa's population currently resides, and where overall fertility rates remain high. Indeed, East and West Africa may quadruple their 1980 population by 2025.

Given the current perspective of population, in Tropical Africa the impacts of society-climates system have been very disastrous. This is especially true of the savanna, arid and semi-arid areas which have relatively fragile ecosystems and which are associated with soil moisture deficits in most parts of the year. It is also these areas which suffer erratic rainfall, and which are characterized by high rainfall variability and unreliability. Moreover, it is in these areas that desertification processes are highest.

Unfortunately, not much has been done to statistically relate the recent climatic events to the dynamics of population in Africa. Probably one major reason is the fact that statistical information on population dynamics resulting from climatic hazards are not available. Yet, due to climatic events, there have been large scale movements of population, which considerably affect the population characteristics, politics and the economy of the different countries of Tropical Africa. Following the 1972–1973 droughts, large scale movements of people occurred from the countries located in the Sahelian regions of West Africa. In 1975, about 726,000 and 348,500 people migrated respectively from Burkina Faso and Mali to Ivory Coast. These figures form considerable proportions of the population of the countries of origin, 6,4 million and 6,0 million, respectively, for the Burkina Faso and Mali (Best and Blij 1980). Similar large scale movements, most of which were free across the border were recorded for other countries like the Niger Republic. Such movements result from the fact that the migrants wanted to better their living conditions, although their economic and social costs were substantial. Immigrants represented more than one-fifth of the total population in Ivory Coast in 1975. The inflow of these migrants had serious socio-economic consequences in the receiving countries including reductions in educational levels and increasing crime rates.

10. Energy system

The need for energy is fundamental to the development of the national economies of Tropical Africa. Energy is needed for cooking, transportation and industries and to counteract the adverse consequences of society-climate systems. In turn, energy systems is Tropical Africa are highly dependent on the variations in the characteristics of the society-climate systems. The major energy source in the region include coal, wood, petroleum and gas, and water. Solar energy and wind are other sources which remain largely untapped although the potential for wind power is rather low.

By comparison with other continents, Tropical Africa is poorly endowed with coal while petroleum is currently being exploited in highly commercial quantities in Nigeria and Gabon. In contrast to coal and petroleum, both water and wood constitute major sources of power and are widely used sources of energy in Tropical Africa.

Probably the greatest effects of the society-climate systems on the energy systems of Tropical Africa are those arising from (a) variations in precipitation and evaporation, which in turn result in variations in the water balance characteristics, (b) changes in temperature, and (c) changes in energy demand due to variations in population and settlement characteristics.

The significance of climate as a factor of energy systems can directly be noted in many parts of Tropical Africa which receive high amounts of rainfall and which in turn provides the basis for construction of hydroelectric power generation, not only in these regions of high rainfall, but also in other parts of the region along which the rivers flow. Because of problems arising from the spatio-temporal variations and change in the precipitation characteristics of Tropical Africa, particularly in recent years, many difficulties have been encountered in the supply of hydroelectricity. Thus, the recent droughts have caused considerable power failures in many parts and have shown the need to be concerned with water storage, water conservation and proper management. Variations in evaporation have also significantly influenced river runoffs and discharges and these in turn have influenced hydroelectric power supply.

Indirectly, solar radiation significantly influences the supply of hydroelectric power because it is an important element influencing both precipitation and evaporation. It is a significant factor in evaporation which in turn results in condensation and precipitation.

Changes in the demand for energy due to changes in population and settlement characteristics also significantly influence and are influenced by the society-climate systems in Tropical Africa. There is always the inverse relationship between space heating requirements and environmental temperatures. The heating characteristics of the indoor environment in turn depends upon solar radiation availability for absorption, temperature, wind and humidity. If the outside temperatures are low, the wall and room temperatures will fall and the demand for fuel and power will be for heating. The reverse is the case when the outside temperatures are high. These two situations which are usually experienced in Tropical Africa, in both cases, increase the demand for energy usage.

11. Society-climate systems and other environmental problems

All nations of Tropical Africa face environmental problems resulting from the impacts of variations in weather and climate. Probably the most important of these problems are those related to (a) violent storms, (b) flooding, and (c) droughts and desertification.

Violent storms in Tropical Africa include hurricanes, hailstorms, lightning, line squalls and thunderstorms. Although hurricanes are not common in Tropical Africa, they are characteristic of the coasts of East Africa, including the island of Malagassy Republic. In economic terms, they are by far the most destructive weather phenomena on a global scale.

Hailstorms can sometimes be significant as weather disasters in Tropical Africa, particularly when accompanied by thunder and lightning. As noted by Samson and Gichniya (1971), hailstorm is a common occurrence in the Kericho area of Western Kenya. During the 66 months of July 1963 to December 1968, there were only two months when no hailstorms were reported. Many of these storms, however, affected only a limited area although the accumulated damages caused by them in a year are sometimes considerable.

Floods are among the other natural disasters which cause considerable losses and damages in many parts of Tropical Africa. The severity of floods, however, varies from place to place, but evidence from many countries indicates that floods sometimes cause greater losses of lives and property in many nations in the region than any other disaster.

Probably the most widespread and most devastating events resulting from society-climate systems in Tropical Africa are droughts and desertification. Both of them have been responsible for many tragic disasters and, particularly in recent years, the tragic human impacts of droughts and desertification in the region have repeatedly drawn the world's attention for the need to be concerned with climatic events. For the Sahelin countries alone, the cost of covering just the food deficits for one year amounts to over 500 million dollars. The realization of the significance of droughts and desertification in Africa has led to active steps being taken to combat them, such as the establishment of the Agrhmet Programme, formed under the auspices of the Permanent Inter-State Committee for Drought Control in the Sahel.

12. Humankind and climate change: The Tropical African dilemma

In recent years, the impact of the substantial increases in the atmospheric concentration of the greenhouse gases has become a great concern to government and people of some of the countries of Tropical Africa. Indeed, it is generally agreed that if the trends in the production of the greenhouse gases should continue until about the middle of the 21st century, substantial implications on environmental processes and their consequent impacts on socio-economic and socio-cultural activities may result.

Tropical Africa is no doubt one of the areas that would be vulnerable to global warming and climatic change and the associated sea level rise. In this region, global warming and sea level rise may compound the serious problems of imbalance between resources, population and consumption, and life.

In addition to rainfall variations and variabilities which have particularly characterized Tropical Africa in recent years, several changes and variations have occurred in the temperatures of Tropical Africa. Over the past three to four decades, temperatures have increased by about 1 to 1.5°C in many cities in Tropical Africa although the changes and variations vary in time and space.

When viewed from the past and present temperature trends, it may reasonably be assumed that with the expected climate changes, the tropical rain belts may experience increased precipitation and increased evaporation although the relative increase would vary from place to place. In contrast to humid areas of the tropical rain belts, the sub-humid and semi-arid areas would have the tendency for a decrease of precipitation. This decrease, coupled with increased temperature and evaporation rates would reduce soil moisture. Indeed, future climate change could worsen the current problems resulting from inter-annual and seasonal climatic variabilities and their impacts on the environment and ecosystems.

On its own, the sea level rise which would accompany global warming, would be significant in several ways. Sea level rise on its own would result in (a) submergence and inundation of the coastal low lands and the island areas of Tropical Africa, (b) increase in the salinity of the estuarine areas, and (c) increase in the coastal erosion, flooding and storm damage.

Other aspects of the society-environmental systems in Tropical Africa would be affected by the expected global warming and climate change accompanied by the sea level rise. Both droughts and desertification would be significantly affected, particularly in parts of the tropical savanna where an increase in evaporation rates is likely to be accompanied by a decrease in precipitation. Moreover, because of the significance of human interference and the projected increase in population pressure on land, drought and desertification may further be intensified except measures are taken to avert or reduce the consequences of climate change (Ibe and Ojo 1990; Alusa and Ogallo 1990).

Both agriculture and water resources are likely to be affected by the expected global climate change. Crop and animal farming are directly or indirectly likely to be affected by (a) increased rainfall and rising temperatures, (b) increased rainfall variability, (c) changes in agroclimatic zones and shifts in agroecological zones, and (d) rise in sea level which is expected to accompany the change in climate (Ibe and Ojo 1990).

The characteristics of climate and climate change would influence (i) the potential growing season and changes in plant growth rates, (ii) crop yield and variability in yield, (iii) level of crop certainty, (iv) yield quality, and (v) sensitivity of crops to differing levels of application of fertilizers, pesticides and herbicides. Climate change in Tropical Africa, which is expected to result in greater variability of rainfall, with more frequent floods and droughts, would lead to more frequent changes in agroclimatic zones while more frequent droughts and long dry periods could lead to decreased food and livestock production causing famine due to crop failure, and much suffering and losses of human and animal lives.

As far as implications on water resources are concerned, it may be emphasized that water supply systems which depend mainly upon precipitation that are strongly concentrated in a rainy season, would likely experience greater water stress and its socio-economic implications. Similarly, increased precipitation in humid areas could increase flood and erosion and water management would require solutions to the problems created by these hydrometeorological events. Lakes and engineered reservoir systems may also be affected by decreased precipitation which may have serious implications as there may be water shortages to maintain the needed storage capacity for later release during periods of deficiency.

Climate change and sea level rise may also have serious impacts on other aspects of energy resources and utilization and many other socio-economic and socio-cultural systems. The implications of climate change on energy would result from the possible impacts of (a) increase in temperature, (b) sea level rise, (c) increase in precipitation in the humid areas, (d) decrease in precipitation in the sub-humid and semi-arid areas, and (e) increase in evaporation throughout Tropical Africa (Ibe and Ojo 1990; Ojo 1990). Because many areas would experience greater drought stress and desertification, there may be greater stress on fuel wood. This fact, coupled with the submergence of the coastal areas which would result from a combination of sea level rise, greater change of storm surges and rising peak runoffs, would decrease the areal extent of forests harvested for fuel wood. Submergence of the coastal areas would also be significant by making the onshore development of petroleum industry more difficulty and more expensive. The increase in drought stress could also lead to a decrease in the supply of hydroelectricity and an increase in the accompanying socio-economic problems.

13. Strategies

The above discussion shows that there are many consequences associated with society-climate systems. No doubt, both now and in the future, climate will continue to have significant impacts on the society, directly through the perceptible effects on humanity, plants and animals and indirectly through the complex effects arising from the numerous interrelationships and interplays between the human society and climate-environmental systems.

Similarly, the human society as a factor, will continue to influence society-climate systems in Tropical Africa, particularly through the use and misuse of the resources available in the region. Deforestation, leading to desertification is very worrying and has created much concern for the government and people of the nations in Tropical Africa. More land has been cleared for cultivation in the last hundred years than in the preceding human history, while the current consumption of fossil fuels in the region has alarmingly increased particularly because of the rapid increase in population.

The consequences and the uncertainties associated with the society-climate systems are now realities that humanity has to face. Many nations in Tropical Africa are particularly vulnerable to the consequences of the impacts of the society-climate systems, while some others would continue to be threatened. Thus, it is necessary that actions be taken in response to the present and potential impacts of climates and climate change. In planning to tackle the problems posed by the consequences of society-climate systems, it is important to plan for both the long-term and the short-term implications of climate variations and climate change resulting from the interplay of society-climate systems. Two categories of responses may be noted in tackling these problems.

First, measures must be taken, if possible, to avert some of the adverse consequences of society-climate systems. Secondly, measures must be taken to mitigate any other impacts resulting from the systems. Efforts should be made to reduce the demand for and use of fossil fuels, especially by conserving and using alternative sources of energy. Reducing or banning bush burning, a common practice in many parts of Tropical Africa, which has considerably increased in recent years because of the pressure to clear more land for agricultural use, will curtail many of the adverse consequences of the society-climate systems. Improvement in forest management and the exploitation and use of land and forest resources are also measures which can significantly reduce as well as avert the adverse consequences of society-climate systems. Increasing biomass production, for example, through afforestation would also avert or reduce the consequences.

Examples of the mitigation measures are development of strategies which can be used to protect the heavily built-up areas, particularly those with high value installations and areas with high population densities where the option of relocation may not be an economic proposition. Relocation of existing settlements and the enforcement of planning regulations could also be significant as preventive measures. Measures which will increase resilience to impacts of society-climate systems include improvement in water management and conservation methods, agrotechnology and effective landuse policies.

14. Conclusion

Many other measures can also be used to avert or mitigate the consequences of society-climate systems. In particular, it is important to note that considerable improvement is needed in (a) environmental monitoring and warning systems, (b) providing and applying improved data on society-climate systems, (c) providing public information, and (d) transfer and effective adaptation and use of technology. A basic problem in many countries of Tropical Africa is the inadequate availability of data, and when available, they can only be found in a variety of sources not always known to scientists who study society-climate systems. Part of the data are scattered in various libraries, record offices and other institutions not known to the scientists, and in many cases the data are hopelessly inaccurate and are of no use for meaningful research and impact studies. There are also problems associated with administrative bureaucracy which have long been recognized but for which no solution has been found.

Added to the problems associated with the characteristics of the sources of data are problems related to inadequate network of meteorological and hydroclimatic stations and coverage of the data on society-climate systems. It may also be noted that it is important to ensure the participation of the population in an organized manner, in the preparation, implementation and evaluation of any programmes aimed at combating the impacts of society-climate systems. It is also important to consider the cultural heritage and leanings of the different groups of people.

The need to transfer and effectively adapt and apply science and technology cannot be overemphasized. This is particularly important in activities related to agriculture, water resources, energy resources, fisheries, forestry and landuse planning.

No doubt, the influence of humans has shown for the first time that "one species, humanity, has the ability to alter the global environment within the lifetime of an individual" (Henderson-Sellers 1991). The fate of the global environment is in the hands of the same species, humanity. Indeed, the fate of Africa's future is in the hands of Africans.

References

Alusa, A.L. and Ogallo, L. (1990): Implications of climate change on the East African region. Executive Summary of Findings, UNEP, 7 pp.

Best, A.C. and de Blij, H.J. (1980): *African survey*. John Wiley and Sons, New York, 626 pp.

ECA (1983): ECA and Africa's development, 1983–2008. Preliminary perspective study. Economic Commission for Africa, Addis Ababa, 103 pp.

Henderson-Sellers, A. (1991): Greenhouse gas and global change: A personal view of the pitfalls of interdisciplinary research. Inaugural Lecture, Macquarie University, North Ryde Australia.

Ibe, A.C. and Ojo, O. (1990): *Implications of expected climate change in the West and Central African region. An Overview*. UNEP, 52 pp.

Landsberg, H.E. (1975): The definition and determination of climatic changes, fluctuations and outlook. In: *Atmospheric quality and climate change*, ed. by R.J. Kopec. Studies in Geography, Vol. 9, Chapel Hill, 52–62.

Ojo, O. (1990): Sociocultural implications of climate change and sea level rise in the West and Central African regions. In: *Changing climate and the coasts*. Vol. 2, *West Africa, the Americas, the Mediterranean and the Rest of Europe*. Report to the IPCC on Climate Change, 103–111.

O. Ojo
Department of Geography and Planning
University of Lagos
Lagos
Nigeria

3.2. CLIMATE AND LIFE IN THE CARIBBEAN BASIN

LAWRENCE C. NKEMDIRIM

1. Geographical setting

The Caribbean Basin is contained roughly between latitudes 10 to 27°N and longitudes 57 to 87°W. The basin which derives its name from the word "Carib", name of one of the ancestral Indian groups of the region, is an archipelago curving 4,000 km from Florida in the north to Venezuela in the south (Figure 1). The west side faces Central America. The northwest boundary is the Gulf of Mexico which is itself separated from the Caribbean Sea by a line running between the Yucatan channel and Florida Keys. To the east, the basin is separated from the Atlantic Ocean by an arc of islands starting with the Bahamas to the north and ending with Grenada and Trinidad-Tobago in the south. The main chain of islands is the Greater Antilles comprising Cuba, Hispaniola (the Dominican Republic and Haiti), Jamaica and Puerto Rico. The narrower chain is the Lesser Antilles, a collection of smaller islands stretching from the Virgin Islands in the north to Grenada in the south. The Lesser Antilles is further divided into windward islands which lie south of latitude 15° and leeward islands which lie north of the line. The islands of the Lesser Antilles and the eastern Caribbean are used synonymously in this chapter. The Caribbean and West Indies are also used interchangeably to refer to all the islands in both the Greater and Lesser Antilles. In other contexts, these two terms, Caribbean and West Indies, also include all countries of Central America as well as Venezuela, Surinam, Guyana and Colombia on the South American mainland and the islands that form its north coast.

Caribbean islands vary enormously in size. Cuba, the largest island, has an area of 145,000 km^2 and Montessarat, one of the smallest, is only 98 km^2 (Table 1). The predominant theme expressed in the Caribbean is smallness. Small size limits the occurrence of any significant variety and quantity of mineral resources, contributes to islandwide vulnerability to severe storms, droughts and floods and restricts the size and viability of drainage basins and

M. Yoshino et al. (eds.), Climates and Societies – A Climatological Perspective, 177–201.

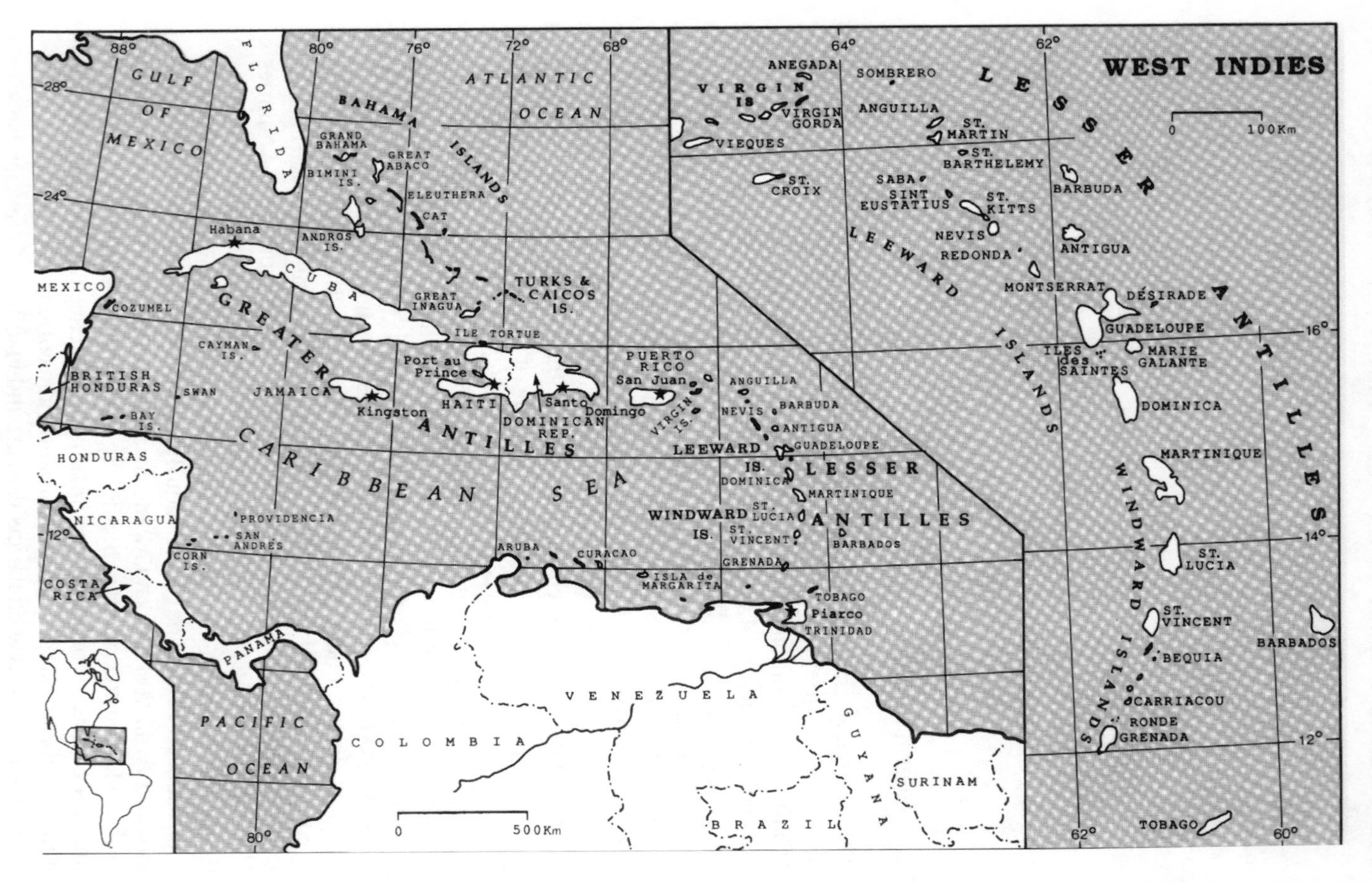

Figure 1. The Caribbean. The whole region contained in this map is sometimes described as the Wider Caribbean. This chapter only deals with

Table 1. Selected statistical data on countries of Insular Caribbean.

Country	Size (km^2)	Population ('000)	% Annual population increase (10^6)	Gross domestic products US $	Arable land (km^2)	Forest and wood land (km^2)	Irrigated land (km^2)
Antigua	442	66	1.4	188	–	–	–
Bahamas	13,935	175	3.6	2,930	70	3,240	–
Barbados	431	238	0.7	311	330	30	–
Cuba	114,524	8,569	1.7	–	25,400	19,300	10,200
Dominica	751	78	1.1	102	70	310	–
Dominican Rep.	48,734	4,006	3.0	2,923	11,100	6,290	1,780
Grenada	344	94	0.4	41	50	30	–
Guadeloupe	1,779	325	1.6	234	50	110	10
Haiti	27,750	4,330	1.6	733	5,502	990	700
Jamaica	10,991	1,849	1.6	2,807	2,207	3,020	330
Martinique	1,102	325	1.5	1,365	110	280	50
Puerto Rico	8,897	2,712	2.8	8,208	700	1,870	390
St. Vincent and the Grenadines	388	87	1.2	136	130	140	10
Trinidad and Tobago	5,128	941	1.1	1,927	700	2,270	210

Source: United Nations Statistical Year Book 1987; FAO Report Resources for Food and Agriculture in Latin America and the Caribbean, Paper 8, 1986.

ground water aquifers. Island economies are small and fragile. The oceans tend to reinforce smallness and the sense of isolation. Unlike the south Pacific where the sea is seen as a binding force, in the Caribbean, the ocean is often viewed as a divider.

2. Physical basis for human-climate interaction

Caribbean islands are rugged and mountainous. Puerto Rico is typical. Here over half the land area lies above 300 m a.s.l. Forty percent is occupied by mountains and 35% by hills. Only 25% is flat land. Slopes especially in the hilly country are steep or very steep. Forty-five degrees of slope is normal.

A major feature of the Greater Antilles is an east-west nucleus of mountains and hills which runs like a spine through the centre of the country. The Cordillera Central in Puerto Rico, the Cord Central in the Dominican Republic and the Central Inlier-Blue Mountain complex in Jamaica, are dominant themes in the physiography and climate of the region. The windward sides of the spine are usually wet, the leeward are drier.

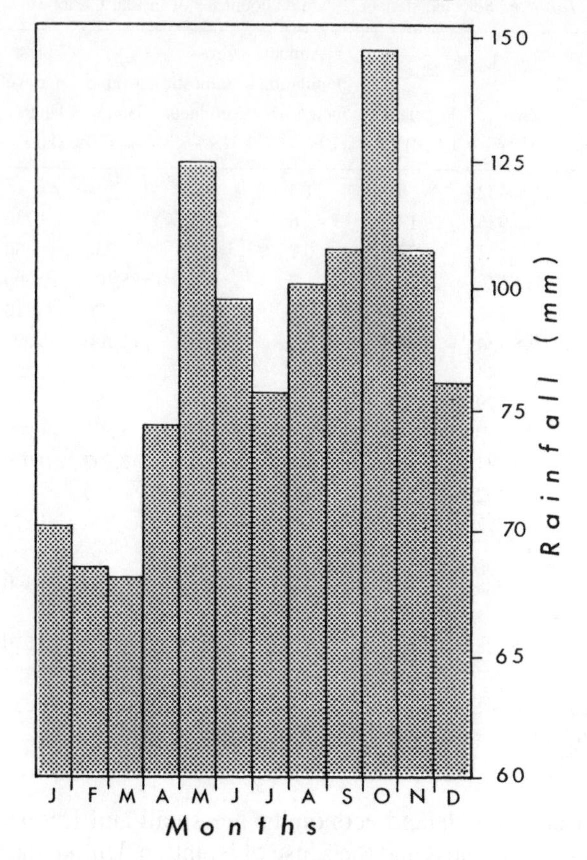

Figure 2. Mean monthly rainfall for Jamaica. The pattern revealed is typical for all the islands.

The Caribbean is tectonically unstable. Earthquakes occur in Cuba and Jamaica. Evidence of recent geothermicity is found in many islands of the Lesser Antilles. The most recent volcanic material is present on higher grounds in the Greater Antilles but is the predominant rock type everywhere in the Lesser Antilles.

3. Climate

Latitude, topography and oceanic influence are the major forces which drive weather and climate in the Caribbean. The West Indies lies between the semi permanent subtropical high pressure belt centred around latitude 30°N (the Azores-Bermuda High) and the Equatorial Low. Winds prevail from the northeast most of the year. The strength, depth and direction of the trades vary

Table 2. Mean climatological data for selected Caribbean locations. Precipitation (mm), temp. (°C), vapor pressures (mb), windspeed (ms^{-1}), radiation (langleys d^{-1}), evapotranspiration (mm).

Country: Trinidad and Tobago; Station: Piarco Airport
Latitude: 10.37; Longitude: −61.21; Elevation: 12 m

J	F	M	A	M	J	J	A	S	O	N	D	Year	
Precipitation	74	55	30	49	116	265	255	240	183	161	205	144	1777
Temp. Average	24.5	24.7	25.4	26.3	26.6	26.1	25.9	26.1	26.2	25.9	25.4	24.8	25.7
Temp. Mean Max	29.2	29.7	30.4	31.3	30.7	30.0	29.8	30.5	30.9	30.3	30.1	29.5	30.2
Temp. Mean Min	19.8	19.7	20.4	21.3	22.5	22.5	22.0	21.6	21.5	21.4	20.7	20.1	21.1
Vapor Press.	24.4	24.4	24.0	24.9	27.6	28.4	28.3	28.2	28.3	28.2	27.7	26.3	26.7
Wind Speed 2 m	1.5	1.6	1.6	1.6	1.5	1.2	1.2	1.2	1.2	1.2	1.2	1.3	1.4
Sunshine %	64	66	67	65	59	50	5557	60	57	56	56	59	
Total Radiation	422	466	504	511	485	444	464	475	481	445	407	386	457
Evapotranspiration	109	111	139	142	138	119	125	130	126	120	103	99	1461

Country: Cuba; Station: Cienfeugos
Latitude: 22.09; Longitude: −80.24; Elevation: 39 m

J	F	M	A	M	J	J	A	S	O	N	D	Year	
Precipitation	17	25	32	44	115	146	117	154	166	154	39	22	1031
Temp. Average	22.2	22.2	23.3	24.4	25.6	26.7	27.2	27.2	26.7	26.7	23.8	22.8	24.9
Temp. Mean Max	27.2	27.8	28.9	29.4	30.6	31.7	32.2	32.2	31.7	31.1	28.3	27.9	29.9
Temp. Mean Min	17.2	16.7	17.8	19.4	20.6	21.7	22.2	22.2	21.7	21.7	19.4	17.8	19.9
Vapour Press.	19.5	19.0	20.3	21.6	24.3	27.0	27.0	27.0	27.3	27.7	22.4	20.8	23.7
Wind Speed 2 m	2.5	2.5	2.6	2.7	2.0	2.0	2.0	2.0	2.0	2.0	2.5	2	2.3
Sunshine %	70	67	66	66	65	57	6160	64	64	59	65	64	
Total Radiation	342	405	456	499	505	506	518	532	476	420	347	330	445
Evapotranspiration	91	97	131	146	153	147	157	152	134	121	93	85	1507

Country: Dominican Republic; Station: La Vega
Latitude: 19.13; Longitude: −70.32; Elevation: 97 m.

J	F	M	A	M	J	J	A	S	O	N	D	Year	
Precipitation	97	71	89	128	197	97	108	108	125	137	160	106	1423
Temp. Average	23.8	24.1	25.2	26.2	26.8	27.5	27.7	28.1	28.0	27.3	25.8	24.3	26.2
Temp. Mean Max	28.4	29.0	30.5	31.3	31.7	32.4	32.8	33.2	33.1	32.3	30.2	28.6	31.1
Temp. Mean Min	19.0	17.4	18.1	18.8	19.9	20.5	22.7	22.8	22.8	20.2	21.7	19.9	20.3
Vapour Press.	22.9	22.7	24.9	24.9	27.9	27.2	27.2	28.3	28.6	27.4	25.9	24.1	26.0
Wind Speed 2 m	2.2	2.2	2.3	2.3	2.1	2.5	2.5	2.2	2.2	2.0	1.9	1.8	2.2
Sunshine %	59	59	58	50	46	59	6255	63	66	56	50	57	
Total Radiation	352	399	447	454	452	506	515	477	479	441	356	312	433
Evapotranspiration	94	100	132	142	140	156	165	154	145	135	99	85	1547

Source: FAO: Agroclimatological Data, Latin America and the Caribbean, Rome, 1985.

with the season. In winter, they are weaker and have a greater northerly component. The height of the trade wind inversion is lower which tends to inhibit the growth of cumulus clouds. In summer, the trades are broader, stronger and more persistent (Table 2). The inversion is weaker and occasional and its base, when present, is higher. Also in the summer, the Intertropical Convergence Zone (ITCZ) sweeps through the region. The cloud band associated with its presence is largely responsible for the summer bias of the annual distribution of rainfall (Figure 2). The timing of maximum rainfall varies with latitude, earlier in the south and later in the north. In addition, the rainfall pattern displays two maxima, one around May and the other in October or November. In general, the earlier peak is the smaller of the two. The latter peak coincides with the hurricane season and occurs during a period of maximum sea surface temperatures.

Within individual islands, strong spatial gradients occur due to topographical alignment (Figure 3). A north-south transect taken along the eastern third of Jamaica illustrates a pattern which is repeated in almost all the islands. In this example, rainfall increases evenly from 2,000 mm/yr on the north coast to about 8,000 mm near the crest of the Blue Mountains and from there drops sharply to less than 1,500 mm on the south coast.

Mean annual temperature is between 25 and 26° everywhere except in high elevation where it can be considerably lower. Mean maxima hover around the 30° mark and minima about 20°. Temperature range increases northwards. July and August are the warmest months (Table 2) and lag the period of maximum insolation by about 1 to 2 months. Vapour pressure ranges from a maximum of 30 mb on the coast in the summer months to 20 mb in the interior during the winter.

Potential evapotranspiration is high throughout the year consequently, humidity is a function of rainfall and its distribution. On the windward parts the ratio of potential evapotranspiration to precipitation (PET/P) is lower than 1 and decreases with increasing elevation to the crest of the mountains. On the lee side PET/P is higher than 1. On parts of the lee coast, the ratio may exceed 2. Indeed the spine of each island approximates the unity line for PET/P.

Occasionally, severe storms occur in the region. Disturbances may begin as minor waves in the easterlies and subsequently deepen into tropical storms with winds between 17 and 32 ms^{-1}. Some reach hurricane strength (winds in excess of 32 ms^{-1}). Severe storms begin in May. Frequencies peak in September and October (Table 3).

4. Vegetation

Vegetation in the Caribbean is a fine weave of loosely linked biological sub communities (Brown 1982). Floristically, there are considerable variations in genera and specie composition from island to island. The older islands of

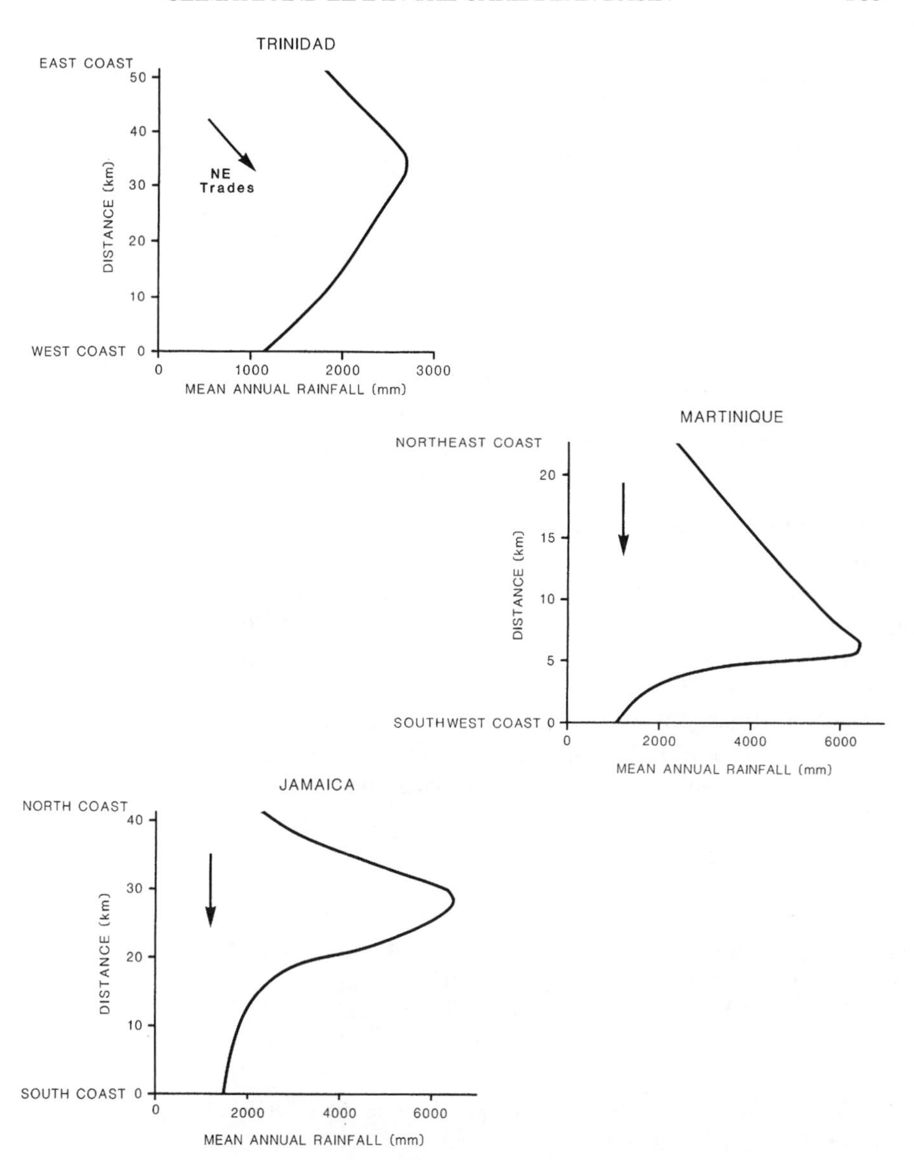

Figure 3. The spatial variations of mean annual rainfall along the mean path of the prevailing winds in several Caribbean islands. This type of gradient is found everywhere the spine lies across the path of the prevailing trades. *Source*: Portig 1974.

the Greater Antilles have more diversity in community composition than the younger ones of the Lesser Antilles. For each island the degree of endemism of plant genera/species tends to be higher in the mountains than elsewhere. Human occupance has played a very large role in modifying vegetation to

Table 3. Percentage monthly frequency of Atlantic tropical storms and hurricanes from 1881 to 1972.

Feb	Mar	Apr	May	June	July	Aug	Sept	Oct	Nov	Dec	
Hurricanes	0	1	0	3	21	30	117	153	76	14	2
Percent	0	0.2	0	0.7	5.0	7.2	28.1	36.7	18.2	3.4	0.5
Storms	1	0	0	8	26	24	4585	80	18	2	
Percent	0.3	0	0	2.8	9.0	8.3	15.6	29.4	27.7	6.2	0.7
Total	1	1	0	11	47	54	162	238	156	32	4
Percent	0.1	0.1	0	1.6	6.7	7.6	22.9	33.0	22.1	4.5	0.6

Source: Alaka, M.A. (1976).

the point where the pre-Columbian mosaic can only be reconstructed through proxy data.

Tropical broad leafed rain-forest is present in hot humid climates (rainfall in excess of 2,000 mm/year and 10 months in which PET/P $<$ 1). The rain-forest is evergreen with as many as four storeys of canopy bound together by woody vines. Epiphytes of ferns, orchid and bromeliads occupy the space between trunks. In the wet dry climate areas (rainfall between 1,500 and 2,000 mm/yr and up to 5 months in which PET/P $>$ 1) lowland vegetation is of the seasonal rain-forest type, is partly deciduous, and about 1/3 of its composition is evergreen. Single canopies are the norm. Undergrowth of fern and other epiphytes still occur but in quantities less profuse than found in the true rain-forest. These forests have been disturbed even more extensively than the broad leafed rain-forest. Timber and fuel wood harvests, urbanization and agriculture, have turned some of them into drier tropical forests with reduced specie abundance, lower epiphyte composition and smaller biomass production. Where the dry season extends from 6 to 7 months and rainfall is between 1,500 and 1,000 mm, dry forest is the typical vegetation. Cactus thorn forest occurs in drier areas.

Mountain vegetation displays considerable differences above the 1,000 m level. In the Blue Mountains of Jamaica, Grubb and Tanner (1976) found that the lower limit of a montane or mist forest is reached at an elevation of 1,370 m. This forest is similar in appearance to the tropical rain-forest due, in part, to higher rainfall and lower temperature. There is a much greater preponderance of ferns but because of the drop in temperature, at least in part, multiple storied layering is replaced by a single story. Towards the summit of the highest peaks, elfin forest gives way to more open plant formation. In most of the Lesser Antilles, these stands are similar to the treeless moorlands of the mid latitudes (Watts 1987).

5. Soils

The diversity of rock types, vegetation, climate and micro-climates produce soil types which vary tremendously within islands and among them. Pico (1974) identified 352 soil types in Puerto Rico's 9,000 km^2. Superimposed on this complexity, however, are some common qualities. the soils are old and deeply weathered. Soil depths of up to 20 m are possible in places. The average is 10 m. Infiltration rates tend to be high. Rates of solution and chemical reaction are fast. Because of this, most of the nutrients is stored in vegetation and in a shallow layer no deeper than a few centimeters of the soil especially in the wettest areas. Exchange capacities of ions are enhanced. Silica is rapidly mobilized and only extremely stable elements such as quartz and kaolinite are left in the uppermost soil layers. These processes contribute to the formation of the three major soil groups found in the West Indies-ferralitic soils (latosols), tropical podsols and lateritic soils. Ferralitic soils are deep, usually more than 10 m, heavily leached and highly acidic, and have an accumulation of red and yellow clays. Usually found under a rain-forest, they are nutrient deficient. Podsols are found in humid climates with a distinct dry season. They are less weathered and less acidic. Where rainfall is lower than 2,000 mm/yr podsols tend to be more fertile. Laterites occur when pans of iron and calcic compounds are hardened to almost concrete-like consistency by alternating severe rainfall, desiccation, and oxidation. In grassland areas where there is a distinct lengthy dry season (5 to 6 months) lateritic soils are nutrient deficient.

Volcanic soils occur in most of the islands of the Lesser Antilles. Ash layers from past volcanic events occur widely. Soils here tend to be younger.

6. The people

Before the arrival of the first colonising Europeans, the Caribbean was home to the Carib and Arawak Indian nations. Estimates of original aboriginal population range from a few hundred thousand to six million. Within 25 years of Spanish colonization, native population had dwindled to less than 10,000. Forced labour in gold mines and sugar cane plantation, and disease (small pox) were factors in their demise (Sauer 1966). Today, persons of pure Indian stock are virtually non-existent.

Blacks of African descent are the majority population group in the islands except in Cuba and Puerto Rico. African slaves were introduced to the islands in 1505 both to replace the dying aboriginal labour force and increase the production of sugar, and later, hides and skins. Whites are dominant in Cuba (~70%) and Puerto Rico (~72%) but constitute small to very small minorities elsewhere. The racial composition is about 23% white in the Dominican Republic and less than 2% in Jamaica. Whites are mostly of south European ancestry. In the mid nineteenth century, immigration from Asia, mainly south Indian and Chinese, introduced an Oriental presence in the Caribbean. Most of the new immigrants arrived in Trinidad and Tobago as indentured labour. Now, people of south Indian extraction comprise about 40% of the population of this two-island nation. Chinese are an important minority in Jamaica and Cuba. Social interaction across racial lines make categorization based on race approximate at best. Large numbers of people are of mixed ancestry.

Several reasons have been advanced for the population mix within insular Caribbean (Watts 1987; Jones 1971). But none of these appear to be more important than climate. The West African and South Asian climates are almost identical to those found in the West Indies. Southern Europe in the summer has a climate not unlike the Caribbean including high humidities. In contrast people of north and northwest European descent are small in number even though opportunities for a larger presence during colonial time were not markedly different from those for other Europeans. Indeed, the ability to survive under marginal conditions is believed to be an important factor in black dominance of the islands (Jones 1971).

7. Climate-human interactions

Occasionally, climate-human interaction is direct. More often, the connection between the two systems is indirect. The web of processes through which linkages occur can be complex; so complex at times that the climate drive on life systems and vice versa may not be immediately apparent.

7.1. *Land use*

Due in part to the ruggedness of the interior, human occupance in insular Caribbean favours the coast; the leeward coast and adjoining plains are preferred. Although rainfall here tends to be variable and aggressive, these parts are drier. Their soils tend to be richer and better drained. Water for irrigation is either available or potentially so. Forests are open. The original peoples who also favoured coastal and near coastal communities adapted their agriculture to both the environmental opportunities offered and the challenges posed. The staple was manoic, a drought resistant and nutrient non-demanding crop which they grew in small plots carved out inside the forest where they were protected against destructive winds including hurricanes. As a response to variability in rainfall and wetness frequency, the harvest time table was staggered. The taking of crops out of the ground as needed was also used as a storage device.

The Caribs and Arawaks planted their crops on mounds some as high as one-half meter above the ground. This practice produced several economic and environment conserving benefits. It protected the crop from physical damage during a flood, controlled runoff and checked erosion. Sediments trapped between mounds were recycled to enrich the soil while the water dammed by them replenished soil moisture between rainfall events. This water also recharged groundwater and helped regulate streamflow which would otherwise be erratic given the variability and aggressiveness of precipitation. When hillslopes were farmed, plots were smaller than in flatter land. Cultivation period was shortened to two years or less after which the land was abandoned to subsequent recolonization by native plant communities. It is clear that the Indians were environmentally sensitive and that this sensitivity was an important factor in sustaining the high population density supported on those fragile lands.

The settlement pattern in post Columbian times is much the same as they were before Europeans arrived. Cities and towns had flourished along the coast and sheltered valleys. The drier coastland is still favoured. Indeed, the 1,500 mm/yr isohytes appears to form the outer boundary of the region within which most major cities are located. Although the reasons for the choice of where to live might have been similar, the land management practices applied by the pre and post Columbian societies contrasted sharply. The results are also markedly different.

The small conuco which took up the bulk of the farmed land in Indian times has now been replaced by large plantations. In Jamaica, farms larger than 50 hectares make up 50% of the arable land but constitute less than 1% of the holdings while those 2 hectares and under account for only 13% of the cultivated area and 70% of the farm total. This pattern which is repeated in all the islands (O'Loughlin 1961) has remained virtually unchanged today despite attempts at land reform (Dale 1977). Agriculture both large and small scale as practised in the modern Caribbean require the removal of the primary forest and this removal is in most cases permanent. The small holdings are cropped more or less continuously and many are on land considered marginal. Cattle and other stocks are grazed and stock density can be unacceptably high.

Penman (1961) and Molion (1975) have shown that evapotranspiration from the forest is a large component of local rainfall in the humid tropics. Crops especially when seasonal, do not fully replace all the water vapour lost when a forest is removed. Nor does cropland compensate adequately for the original aerodynamic roughness which promoted the development of deep mixing layers and subsequent growth of cumulus rain bearing clouds. Nkemdirim (1990) speculated that the trend towards an increase in seasonal droughts in the eastern Caribbean might be due, at least in part, to deforestation. The islands of the Lesser Antilles have been virtually stripped of all their forest (UNEP 1982). Lowenthal (1960) estimated that forest land in the islands amounts to less than 0.25 hectares per capita. On steeply sloping land, deforestation has gone too far to be sustainable.

Eyre (1988) linked the climatic change occurring in parts of Jamaica to the island's plantation history. Prior to the late 18th century, the Fall and Hope River basins were entirely covered by rain-forest. This landscape was greatly modified during the coffee boom of the early 1800s when British, Jamaican Creole and German planters replaced much of the native forest with Coffeea Arabica, even on the steepest slopes. The two basins now include only remnants of several types of forest, all of them seriously degraded. At present, wynne grass occupies large tracts of preciously forested hillslopes. Rainfall has declined (NOAA 1979), and the prospect of recolonization of the hillslope by humid rain-forest appears very remote even in abandoned land. Eyre (1986) believes that we might be witnessing the early stages of desertification of a once humid rain-forest. Given the size of documented evidence of overgrazing, logging and other forms of deforestation in the region, and also given that the major island groups lie close to the southern edge of the Bermuda-Azores High, some investigators including Nkemdirim (1988) have speculated that the type of change observed by Eyre may not be inconsistent with the albedo driven Charney effect (Charney 1975).

A closed forest is an effective wind break. Earlier we recalled how the original peoples used the forest to protect against wind damage. It does not appear that agricultural practice in the Caribbean has benefitted from the history of wind damage on the islands. In 1963, for example, Hurricane Flora cause very extensive damage in deforested parts of Cuba while other areas which were protected by a forest or were in a forest suffered insignificant losses. Data supplied by Granger (1988a) indicate that in the eastern Caribbean trends in hurricane damage parallel those in deforestation.

Deforestation is also partly responsible for the severe soil erosion being experienced in the islands. Lal (1983) has demonstrated the relationship between heavy rainfall, high temperature and soil erosion in the tropics following deforestation. In Jamaica, the Yallahs valley has been declared a protected area subject to very stringent soil conservation programmes. This is the direct result of the loss of up to 100% of the topsoil and 75% of the subsoil in this coffee growing region (Clarke and Hodgkiss 1974). Similar programmes are in place in the Chibarazo valley in Barbados where extensive gullying is threatening the viability of the sugar cane and banana industries. Mudslides have dammed sections of the main stream. Extensive damage is also found in parts of Cuba and in the Cordillera central of Puerto Rico. In Haiti, the erosion problem is compounded by lack of education. In a country where 90 to 95% of the rural population is illiterate, even the most rudimentary erosion control is unknown (Watts 1987).

7.2. *Hurricanes*

Of all the natural hazards faced by insular Caribbean, the hurricane is by far the most destructive. Hurricane is the modern form of "huracan" used by the Arawaks to describe the severe storms with which they were all too familiar. Huracan also appears in Quiche (a Central American group) cosmogony and theology as the god of tempests (Arenas 1988). In the mythology of certain

Antillean groups the god of tempests controlled the intensity and frequency of the storm. Spanish sailors first experienced a hurricane in 1494. Fernandez de Oviedo detailed the chaos and disaster caused by the storm in Hispaniola in 1508. Poey (1856) provided an excellent chronology of Caribbean hurricanes. Reading (1990) analysed the spatial and decadal frequency of West Indian cyclones from 1500 to present.

Between 1871 and 1890, tropical cyclones averaged 4.9 per year and each lasted 8 days on the average. From 1910 to 1930, hurricanes averaged 3.5 annually. The frequency rose to 6.0 between 1944 and 1980. Since 1960, however, average numbers have dropped but intensity has increased (Granger 1989). The 1995 Hurrican season turned out to be one of the most active. Up to September 20, 13 storms had developed in the Atlantic and the Caribbean Sea.

Islands in the path of the hurricane are literarily laid waste. The toll on human life can be very high. *Flora* (1963) killed 7,000 people in the Caribbean. *Donna* (1960) took 114 lives in Barbados and Anguilla and 56 died in Dominica from *David* (1979). Financial losses are also high. The Jamaican government estimated that the island lost between 2 and 5 billion US dollars from damage inflicted by *Gilbert* (1988).

The underclass and marginalized persons are the first and worst victims of hurricanes. Granger (1988a) provides the following description of the chaos generated by *David* in Dominica.

> David destroyed or badly damaged 75% of all buildings in Dominica together with the only deep-water harbour and its warehouse and waterfront buildings. Roofs were blown off, galvanized iron sheets and glass were strewn everywhere (and deadly as missiles), trees were uprooted, and flooding was rampant. Mothers with children roamed the streets in terror looking for shelter.

Wind and storm surges are the primary vehicles for damage in coastal communities. In the cities and hinterland, the agents are wind and floods. The most severe floods in the Caribbean are hurricane related so much so that Arenas (1988) advocated the separation of flood populations on the basis of storm type for extreme value analysis. He showed that hurricane type floods form an annual maxima series composed of instantaneous peak discharge three or more times the size of the peak in non-hurricane series for equivalent recurrence intervals.

Because the islands are so small, damage is usually island wide, swift and virtually simultaneous. As a result, help from within is not an effective option. Emergency foreign relief is normal but longer term reconstruction can be elusive and if available does not always come without strings. Also, some of the strategies designed to repair damage may cause social instability and political change as was the case in Dominica following hurricane *David*

(Granger 1988a). They may also provide a powerful push for rural-urban migration. A common recommendation made following a storm is economic diversification. Preferred methods through which this may be achieved include small scale manufacturing, expansion of the tourist industry, and public and private sector financing. However, these types of activities are either urban based or capable of creating new cities where none existed. Consequently, they tend to exacerbate rural out migration and all the problems associated with it.

The high frequency of hurricanes robs those who are chronically impacted by it of a sense of permanence to life. Fresh start is a recurring theme. But a fresh start hardly takes root before the next one begins, and hope for the ever hopeful turns to despair.

Every country in the Caribbean has an office of emergency preparedness or its equivalent. However, there does not seem to be a well thought out vertically and longitudinally integrated policy agenda for dealing with hurricane driven danger and damage. Building codes designed to mitigate the impact of severe storms exist but are virtually uneforceable among the vulnerable marginised society in squatter settlements and shanty towns. Research ought to be directed towards technology consistent with what those communities can understand, afford and implement. Storm shelters and speedy evacuation of target zones are attractive initial reactive policy options especially given the considerable lead time made possible by improved forecasts and tracking. Compensation following a storm should be designed to restore victims to their prestorm standing. If improvements must be made they should be based on very careful identification and analysis of the social and economic processes that flowed from similar experiences in disaster impacted areas both in the Caribbean and similar environments overseas. Because it is 100 percent certain that hurricane damage will occur consideration should be given to the creation of a significant Caribbean regional disaster fund financed through a special hurricane tax and foreign donations. The newly formed Caribbean Disaster Emergency Response Agency (CDERA) headquartered in Barbados is a welcome development.

7.3. *Urbanization*

Rapid urban growth has been a major feature of the Caribbean since 1900. It accelerated after World War II and peaked between 1950 and 1970. At its height annual rates of growth averaged between five and seven per cent. Now they average about 3.5%. City growth is exceedingly high compared to natural increases which has fallen to less than one percent in some islands (Table 1). About 75% of the population live in cities of 20,000 or more and nearly 25% reside in urban areas larger than 50,000. Each of the two largest cities, Habana and Santo Domingo has a population of more than 1.5 million. The biggest city in every island is a sea port, the nation's capital, and a service and distribution centre. Almost all the white collar jobs outside tourism are based there. Employment opportunities, real and imaged, are major pull factors. Environmental deterioration including soil degradation and erosion, and recently, climatic variability, are also pushing some of the younger farmers away from their small rural holdings into the city.

The city has not been exactly heaven for the rural immigrant. There is a housing crunch in every major city. Numbers of both public and private sector accommodations are grossly inadequate (UNEP 1985). Also what is affordable is more likely to be substandard than not. The results are overcrowding and the growth of shanty and squatter towns within the inner city and in the gullies of surrounding hills. In these areas, density can reach 800 persons per hectare more than three times the number found in overcrowded middle class neighbourhoods (Clarke 1973). Infrastructure including sewerage and storm drainage systems if they exist are less than adequate. The open gutters which run through some communities are so clogged with garbage that they impede rather than enhance drainage.

These communities are disaster prone (Troy 1978; McPherson and Saarinen 1977). West Kingston and Kingston's squatter towns have been flooded or blown out by hurricanes several times in the last twenty years (Lewis 1991). In these communities, natural hazard is as much human made as it is an act of God. Persons on the receiving end might not even make that distinction. To some, poverty and marginalization are also divine driven.

Flood frequency in cities is significantly a function of the degree of urbanization. In general, the greater the urbanization the higher the frequency of floods and the smaller the recurrence interval for floods of a given magnitude. Nkemdirim (1988) has shown that peaks have risen and lag time reduced for floods generated by unit rainfall as a result of urban growth in Kingston and Montego Bay. Indeed, most Caribbean cities experience a high frequency of "nuisance" floods; floods which though not severe enough to cause acute or chronic economic damage, snarl traffic, inundate floors of dwellings, courtyards and gardens, and pollute streams.

Granger (1988b) studied the hydroclimatological consequences of increased urbanization in Trinidad during the oil boom of the nineteen seventies and early eighties. He found that (a) runoff coefficient has risen significantly, (b) evaporativity has increased, (c) soil moisture has decreased, and (d) instantaneous peak stream runoff has increased. He also provided evidence of decreased aquatic habitats in some wetlands. He concluded that these changes have begun to place constraints on agricultural production and productivity in the impacted areas. They are also degrading the general health of the environment.

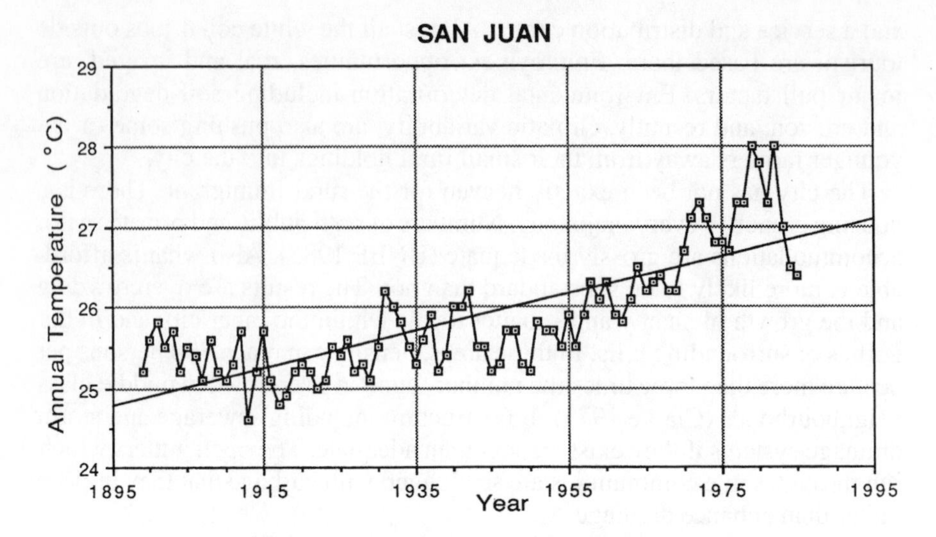

Figure 4. Time series of mean annual temperature for San Juan, Puerto Rico.

Table 4. Mean decadal temperature change in selected Caribbean cities.

City	Period covered	Mean decadal temperature change	Standard deviation
San Juan	1901–1986	0.22 °C	0.7 °C
Kingston	1901–1987	0.16 °C	0.3 °C
Port-au-Prince	1883–1967	0.17 °C	0.5 °C
Piarco	1946–1987	0.16 °C	0.4 °C

7.4. *Urban heat islands*

The tropical environment is by definition hot. The addition of anthropogenic forcing that could make that environment even warmer is undesirable especially in light of the persistent high humidity. Except in higher altitudes, insular Caribbean has an annual mean discomfort index in the 75 range. In mid summer the index can and often exceeds 80 at which level everyone is uncomfortable (Kamiyama 1961). High effective temperatures cause physical discomfort for those not fully adapted. The performance of tasks requiring sustained mental concentration may also be affected (WMO 1987).

Urban heat islands are as pervasive in the tropics as they are in the mid-latitudes. Jauregui (1991) showed that heat island intensities in tropical cities are correlated with population in much the same way as they do in mid-latitude cities. Absolute values, however, are lower. There are no reliable figures on urban heat islands in the Caribbean. Granger (1989) indicated that a significant portion of the upward revision of mean temperatures observed in this region since 1930 could be attributed to urban heat islands. This speculation is supported by the trend in the mean annual and seasonal temperature series for San Juan, Puerto Rico (Figure 4). The trend is unmistakable and consistent with similar patterns for Kingston, Habana and Piarco (Table 4). Although the correlation coefficient between decadal changes in mean annual temperature and corresponding values of population change was only 0.35 for the whole region, the statistic is significant at the 99% confidence level, suggesting that urban forcing has had a measurable role in elevating mean temperatures in the islands' cities. The series for rural locations were not as long. But analysis of temperature data for small towns including Negril, Mandeville, Fajarado and Mayaguez Ceer shows that temperature has remained trend free. Mean temperature in Guatemala city actually declined.

The impact of these fairly sizeable temperature gains amounting to between 1.5 and 2°C over the century (even if they are only partially due to urbanization) have yet to be measured and should be. They might have stimulated convective activity and increased wetness frequency and perhaps rainfall amounts either within the city or downwind. They could have played a role in the increased frequency of floods especially those of the nuisance type. They might have moderated a trend towards a deforestation driven drought. They certainly have contributed to the thermal discomfort felt in the islands' cities and forced up demand for energy is a region which outside Trinidad and Tobago, depends heavily on imported fuel for electricity (UN 1985).

8. The future

Looking into the crystal ball to predict a future for the Caribbean especially in the context of climatic forcing is not an enviable task. The Caribbean society is as integrated with the wider off shore community as it is isolated from it. External political and economic forces could well determine how well the islands are able to adjust to changing environmental futures, and sustain their own economic and social viability. The contribution of climate change to this determination could be puny. It is not an exaggeration to state that the Caribbean has seen better days economically. In 1988, the economic crisis in the West Indies took a dramatic turn for the worse (UN 1989). For the first time since the 1981–1983 recession, the region's per capita product declined and now stands at a level equivalent to 1978. In general, these countries have island economies of relatively small dimension and rely on a small number of commodities mainly agricultural and some minerals. Caribbean countries customarily show a considerable deficit in their merchandise trade but this is generally offset by net income received from tourism. Since both pillars of their economy, agriculture and tourism, depend on climate, the future of the islands could be vulnerable to climatic change; local, regional or global.

The foremost environmental question of our time is the greenhouse effect due to the large growth of greenhouse gases (GHGs) in the atmosphere. Following a thorough review of available research, the Intergovernmental Panel on Climatic Change (IPCC) concluded that if current trends in GHG emission persist, warming of 0.3°C/decade could occur through the next century. Lower values in the 0.5 to 1°C range per century are predicted for the tropics including the Caribbean (WMO 1990). The estimates for the low latitudes could well be low. Hard data collected during the abnormally warm years of 1987 and 1988 show that mean annual temperature anomalies in excess of 2°C were as common in the Tropics as they were in Central North America and southern Europe (WMO-UNEP 1990). Nkemdirim (1990a) has shown that the eighties was the warmest decade in the West Indies. This is consistent with global data for the Northern Hemisphere (Hansen et al. 1989). The average annual warming experienced in the region during the eighties was 0.3°C. In the eastern Caribbean warming at 0.7°C was even higher. Although the anomaly was within one standard deviation of the normal in much of the Caribbean, in the eastern Caribbean it fell between 2 and 3 standard deviations.

Whether the strength of these signal is sufficient to indicate greenhouse forcing is uncertain. More studies must be conducted using methodology similar to those recommend by the WMO Working Group on detection of climate change (WMO-UNEP 1991) before drawing a firm conclusion. The search for evidence for anthropogenic forcing will be further complicated by urbanization, the strong ENSO events of the eighties and nineties, and the fact that the pattern of warming was not totally consistent with General Circulation Model (GCM) predictions, e.g. the largest gains in temperature occurred in the summer rather than in the winter.

The rainfall profile for the future is even less certain. GCMs disagree, often significantly, on both the magnitude and direction of change. If a consensus has emerged, it is that rainfall volumes in the Caribbean will not be impacted by GHGs. The run of dry summers evident in Figure 5 is not substantially different or more severe than what was observed earlier in the century.

Let us assume that rainfall remains constant while temperature rises. Such a combination could cause agricultural drought especially in the less humid parts because of higher rates of evapotranspiration and soil moisture depletion. It is not helpful that the temperature rise observed at this point is larger in the summer, the major growing season, and when the majority of the rainfall is received. Should the decline in summer precipitation worsen or even persist, agricultural drought will be further aggravated. It is not too early to address these scenarios and to put together a variety of matching policy options which may be tested as climatic opportunities unfold.

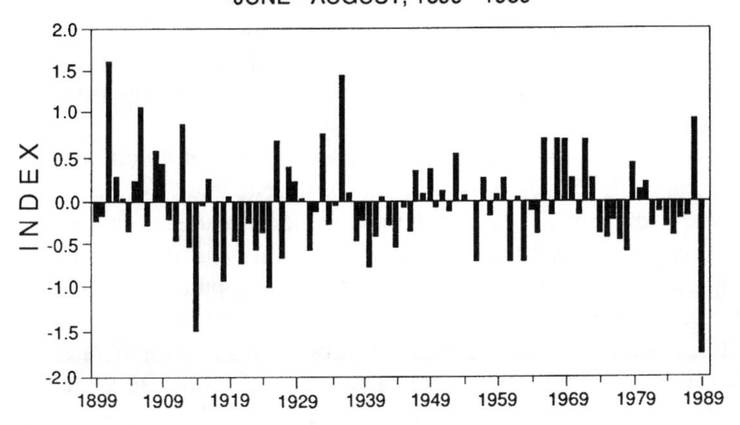

Figure 5. Caribbean precipitation index 1899–1989. *Source*: National Climatic Data Centre.

Parry (1991) has suggested a fine-tuning of the crop calendar to take advantage of suitable moisture and temperature combinations that may become available. Drought resistant crops may be substituted where markets and palates permit. Research should be encouraged into drought tolerant variants of current products. Based on past performance, the preferred option may be irrigation.

Table 1 contains data on irrigation usage in various Caribbean countries. In Puerto Rico and Martinique, the proportion of arable land under irrigation is 56%; in Cuba it is 40%. However, for the region as a whole, the figures indicate that there is considerable room for expansion. The Food and Agricultural Organisation of the United Nations (FAO) estimates that only 10% of the potential cropland area is currently irrigated. The current annual growth rate is only 3.2% down from the five to six per cent level achieved between 1966 and 1977 when irrigation expanded by more than 50% (FAO 1986).

The waning of enthusiasm for irrigation derives from many factors. Considerable doubt has been cast on the real economic viability of many projects because of their heavy subsidization by the State (UN 1985). Increasingly, it has been recognized that the transport of water through the building of physical infrastructures is not sufficient in itself to obtain the anticipated rise in both production and productivity. Waterlogging and salinisation have damaged previously profitable agricultural lands some so severely that they have been taken out of production. Also, there is resentment that most of the water diverted for irrigation targets the large export crop oriented plantations rather than the smaller holdings which produce the bulk of the domestically consumed food.

But those reasons are not sufficient for abandoning the irrigation option. Instead, they should provide a framework for a reformulation of an irrigation policy designed to achieve some of the results that eluded the previous practices as well as meet the challenges of a potential change in climate. Essential components of a revised policy should include (a) a reduction in the size of engineering structures required to convey water, (b) ensuring that water is supplied to the smaller farms, (c) sustaining the productivity of the land irrigated, and (d) reducing the amount of bureaucracy. Nkemdirim (1990b) has advocated the use of microdams and flood management schemes as a tool not only for expanding irrigation and rural electrification and development, but also as a strategy for stabilizing rural population in the Caribbean. Local control, preferably under co-operative management has always been attractive. Postel (1989) has advanced similar views. Groundwater resources, plentifully in the heavily fractured karstic aquifers of the Greater Antilles and under utilized at present (Jamaica Underground Water Authority 1985) should be more widely used for irrigation.

But irrigation expansion cannot be accomplished without external funding. Recently, faced with huge Third World debts, foreign banks and foreign aid donors have been reluctant to invest in water resources projects. Falling commodity prices have also reduced the attractiveness of investment in commercial agriculture. By 1988, the total capital outlay devoted to water resource projects including irrigation had declined to its 1960 level (Brown and Young 1990).

The end of communism in Eastern Europe will have far reaching consequences on the Caribbean, two of which could be immediate. Cuba is a client state of the Soviet Union. Cuban agriculture has prospered under the preferential treatment accorded its products by the Soviet Union. Cuba in turn has helped to promote scientific agriculture in some parts of the Caribbean Basin (Dale 1977). Faced with new economic and geopolitical realities a redesigned soviet policy is unlikely to favour Cuba and the Caribbean as did the previous one. The opening of Eastern Europe to western capital will also drain investment away from the Third to the Second World.[1]

The Caribbean is a significant supplier of tropical commodities, such as bananas, mangoes, coffee and a variety of tropical vegetables, to North America and Western Europe. Apart from tourism, these products are for most of the insular Caribbean the major source for foreign exchange. Given that temperature change is predicted to amplify polewards under a 2X CO_2 climate, southern North America and southern Europe may receive enough warmth to grow crops which hitherto were exclusively tropical. With their technological

[1] Since this paragraph was written, Russia has ended most of the preferential treatment accorded to Cuba. The International Monetary Fund has approved multi-billion dollar loans for Eastern Europe, especially Russia, and the share of foreign aid directed to the Caribbean has declined by more than 50% since 1990.

superiority, the two areas could challenge Caribbean and tropical monopoly in these products.

The other pillar of the economy, tourism, will also face competition from the same regions. GCMs predict that under a double CO_2 climate temperature gains in the mid-latitudes will show a decidedly mid-winter bias. This will affect the West Indies in two ways. Many North Americans and Europeans, the major consumers of Caribbean tourism, may not feel a need to escape to a warmer climate for a winter weather break. And if they decide that a vacation is needed, the now warmer beaches closer to home might be financially more attractive. In both cases, the West Indian tourist trade whose high season is the winter could be devastated.

The beach is also threatened even more directly on the home front. Under the business-as-usual scenario, the IPCC estimates that sea level could rise by as much as 0.6 m by the end of the 21st century. Based on estimates by Bruun (1962) a change of this magnitude could result in a shoreline retreat of up to 100 m. Beach front property, a major attraction within the tourist industry, could be wasted. Granger (1989) points out that in many of the islands of the Lesser Antilles most beaches are not much more than 30 m wide at high tide. Salt water intrusion into coastal aquifers and salinisation of coastal agricultural lands are possibilities. Also, coral reefs could vanish because coral populations will not tolerate water temperatures in excess of 30°C.

The prognosis on tropical storms and hurricanes is not reassuring. Recent hurricane history both in the Caribbean and the North Atlantic, and also in the Eastern Pacific seems to underscore the worst fears, namely, a trend towards fiercer storms. The number of hurricanes may have dropped slightly since 1960 (Reading 1990) but their intensity has risen. The recurrence intervals tor categories 4 and 5 storms are estimated to be about 30% lower now than they were 30 years ago. Emmanuel (1987) has used a simple Carnot cycle model to estimate the effect of greenhouse warming on the maximum intensity of tropical cyclones. Based on the sea surface temperature changes expected in a doubled CO_2 climate, he estimated that the minimum sustainable sea level pressure could drop to 800 mb, down from the present day value of 880 mb. The corresponding maximum windspeed would rise from $\sim$280 to 350 km h^{-1} and the destructive potential could increase by between 40 and 50%. In light of what has been said about human suffering in coastal urban communities, the net effect of this combination of sea level rise and fiercer storms is better imagined than described.

9. Concluding remarks

The outlook for the Caribbean for the 21st century is not the rosiest. The potential change of climate, natural or human driven, local, regional or global, and the environmental impact of that change will require new thinking on how best resources can be managed to sustain and as many have come to expect, improve life. Although there is a rich history of environmental conservation left behind by the pre-columbian Indians, new realities in the West Indies will not accommodate those very intelligent practices. This means that new methods must be found to effectively address modern and future problems and in the context of this chapter those that flow from climatic change and variability. The task is not impossible given the rich diversity of natural and human resources available in the region. The secret lies, perhaps, in resource pooling at a regional level.

Several attempts at regional cooperation have been instituted in the past and some like the Caribbean Community and Common Market (CARICOM), the Central American Common Market (CACM) and the Economic Commission for Latin American and the Caribbean (ECLAC) are still in existence. Dale (1988) provides an excellent review of the history of Caribbean cooperation and identified the obstacles, most of them political, to long term integration. He recommends the Association of South East Asian Nations (ASEAN) as a useful model for Caribbean cooperation. The Bangkok declaration of August 1967 agreed to strengthen regional cohesion and self reliance among the five signatory nations – Indonesia, Malaysia, the Philippines, Singapore and Thailand – and to emphasize economic, social and cultural cooperation and development among them in national plans. This apparently inward looking declaration has been very profitable in mobilizing resources in a region where different languages are spoken, where different political systems are practised and where the colonial legacies left behind are as varied as the peoples themselves. At a 1990 meeting in Bangkok, ASEAN which now includes Brunei Darussalam issued an action policy on GHG driven climate change.

The larger Caribbean has a land area of more than 5 million km^2 and a population of 183 million. Integration if successful, will enable individual countries to overcome some of the vulnerabilities associated with smallness. It will also reduce the dependence of these countries on external economic forcing and hopefully improve their collective bargaining strength in international trade arrangements. Hopefully, the new economic and social order arising from enhanced regional cooperation could be used to promote a viable and sustaining environment. UNEP (1982) has made several recommendations on strategies for action including suggestions on the conduct of environmental assessment and management. These recommendations rely very heavily on regional cooperation (Regional Coordinating Unit) as well as on national focussing (National Focal Point). Collymore (1988) has called for a revised agenda in disaster studies with a view to addressing the range of options available for hazard mitigation and the social and economic processes and results which flow from them. The time to do the calibration and perform the sensitivity tests required to measure the effectiveness of the options is now. However, it may be naive to think that governments and peoples of this region do not know what must be done to advance their economy and promote environmental health at the same time. I believe they know. I suspect that the framework within which this can be done is not in place and most importantly that reliable funding is not assured.

Unlike North America and Europe where there are potential winners and losers from a global climate change, there are no winners in the Carribbean. Caribbean contribution to global warming is puny, but as we have seen the impacts of greenhouse warming on the region will be devastating. Little wonder that a number of West Indian nations joined a group of small islands at the second world climate conference in Geneva (1989) in calling for severe reductions in GHGs and for financial help to their communities to enable them cope with the impacts of climatic change. In the end the future of the Caribbean is as much a regional issue as it is global.

References

Alaka, M.A. (1974): Climatology of Atlantic tropical storms and hurricanes. In: *World Survey of Climatology*, Vol. 12. *Climates of Central and South America*, ed. by H.E. Landsberg. Elsevier, Amsterdam.

Arenas, A.D. (1988): A new approach in surface hydrology for severe storms data treatment in the Caribbean basin. In: *The tropical environment*, ed. by L.C. Nkemdirim. Proceedings of the International Symposium of the Physical and Human Resources of the Tropics, Kingston, Jamaica, 1987, 27–31.

Brown, K.S. (1982): Paleoecology and regional patterns of evolution in neotropical forest butterflies. In: *Biological diversity of the tropics*, ed. by G.T. Prance. Columbia University Press, New York, 255–308.

Brown, L.R. and Young, J.E. (1990): *Feeding the world in the nineties. State of the World, 1990*. World Watch Institute, Norton, New York, 59–78.

Brunn, P. (1962): Sea level rise as a cause of shore erosion. Amer. Soc. Civil. Engrs. Proceedings, *J. Waterways and Harbours Division*, 88, 117–130.

Charney, J. (1975): Dynamics of deserts and drought in the Sahel. *Quart. J. Roy. Meteor. Soc.*, 101, 193–202.

Clarke, C.G. (1973): Ecological aspects of population growth in Kingston, Jamaica. In: *Geographical analysis for development in Latin America and Caribbean*, ed. by R.P. Momsen, 42–55.

Clarke, C.G. and Hodgkiss, A.G. (1974): *Jamaica in maps*. University of London Press, London.

Collymore, J.M. (1988): Geophysical events and human use systems: a revised agenda. In: *The tropical environment*, ed. by L.C. Nkemdirim. Ibid cit., 85–91.

Dale, E.H. (1977): Spotlight on the Caribbean, a microcosm of the Third World. *Regina Geographical Studies*, 2, 95 pp.

Dale, E.H. (1988): Integration of all the Caribbean basin states, an elusive phenomenon. In: *The tropical environment*, ed. by L.C. Nkemdirim. Ibid cit., 146–157.

Emmanuel, K.A. (1987): The dependence of hurricane intensity on climate. *Nature*, 325, 483–485.

Food and Agricultural Organisation (FAO) (1986): Report on natural resources for food and agriculture in Latin America and the Caribbean. FAO Environment and Energy Paper, 8, Rome, 102 pp.

Eyre, L.A. (1986): Vegetation change and desertification in the Caribbean; Parameterisation of land surface characteristics. In: *Use of satellite data in climate studies*. European Space Agency, Paris, 509–514.

Granger, O.E. (1988a): Geophysical events and social change in the eastern Caribbean. In: *The tropical environment*, ed. by L.C. Nkemdirim (ed.). Ibid. cit., 78–84.

Granger, O.E. (1988b): Hydroclimatological constraints on urban-industrial development in a tropical, wet-dry regime. In: *The tropical environment*, ed. by L.C. Nkemdirim. Ibid cit., 32–35.

Granger, O.E. (1989): Implications for Caribbean societies of climate change, sea-level rise and shifts in storm patterns. In: *Proceedings of the Second North American Conference on Preparing for Climate Change*. Climate Institute, Washington, DC, 422–430.

Grubb, P.J. and Tanner, V.J. (1976): The montane forests and soils of Jamaica: a reassessment. *Journal of the Arnold Arboretum*, 57, 313–368.

Hansen, J., Rind, D., Delgeimio, A., Lacis, A., Lebedeff, S., Prather, M., Ruedy, R. and Karl, T. (1989): Regional greenhouse climate effects. In: *Proceedings of the Second North American Conference on Preparing for Climate Change*. Climate Institute, Washington, DC, 68–81.

Jamaica Underground Water Authority (1985): Water resources development master plan project, Jamaica. Jamaican Ministry of Agriculture.

Jones, C.L. (1971): *Caribbean backgrounds and prospects*. Kennikat Press, London, 354 pp.

Jauregui, E. (1991): Aspects of monitoring local/regional climate change in a tropical region. Preprints, International Conference on Climatic Impacts on the Environment and Society, Tsukuba, Japan, Feb. 1991.

Kamiyama, K. (1961): Sensible climate. *Kishokenkyu Note*, 12, 214–248.

Lal, R. (1983): Soil erosion in the humid tropics with particular reference to agricultural land development and soil management. In: *Hydrology of humid tropical regions*, ed. by R. Keller, I.A.S.H. Publication No. 140, 221–237.

Lewis, D. (1991): The impacts of surface flooding in the Kingston Metropolitan Area. Unpubl. M.Phil. thesis, The University of the West Indies, 180 pp.

Lowenthal, D. (1960): Physical resources. In: *The economy of the West Indies*, ed. by G.E. Cumper. United Printers, 48–94.

McPherson, H.M. and Saarinen, T.F. (1977): Flood plain dweller's perception of flood hazard in Tucson, Arizona. *Annals of Regional Science*, 2, 241–247.

Molion, L.C.B. (1975): A climatonomic study of the energy and moisture fluxes of the Amazonas with consideration to deforestation effects. Ph.D. thesis, University of Wisconsin, Madison.

Nkemdirim, L.C. (1988): Deforestation and climate. In: *The tropical environment*. Ibid cit.. 92–105.

Nkemdirim, L.C. (1990a): Is the climate signal in the Caribbean strong enough to disclose a climate change? Paper presented at the IGU Regional Conference, Beijing, August 1990.

Nkemdirim, L.C. (1990b): Water resource development in developing countries in a changing global environment. A contribution to New Vision V Workshop on Growing into Maturity, Olds, Canada, April 27–29, 1990, 37 pp.

NOAA, Climate Assessment Division (1979): A study of the Caribbean Basin drought/food production problems. Washington, DC.

O'Loughlin, C. (1960): Economic structure in the West Indies. In: *The economy of the West Indies*, ed. by C.E. Cumper. United Printers, 95–123.

Parry, M. (1991): The potential impact of climatic change on agriculture. Preprints, International Conference on Climatic Impacts on the Environment and Society. Tsukuba, Japan, Feb. 1991.

Penman, H.L. (1963): Vegetation and hydrology. Tech. Comm. No. 33, Commonwealth Bureau of Soils, Harpenden, U.K., 124 pp.

Pico, R. (1974): *The geography of Puerto Rico*. Aldine, Chicago, 439 pp.
Portig, W.H. (1976): The climate of Central America. In: *World survey of climatology*, Vol. 12, ed. by H.E. Landsberg. Elsevier, Amsterdam, 405–454.
Postel, S. (1989): Water for agriculture: Facing the limits. Worldwatch Paper No. 93, Washington, D.C.
Reading, A.J. (1990): Caribbean tropical storm activity over the past four centuries. *Int. J. Climatol.*, 10, 365–376.
Sauer, C.O. (1966): *The early Spanish Main Berkeley and Los Angeles*. University of California Press.
Tanner, E.V.J. (1981): The decomposition of litter in Jamaica montane forests. *J. Ecol.*, 69, 263–276.
Tory, W. (1979): Anthropology and disaster research. *Disasters*, 3, 43–52.
United Nation (1985): Water resources of Latin America and the Caribbean and their utilisation. Esterdros e Informes de la Cepal, 53, 142 pp.
United Nations (1987): *Statistical Year Book*, New York.
United Nations (1989): *Statistical Year Book*, New York.
United Nations Environment Programme (UNEP) (1982): Development and environment in the wider Caribbean region: A synthesis. UNEP Regional Seas Report and Studies No. 14, 37 pp.
Watts, D. (1987): *The West Indies: Patterns of development culture and environmental change since 1492*. Cambridge University Press, London, 609 pp.
WMO (1987): Climate and human health. World Climate Programme Applications, 16 pp.
WMO (1991): Report of meeting of experts on climate change detection project. WCDP No. 13, Geneva.
WMO-UNEP (1990): Scientific assessment of climate change. Policy maker's summary of the Report on Working Group I to the International Panel on Climate Change. Geneva, 26 pp.
WMO-UNEP (1991): The global change system: climate system monitoring, June 1986–November 1988, London, 70 pp.

Lawrence C. Nkemdirim
Department of Geography
The University of Calgary
Calgary, Alberta
Canada T2N 1N4

3.3. CLIMATE AND SOCIETIES IN SOUTHEAST ASIA

K.U. SIRINANDA

Climate relates to many sectors of human society and economy. Climate itself is variable in time and space as are human affairs. That climate is a potent factor in the development of human affairs cannot be gainsaid. Directly through physiological effects and indirectly through other environmental conditions, climate plays a fundamental role in the entire spectrum of human endeavour, from economic to social to cultural; the problem is, however, to ascertain, much less to evaluate, in tangible or quantitative terms, the exact nature of the relationship. This is primarily because humans react to their environment as they perceive it, and optimum climate is variable with space and time. Human perception is a function, among other things, of culture including all mentifacts, sociofacts and artifacts available to a particular society. Cultural dynamics plays an important role in the evaluation of the opportunities, challenges and constraints of the climatic environment and of the nature of human response.

The relationship between climate and society has been interpreted by different people from different standpoints. Some schools of thought have emphasized the domineering role that either may play whereas others have highlighted the reciprocities. While not wishing to get bogged down in the arguments and counterarguments that have been put forward it is nonetheless necessary to draw attention to some general points emanating from what has been said about humans and climate in the tropics in general and Southeast Asia in particular.

The crucial question is not whether climate influences human behaviour but to establish what and how strong the influence is and whether it is of any real significance. Much of the material advanced to support or refute various theories and viewpoints are speculative at worst and anecdotal at best. Writing on the climatic conditions in Southeast Asia most authors have paid only scant attention to climatic impact on human affairs except to point up obvious, general relationships (Gourou, 1974, 1975, 1982; Jackson 1977; Koteswaram 1974; Nieuwolt 1960, 1975, 1982; Pagney 1984; Yoshino 1984a, b and c).

M. Yoshino et al. (eds.), Climates and Societies – A Climatological Perspective, 203–234.

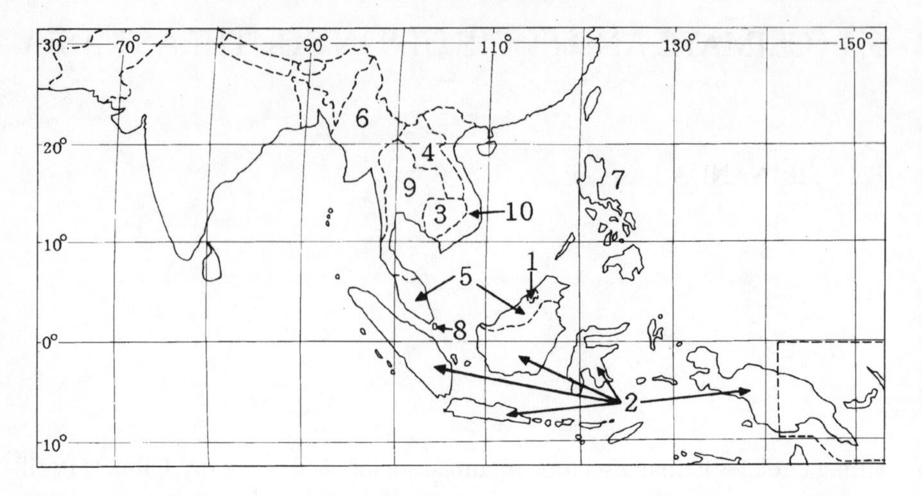

Figure 1. Southeast Asia – location and countries: 1. Brunei; 2. Indonesia; 3. Kampuchea; 4. Laos; 5. Malaysia; 6. Myanmar (Burma); 7. Philippines; 8. Singapore; 9. Thailand; and, 10. Vietnam.

Some environmental scientists, however, have drawn the attention of planners, policy makers and the public at large, to the dangers of disregarding the possible impact of human activities on the climate-society equation in Southeast Asia (David and Lim 1990; Hill and Bray 1978; MacAndrews and Chia 1979). Social scientists too have either explicitly or implicitly marginalized climate's role as a variable in tropical development (Fisher 1964; Hall 1964; Ichimura 1988; Lee 1957; Pelzer 1968; Tilman 1969). This can perhaps be attributed partly to the inherent complexity and subtlety of climate-society interaction and partly as a way of avoiding being classified as subscribers to the 'naive and simplistic' (Biswas 1979, p. 189) ideas of climatic determinism.

The earliest ideas on the nature of human-climate assumed cause-and-effect relationships, i.e. variation in climate is regarded as the causal factor in the variation of human activities. The best known exponent of such a relationship-climatic determinism was Ellsworth Huntington whose prolific writings (1915, 1924, 1926, 1927, 1945) are well known but other writers too have subscribed to it (Bryson and Murray 1977; Fleure 1919; Gourou 1960; Lamb 1982, 1988; Taylor 1957). Although in its undiluted form climatic determinism may be 'naive and simplistic' the fact remains that the 'correlation between climatic regime and economic development is as good as most correlations between non-economic factors and economic development' (Biswas 1979, p. 189). The net result of the reaction to climatic determinism was that climate received only passing reference in most discussions on economic development in the tropics (Brown 1974; Lee 1957; Myrdal 1974) until lately (Biswas 1984; Yoshino 1984a).

1. The region

'Southeast Asia' is a collective term which gained currency during the Second World War to refer to the region 'placed at the cross-roads between the Indian sub-continent, the Far East and the Antipodes' (Tate 1971, p. 3). It really comprises ten states spread over the mainland and in insular locations (Figure 1 and Table 1): Burma, Thailand, Kampuchea, Laos and Vietnam (all mainland nations); Indonesia and the Philippines (insular); Malaysia (both mainland and insular), and the small states of Singapore (an island linked to the mainland by a causeway) and Brunei (a sultanate on the northwest coast of Borneo island). Thus defined, Southeast Asia does not have a uniform climate although tropical maritime monsoonal characteristics prevail over much of the region. Nor does it have a uniform socio-economic landscape. As those knowledgeable about the detailed geography of the region would concur, Southeast Asia has a marked spatial differentiation of climatic conditions (IRRI 1974; Koteswaram 1974; Nieuwolt 1977; Yoshino 1971, 1984a) and an even greater diversity of economies, societies and cultures (Dobby 1950; Fisher 1964, 1969; Gourou 1974, 1982; Hall 1964; Harrison 1986; Osborne 1979; Spencer 1954; Tilman 1969). On a global scale, however, Southeast Asia is considered a regional entity, or more accurately, a unit in diversity (Dobby 1950; Fisher 1964, 1969; Spencer 1954) not only as a zone of transition between the Indian and Pacific Oceans but also due to commonalities in history, economy and culture.

Southeast Asia occupies a latitudinal zone that extends about 35 degrees (roughly 10–25°N) and a longitudinal span of nearly 50 degrees (approximately 90–140°E, Figure 1). This, along with the geographical position and the nature of the physiography, would bring about diversity in the biophysical environment in general and climate and human geography in particular. The land area of the region is about 4.51 million km^2 (mainland −1.96 km^2 and insular −2.55 km^2) (Table 1). In the north, mountain barriers stand between the mainland sector of Southeast Asia and the continental Asia. The bulk of the region is an area of 'geographical fragmentation' (Fisher 1969) forming peninsulas and island arcs. Physiographically the peninsulas and island arcs have mountain backbones or 'exceedingly difficult upland country' (Fisher 1964) running in a north-south or westnorthwest-eastsoutheast direction with lowlands occupying a relatively smaller proportion of the total land area. In between the mountain ranges in the mainland are such great rivers as Irrawady, Menam Chao Praya, Mekong and Red River forming the main arteries of a dense and elaborate drainage network, a direct consequence of rugged terrain and copious rainfall. In the archipelago sector too, high density drainage patterns are a salient characteristic of the physical environment.

In sum, it may be said that the 'cross-roads' or the 'march lands' position of Southeast Asia, the remarkable interpenetration of land and sea, the rugged relief in the north and in between large rivers, the intricate and high

Table 1. Southeast Asia: Area, population and economic status.

Country	Area (km^2)	Nos. (millions; 1989)	Population density (per km^2)	Rural (%)	GNP (per capita) (US$)	Agricultural share of GDP (%)
Brunei	5,765	0.25	45	w	20,000 y	
Indonesia	1,919,300	184.6	95	74	430	21
Kampuchea	181,035	6.8	38	89	x	z
Laos	236,800	3.9	17	84	180	65
Malaysia	329,079	17.4	53	65	1,870	20
Myanmar	696,500	40.8	60	76	x	38
Philippines	299,400	64.9	217	59	630	27
Singapore	620	2.7	4,355	7	9,100	0.4
Thailand	513,517	55.6	108	83	840	17
Vietnam	334,331	66.8	200	81	x	38

w: about 50% of Brunei population can be said to be living in rural areas; however, urban development is essentially in the form of small towns with relatively low-density built-up areas.
x: World Bank (1990) estimates the per capita GNP in these countries to be less than US$ 500, but the actual figures are likely to be close to that for Laos.
y: More than 80% of Brunei's GDP is from the Petroleum sector and services account for much of the balance. Agriculture is negligible.
z: Data not available.
Compiled based on information in Myres (1990), Sirinanda (1984c) and World Bank (1990).

density drainage network, and the details of its regional and local geography have directly or indirectly influenced the climatic environment and the human geography as well as the perception, evaluation and responses of human societies in Southeast Asia towards their climatic environment. The admixture of environmental influences arising out of the peculiarities of the physiography and climate and the external influences due to the 'gateway' role between the Pacific and Atlantic Oceans and the 'causeway' or 'stepping-stone' role between Asia and Australasia have brought about a complex economic, social and cultural diversity (Hall 1964; Fisher 1964; Tilman 1969).

2. Climate

Southeast Asia, as mentioned earlier, straddles the Equator, is entirely within the tropics and is in the monsoon realm. Added to this is its location between the two vast ocean masses and at the interface between the continental masses of Asia and Australia. The vast latitudinal and longitudinal extent together with the interpenetration of land and sea, physiographic complexity and altitude have brought about spatial and temporal patterns of climatic variation (Figures 4a and b). While regional and local differentiation in the climate due to circumstances outlined above are the rule rather than the exception, the area as a whole shares some common climatic characteristics due to the fusion in various degrees of the equatorial/tropical/monsoonal domains.

Broadly, three main climatic types can be identified (Besancenot 1984; Garbell 1947; Koteswaram 1974; Nieuwolt 1981b; Yoshino 1971, 1984a; Watts 1955), each with some or all of the traits of the equatorial/monsoonal conditions: (a) equatorial, rainforest type; (b) tropical evergreen/tropical monsoonal type; and (c) tropical monsoonal with distinct dry season. Besides these three, altitudinal influences have led to the formation of highland climates in certain localities.

Climatic variation is mainly brought about by the spatial and temporal variations of rain producing phenomena and the determinants of the spatial distribution of rainfall. Hence directly more relevant to this chapter would be the zonation based on rainfall distribution as illustrated in Figure 2 (IRRI 1974) and the biological significance of climate (Figure 3) (Besancenot 1984). These macro type features are modified regionally and locally due to the special relationship between land sea and the variation in relief.

Figures 4a and b give the temperature and rainfall distribution pattern. It is apparent that, in general, the region experiences relatively uniform thermal characteristics both in space and time, the major deviations being the result of latitudinal and altitudinal differences. From the point of view of socio-economic activities the thermal environment can be said to be non-limiting although combined with high humidities, in most localities in Southeast Asia except in the highland and hilly areas, the high temperatures (around 30°C) create enervating, steamy, sultry conditions. Intense insolation and high temperatures have often been regarded as the scourge of the tropics; one comes across such expressions as 'lands of the cruel sun' 'regions of debilitation', 'regions of energy sapping, steamy heat', being used to characterise the thermal conditions of tropical lands.

As a whole, Southeast Asia receives an annual rainfall of more than 1500 mm rising in certain areas to over 5000 mm. In the equatorial regions, rainfall is generally well distributed throughout the year. However, due to altitude, orography and windward or leeward effects the spatial distribution is subject to significant variations, and the seasonality of the rain-producing atmospheric systems gives rise to a seasonal variability in wetness. Dry seasons in the normal sense of the term are rare but can be identified in a relative sense and with reference to other geoecololgical factors. However, as one gets away from the equatorial rainforest climatic region towards the margins of the tropics and in rainshadow areas, a distinct seasonality in the rainfall regime would be evident (Figures 4a and b), with over 50% of the total annual rainfall occurring within 2–3 months (Figure 2). Water balance conditions are summarized in Figure 5. While a large part of the archipelago region experiences

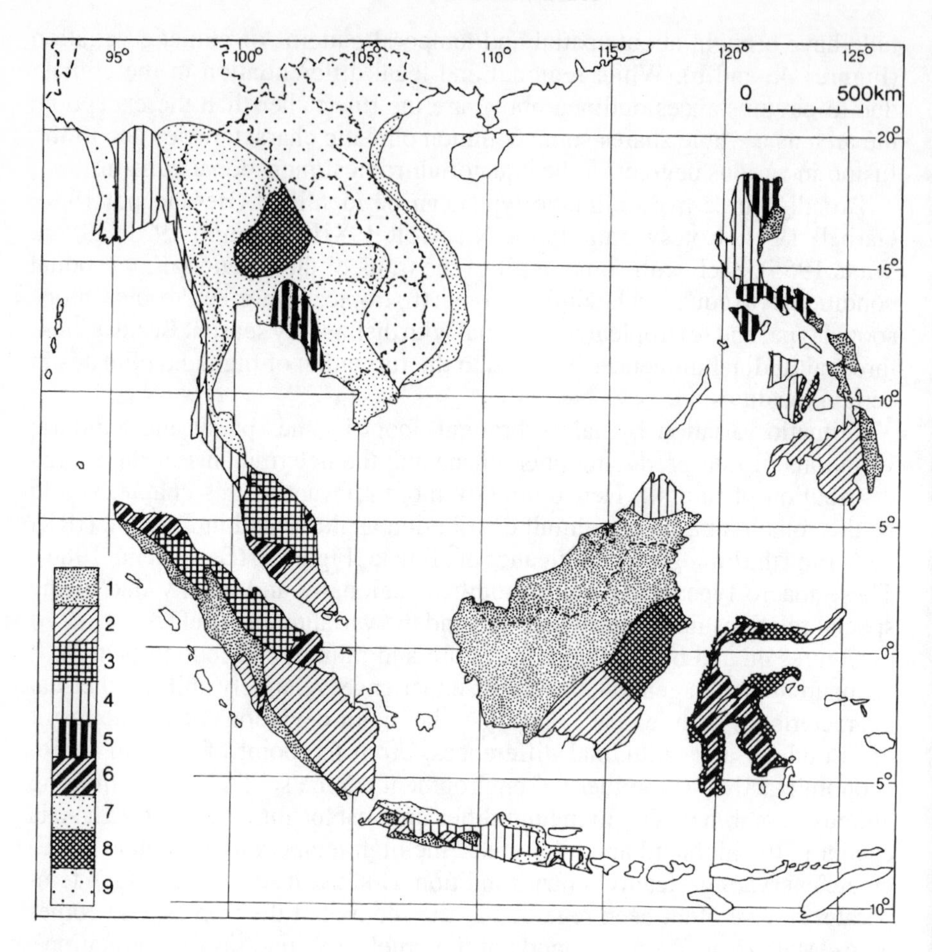

Figure 2. Rainfall zones (after IRRI 1974): 1. more than 9 consecutive wet months; 2. 5–9 consecutive wet months and no pronounced dry season; 3. 5–9 consecutive wet months and secondary rainy season; 4. 5–9 consecutive wet months and pronounced dry season; 5. 5–9 consecutive wet months and pronounced end to rainy season; 6. 2–5 consecutive wet months and no pronounced dry season; 7. 2–5 consecutive wet months with pronounced dry season; and, 8. less than 2 consecutive wet months (wet month: receiving at least 200 mm of rain; pronounced dry season: at least 2–3 months with less than 100 mm of rain).

adequate water supplies, the existence of areas of annual or seasonal water deficits in the mainland areas and in the southern island areas of Indonesia and in the Philippines are of direct relevance to land use patterns.

Temporal variability of rainfall has at times caused problems for water-dependent economic activities and water supplies for domestic and industrial use (Low and Balamurugan 1991; Sirinanda 1976). However, societies in Southeast Asia from time immemorial have resorted to various adaptations and adjustments in their socio-economic life in their efforts to make opti-

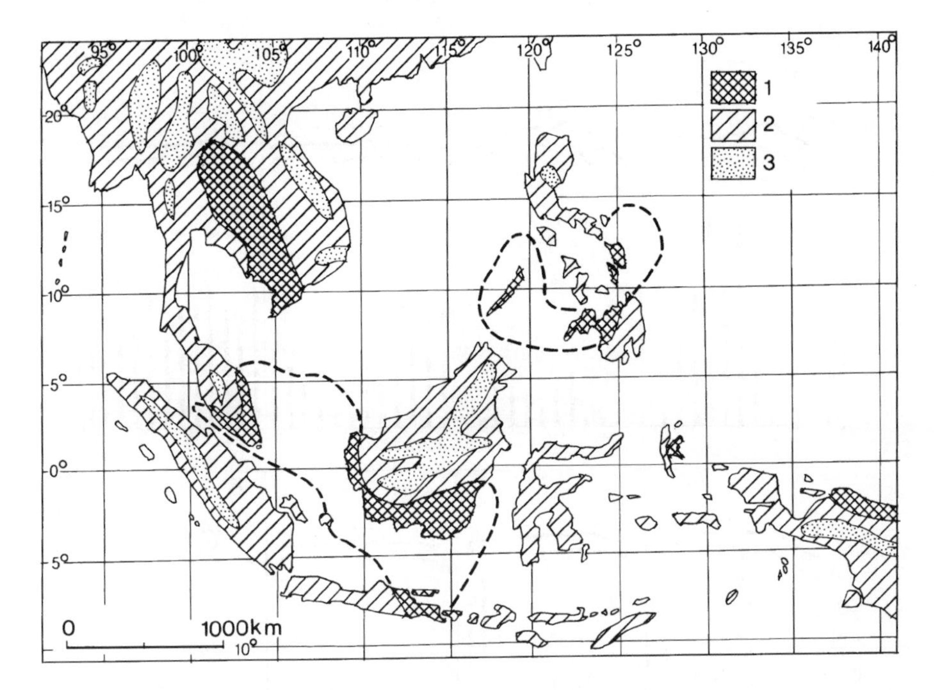

Figure 3. Bioclimatic regions of Southeast Asia (adapted from Besancenot 1984): 1. equatorial climate; 2. tropical climate; and, 3. warm temperate climate.

mum use of what they perceive as the benign aspects of the thermal and hydro-climatic environments and to hedge against the detrimental impact of excessive heat, humidity and, on occasion, hazardous water availability conditions.

Being an arena for the interaction of airstreams and atmospheric systems from different sources and with different properties, Southeast Asia is subject to storms of different magnitudes affecting certain areas more than others. The typhoons in the South China Sea affecting directly the Philippines in particular and the surrounding areas occasionally, the tropical depression in the southeast monsoon airstreams leading to almost annually recurrent floods in Peninsular Malaysia, Thailand and most other lowlying areas of Southeast Asia, the squall-lines in the Straits of Malacca, and thunderstorms in the intermonsoon seasons in much of the area are some of the more obvious weather hazards; these sometimes are considered necessary evils because they are responsible for the abundance of rainfall and of guaranteed water supplies in the areas concerned, the actual and potential damage notwithstanding.

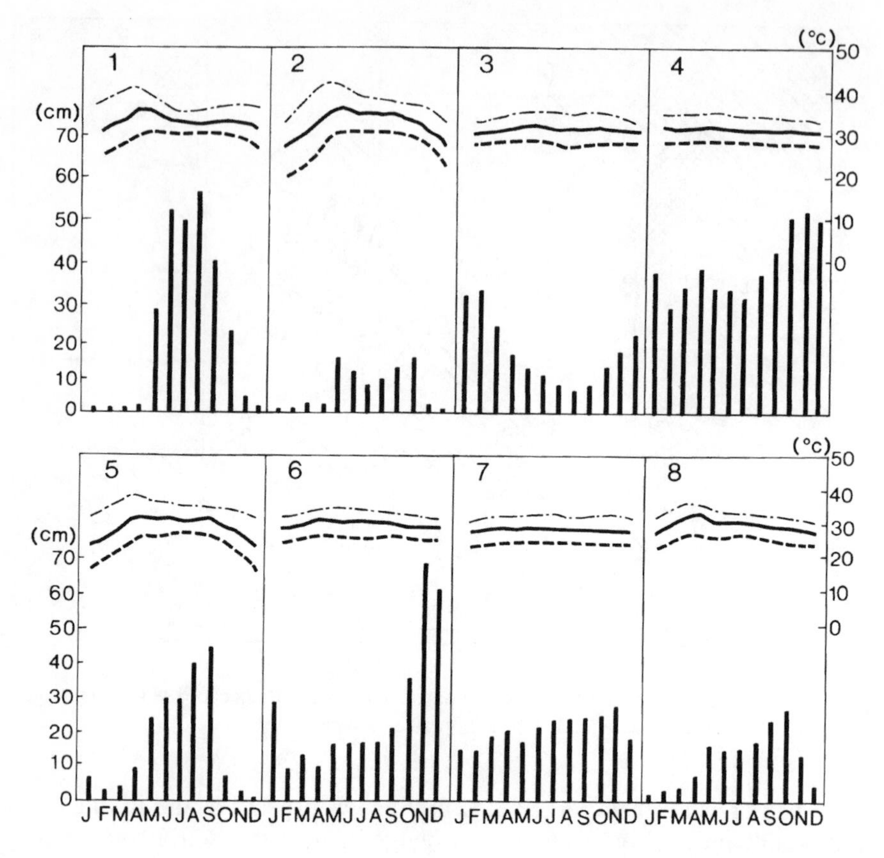

Figure 4. (a) Monthly temperature and rainfall for selected stations (adapted from Koteswaram 1974): 1. Rangoon (Lat. 10°46′ N, Long. 96°11′ E); 2. Mandalay (Lat. 21°59′ N, Long. 96°06′ E); 3. Jakarta (Lat. 06° 11′ S, Long. 106°50′ E); 4. Padang (Lat. 00°56′ S, Long. 100°22′ E); 5. Vientiane (Lat. 17°50′ N, Long. 102°36′ E); 6. Kota-Bharu (Lat. 06°10′ N, Long. 102°17′ E); 7. Malacca (Lat. 02°16′ N, Long. 102°15′ E); and, 8. Phnompenh (Lat. 11°33′ N, Long. 104°51′ E).

3. Climate and society

Southeast Asia as a whole has an abundant supply of the most basic climatic resources, viz., heat, sunshine and moisture, subject to regional and seasonal variations of different magnitudes due to natural environmental factors as well as land use changes brought about by humans. It may be said as a broad generalization that much of the traditional socio-economic life in Southeast Asia follows the natural rhythm of the macro-climatic conditions as expressed in the distributional pattern of the said resources. At the smaller scale, food production, storage and consumption, and settlements and social activities reflect, by and large, the climatic environment, particularly the seasonality and diurnal rhythms of the rainfall distribution pattern. Primitive

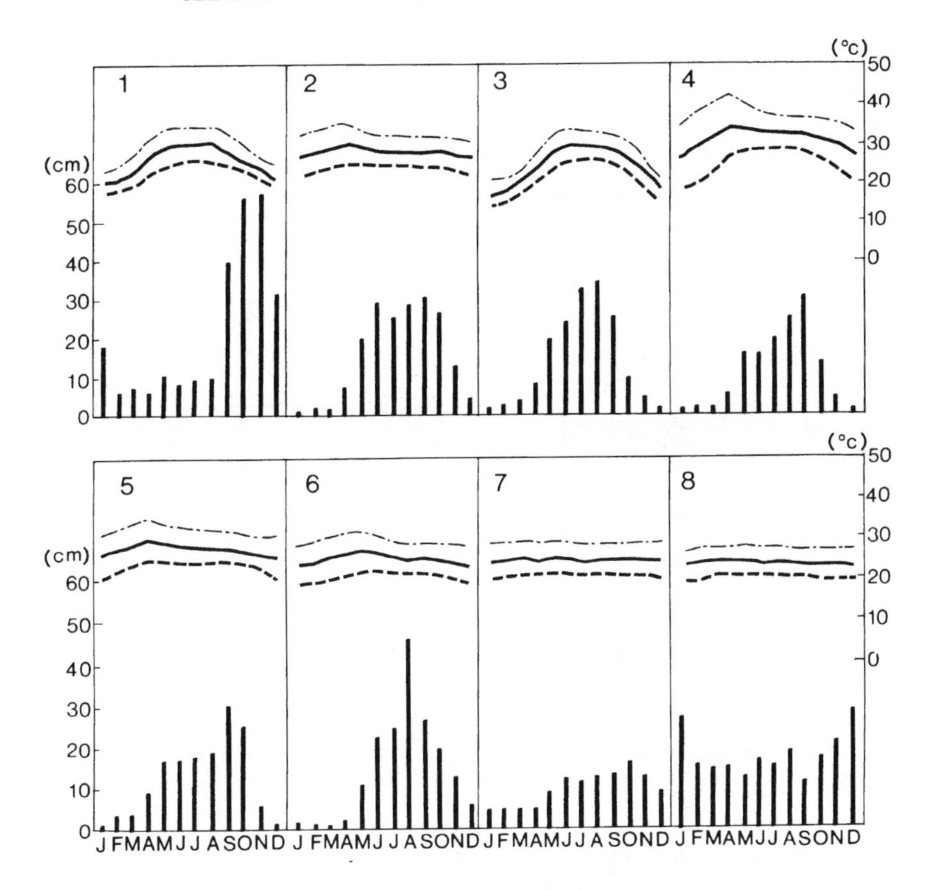

Figure 4. (b) Monthly temperature and rainfall for selected stations (adapted from Koteswaram 1974): 1. Quangtri (Lat. 16°44′ N, Long. 107°11′ E); 2. Saigon (Lat. 106°42′ E); 3. Hanoi (Lat. 21°02′ N, Long. 105°52′ E); 4. Chiangmai (Lat. 18°47′ N, Long. 98°59′ E); 5. Bangkok (Lat. 13°45′ N, Long. 100°28′ E); 6. Manila (Lat. 14°35′ N, Long. 120°59′ E); 7. Zamboanga (Lat. 06°54′ N, 122°05′ E); and, 8. Singapore (Lat. 01°10′ N, Long. 103°50′ E).

hunting and gathering and shifting agricultural economies practised by tribal communities in remote, hilly, interior areas of both the mainland and the archipelago bear testimony to the adaptations to the climatic environment. In order to allow for intense solar radiation, heat and humidity, the traditional communities have used natural housing material and housing designs that allow maximum cooling and ventilation. The village communities in Southeast Asia as a whole have come to terms with their climatic environment by adapting to it and making minor adjustments in their economic and social activities. However, with the growth of large urban centres with high population densities, the dearth of land and the process of modernization with the concomitant increased tempo of economic activities, settlement densities

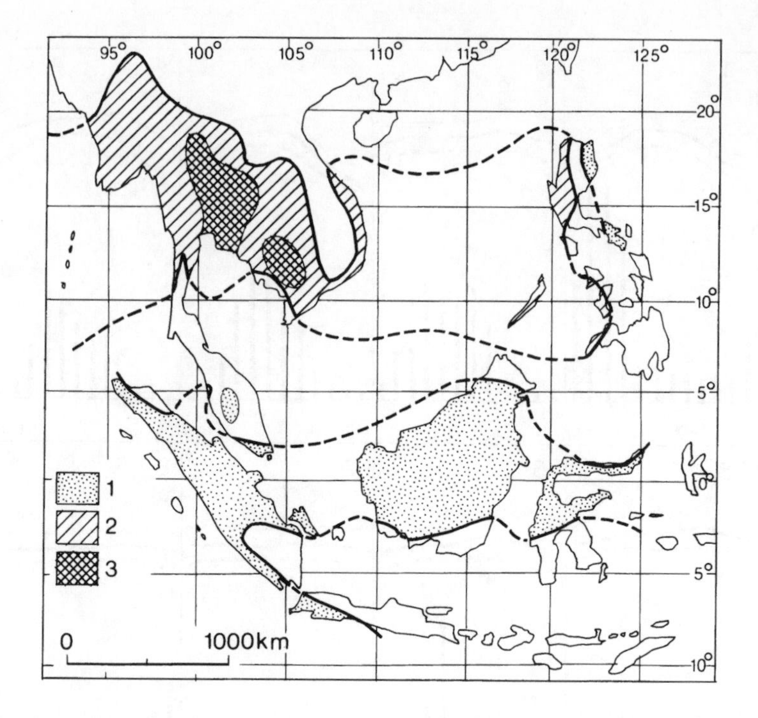

Figure 5. Hydrological regions of Southeast Asia (adapted from Kayane 1971): 1. No annual water deficit; 2. Annual water deficit of over 200 mm; and, 3. No annual water surplus.

and the proliferation of numerous recreational and social activities, weather and climate appear to have been pushed to a back seat in the decision making processes (David and Lim 1990). Indeed, certain patterns of living and life styles in the urbanized areas of Southeast Asia are not at all reflective of the hot, humid climatic conditions. Air-conditioning, refrigeration, 'high-tech' building materials, and building design and architectural developments have all made the creation of desirable indoor climates within the reach of the middle income groups in most Southeast Asian countries. Often, many of the affluent people in Southeast Asia can be said to be 'insulated' to different degrees against hot, humid conditions with the availability of facilities to create comfortable indoor climates artificially in houses, in transport media, in work places or in shopping complexes. The extent to which a person is able to live in bioclimatic comfort in spite of the climatic conditions is often regarded as a measure of that person's standard of living.

4. Demography

With a total population of around 450 million, Southeast Asia lies between two extremely populous regions on Earth: South Asia (1100 million) and East Asia

Table 2. Population growth in Southeast Asia (1600–2000) (in millions).

Country	1600	1800	1950	mid-1970s	mid-1980s	1989	2000 (projected)
Brunei	–	–	–	0.21	0.22	0.25	0.3
Indonesia	9.1	12.3	77	142.9	176.0	184.6	222
Kampuchea	1.2	1.5	4	–	–	6.8	9
Laos	–	–	–	3.3	3.4	3.9	5
Malaysia	1.2	1.5	6	13.1	13.4	17.4	27
Myanmar	3.1	4.6	18	32.6	33.3	40.8	51
Philippines	1.0	2.0	20	46.8	47.9	64.9	86
Singapore	–	–	–	2.3	2.4	2.7	3
Thailand	2.2	2.5	20	45.5	46.4	55.6	66
Vietnam	4.7	7.0	24	52.9	54.1	66.8	86

Compiled on the basis of information in Myers (1990), Rigg (1991) and Sirinanda (1984c).

(1300 million). The average density of about 100 people per km^2 is relatively lower than that for East Asia (130) and South Asia (240). However, the average density is determined, especially in the context of large areas which are heavily inhabited or sparsely populated, by one or more environmental constraints, and the presence of large urban agglomerations with very high population densities (Tables 1 and 2).

Much of the population is concentrated in agriculturally productive river valleys or coastal areas and in urban clusters. Tables 1 and 2 illustrate the wide disparities among the different countries. Even within any one country the distribution of population is markedly uneven. Java, for instance, with a land area of only about a seventh of the total areal extent of Indonesia contains nearly 70% of the country's population, averaging 100 per km^2. Singapore represents the extremely high densities found in urban centres in Southeast Asia; on average over 25% of the total population of Southeast Asia is urban (Chao and Bauer 1987) and much of this live in a dozen or so mega cities. In 1950, Southeast Asia had only 2 cities with over one million people, by 1980 the number increased to 10, and by the year 2000 there will be at least 15 such cities; Bangkok, Manila and Jakarta which are already saddled with very high population numbers are estimated to have more than 10 million each by the year 2000.

The fertile river valleys of Menam Chao Praya, Irrawady, Mekong, and the Red River and the volcanic soil regions of Java, Madura and Bali have been traditional population centres based on wet rice cultivation, whereas the urban clusters display the development of commercial agriculture ('plantation

Table 3. Changes in the economic structure of the ASEAN countries 1960–1989 (%).

Country		1960	1970	1985	1989
Brunei	1	–	–	1	1
	2	–	–	84	84
	3	–	–	15	15
Indonesia	1	50	46	24	21
	2	25	21	36	40
	3	25	33	41	39
Malaysia	1	36	32	21	20
	2	18	25	35	41
	3	46	43	44	39
Philippines	1	26	28	27	27
	2	28	30	32	33
	3	46	42	41	40
Singapore	1	4	2.3	1	0.4
	2	18	30	37	36.6
	3	78	67.7	62	63
Thailand	1	40	30	17	16
	2	19	26	30	33
	3	41	44	53	51

(1: Agriculture, 2: Industry/mining/manufacturing, 3: Services)
(*Source*: World Bank 1987)

agriculture' in particular) and secondary and tertiary activities over the last few decades, especially since the 1970s (Table 3).

The highland terrain in much of the region is forbidding as far as settlement is concerned as shown by the largely empty or sparsely peopled tracts in the interior of Borneo, Sumatra, Irian Jaya, Celebes, Peninsular Malaysia, and the highland regions of Burma, Thailand and Indochina, and have very little economic activity except for low temperature agriculture in certain specific areas and the presence of hill resorts. Furthermore, there are, also, large coastal areas of swamps and marshes which are practically uninhabitable.

The population of Southeast Asia has grown over the last 400 years from a mere 23 million in the 1600 to the present 450 million; during the next decade it is projected to rise to over 500 million (Table 2). As is shown in the section on the peopling of Southeast Asia, the nuclei of early settlements were established with particular reference to favourable land and soil resources and climatic conditions which lend themselves to easy adaptive responses such as rain dependent 'fire' agriculture and hunting and gathering activities.

River valleys with gentle slopes or undulating terrain with a marked seasonality of rainfall and controllable water supply conditions appear to have been preferred. However, as population numbers increased and with developments in technology and modernization of the economies, the element of discrimination seems to have been abandoned, and areas earlier considered to be inhospitable or inappropriate for any productive activities were opened up for human settlement and economic activities. Huge land development, settlement or resettlement projects (irrigated rice plantations, plantation agriculture and urban development), gradually in certain countries but rapidly in others, have flourished, in spite of certain climatic constraints. The failure of the sugar cane project in Perak, Malaysia and some of the transmigration schemes in Indonesia are notable exceptions.

Thus, limits to human settlement in Southeast Asia is set by physiography, drainage and soils. The climate can be said to be non-limiting, the seasonal occurrence of floods and droughts, tropical storms, climate induced pests, diseases and parasites (Monteiro 1962; Ooi 1959) notwithstanding. In other words, climatically speaking, Southeast Asia in general does not have any truly marginal areas compared to certain parts of South Asia, tropical Africa or South America, so much so that Southeast Asia does not feature much in the regional or international media as a climatically serious problem area, the occasional localized damage due to typhoons in the South China Sea being the exception. As has been pointed out by Chang (1968), Gourou (1960, 1974) and Pagney (1984), humid tropical parts of Southeast Asia can easily be identified as a separate entity from the rest of the tropics in terms of the relatively high degree of economic and social development which, according to those authors and results of historical analysis (Fisher 1964; Hall 1964; Harrison 1986; Osborne 1979; Tate 1971; Tilman 1969), is due more to its greater historical intercourse with the external world and not an absolute reflection of inherent characteristics of the climatic environment.

5. Economies

Southeast Asia is traditionally understood as having basically agrarian economies variously described as constituting the essential ingredients of 'fluvial' or 'vegetable' civilizations in the core areas. The main anchor of the economies has been subsistence agriculture although from time immemorial a certain element of commercialism was evident particularly in the coastal areas in relation to the spice trade. Collecting and gathering activities and shifting agriculture were the main activities in the highland areas. The lowland peoples depended on wet rice cultivation or a mixture of sedentary and shifting cultivation. Successive waves of external influences have changed this leading to the expansion of commercial agriculture in selected areas especially the plantation type of market-oriented production and forestry industries.

Table 4. Tropical moist forest cover (km^2), deforestation rate (km^2) and estimate of release of carbon (million tons) in Southeast Asia.

Country	Original forest cover	Present forest cover	Present primary forest cover	Present annual rate of deforestation	Annual rate of carbon release
Indonesia	1220,000	860,000	530,000	12,000	124
Kampuchea	120,000	67,000	20,000	500	5
Laos	110,000	68,000	25,000	1,000	10
Malaysia	305,000	157,000	84,000	4,800	50
Myanmar	500,000	245,000	80,000	8,000	83
Philippines	250,000	50,000	8,000	2,700	28
Thailand	435,000	74,000	22,000	6,000	62
Vietnam	260,000	60,000	14,000	3,500	36

(Adapted from Myers 1990)

Further diversification of the economies has occurred during the past few decades resulting in structural and sectoral transformation in the ASEAN economies in particular (Table 4). This shows that in the bulk of the region the situation today is anything but reflective of the traditional description of the region as having 'vegetable' civilizations. In Burma, Laos, Kampuchea and Vietnam, the description may still be true, although the winds of change are blowing in these countries too. On a global perspective Southeast Asia is still very much an agrarian region with dual-economies (subsistence-oriented and market-oriented) and are classified as developing or underdeveloped as a group. It is customary to classify these countries into 'low-income economies (Laos, Indonesia, Burma, Kampuchea, Vietnam), 'middle-income economies' (Philippines, Thailand, Malaysia) and 'high-income economies' (Singapore, Brunei), (Asian Development Bank 1990; World Bank 1989). However, on the basis of per capita GNP (Table 1), wide intra-regional variations are evident. On the one extreme are Brunei and Singapore with very large per capita GNP; Brunei's wealth being derived almost exclusively from oil and natural gas whereas Singapore, a 'little dragon of Asia', has highly developed secondary and tertiary economic activities. On the other extreme are countries with very low per capita GNP such as Burma, Laos, Kampuchea and Vietnam. In between are Malaysia, Thailand and the Philippines with a progressively lower economic growth and development. It can be said that the difference in levels of economic development achieved by these countries is more due to differences in socio-economic, cultural and political factors rather than to differences in any environmental conditions.

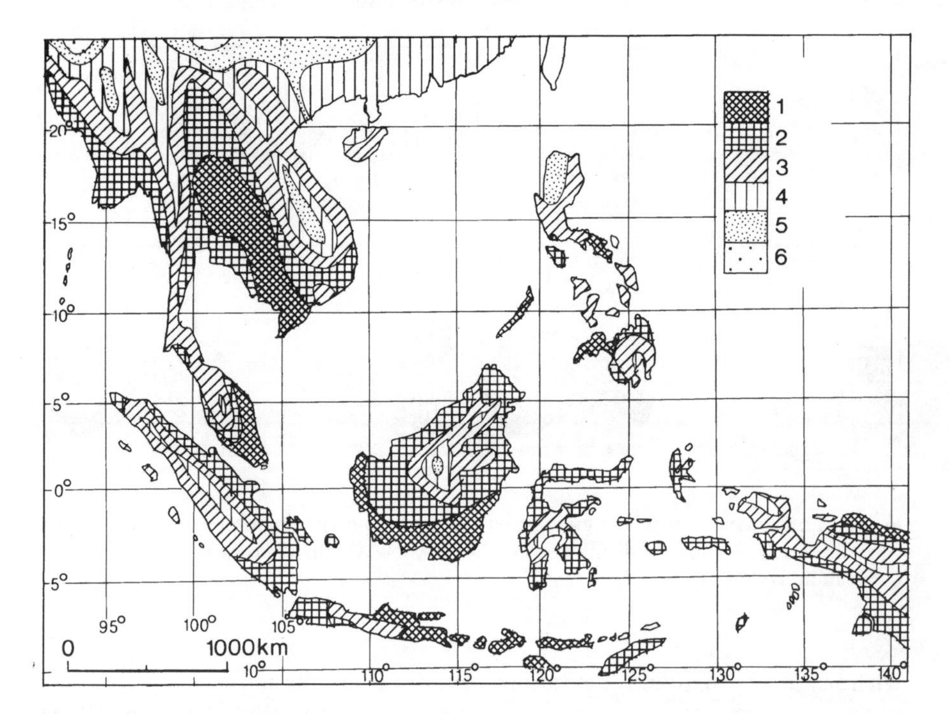

Figure 6. Annual discomfort in Southeast Asia (adapted from Besancenot 1984). Discomfort indices from 0 (absolute comfort) to 36 (absolute discomfort).

6. The humid tropics, Southeast Asia and climate

The tropics in general have long been characterized as negative regions for human settlement. They have been described as 'regions of debilitation' (Fleure 1919), 'regions of low climatic energy' or 'regions of low civilization' (Huntington 1915, 1924). The relative underdevelopment of the tropical regions is generally attributed to its enervating climate and the attendant environmental ills. Although the extreme notions of climatic determinism have largely been disproved, or 'disapproved', the fact remains that climate does act as a constraint or is a source of distress for most tropical areas especially when the human organization to manage the society and the economy is inefficient or illequipped (Biswas 1979; Biswas 1984; Biswas and Biswas 1979; Gourou 1974; Ooi 1959).

Some writers consider the high degree of civilization achieved in Southeast Asia during historical times and the current relatively high level of socio-economic performance of the region as a whole *vis-à-vis* the rest of the tropics as being due to cultural factors of extraneous origin, the main spurs being the influence of the Indian and Chinese domains with less humid tropical climatic conditions (Gourou 1975, 1982; Pagney 1984; Chang 1968),

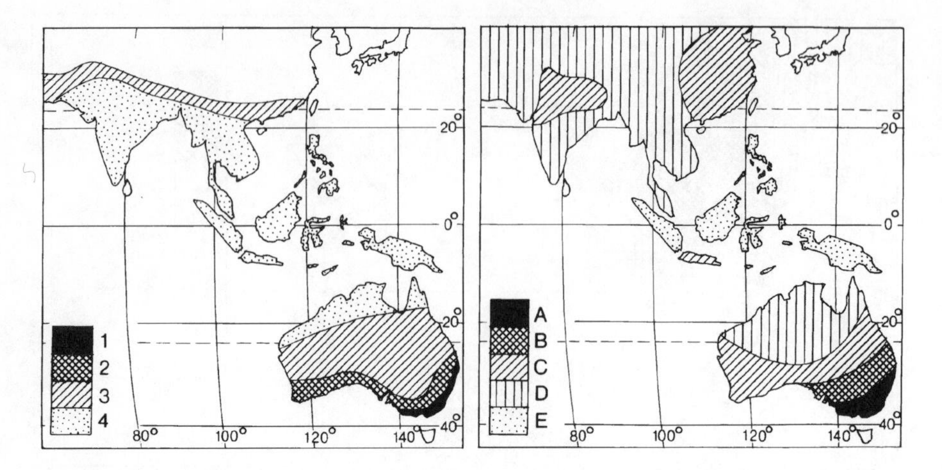

Figure 7. Climatic energy and civilization (adapted from Huntington and Cushing 1934). Climatic energy: 1. High, 2. Medium, 3. Low, 4. Very low; Civilization: A. Very high, B. High, C. Medium, D. Low, E. Very low.

whereas others maintain that humid conditions are not necessarily obstacles to the development of advanced civilizations (Carter 1968; Nestman 1975). Huntington (1915, 1924, 1926, 1927, 1945) and associates (Huntington and Visher 1922; Huntington and Valkenburg 1933; Huntington and Cushing 1934) maintained in their voluminous writings that, what they considered to be, the low level of civilization and development in the tropical world in terms of physical, mental and socio-economic attributes of the inhabitants are a direct or indirect consequence of non-optimal climate, as shown in Figure 6, or low 'climatic energy'. Climatic energy is defined as the 'combined physical and mental activity which might be expected if people's activities depended on climate alone' (Huntington et al. 1933, p. 118). According to this line of thinking a map prepared to show the distribution of climatic energy depicts Southeast Asia as a region of very low climatic energy (Figure 7). Huntington goes on to amplify the 'connection between climate and efficiency' and says that 'the only question is how far this efficiency depended on climate directly and how far indirectly through race, diet, parasitic diseases, hygiene, sanitation and social and political customs' (ibid, p. 124).

Carter (1964) has examined the Huntingtonian assertions about the history of Asia and the cultural status of the tropics. From a 'cultural-historical approach', he rejects the suggestion that the history of Asia is 'a series of climatically determined human pulsations' and that the tropics are 'doomed by the enervating effect of heat and humidity to a perpetually low cultural status'.

Thus the deterministic ideas in their original form have been discarded but still the fundamental role of climate as an environmental variable affecting in different ways and to different degrees the fortunes of human societies has been reemphasized by later writers, especially those who have attempted to reconstruct the impact of climate on society from the historical perspective (Bryson and Murray 1977; Lamb 1982, 1988; Parry 1985) (see Kates 1985; Wigley et al. 1985, for reviews).

7. A historical perspective

The socio-economic history of Southeast Asia is a fascinating saga of how a cross-roads region came to be colonized, settled and influenced by successive waves of migrations of people with different socio-cultural backgrounds. From the earliest times, the socio-economic fabric of Southeast Asia has been fashioned by movements of people whether from coastal to interior areas and vice versa and from outside the region for settlement, trade, conquest or missionary work. Fisher (1964, 1969), Hall (1964), Harrison, (1966), Osborne (1979), Tarling (1966), Tate (1971) and Tilman (1969) sketch the main episodes in this saga, and the material from these works is used extensively in the present paper to outline the socio-economic history of Southeast Asia in relation to the role of the climatic environment. The nature of the climatic environment, particularly the hydroclimatic conditions, seasonality of rainfall and wind systems, have directly or indirectly played a fundamental role. Unfortunately, it is not easy to pinpoint exactly how climate might have influenced human affairs in the region. No historical studies of climatic impact on society have been made compared to, for example, the many studies done for Europe, North America, Australia, and Africa (Chen et al. 1983; Kates et al. 1985; Kellogg and Schware 1982; Leggett 1990; Maunder 1970, 1974, 1986, 1989; Pittock et al. 1978). Even the studies on the possible impact of climate on contemporary socio-economic conditions are very general and descriptive for the most part, and it is generally assumed that the correlation or the covariation of climatic conditions and economic activity necessarily indicate a causality.

Although the beginnings of the human settlements in Southeast Asia are very much shrouded in antiquity, it can be surmised that from early times people in Southeast Asia depended on agricultural and marine resources as basic to their economies and societies. Historically important populations occupied the valleys and deltaic areas or inland irrigated areas which constituted the nuclei of the societies engaged in wet-rice cultivation as the main mode of livelihood. Dependable water supplies, the hot wet climate and the techniques of water control to tide over seasonal excesses or deficits seem to have been the decisive factors. Other smaller groups occupied interior, hilly, remote areas with less developed adaptive technologies to face the hot, humid conditions. Hunting, collecting and/or primitive forms of migrant hillside agriculture were practised. Even today hillside agriculture or shifting cultivation, identified by different regional names (e.g. 'caingin' in the Philippines, 'ray' in Indochina, 'tam rai' in Thailand, 'taungya' in Myanmar and 'ladang' in Indonesia) (Spencer 1966), extends over enormous stretches of elevated terrain throughout Southeast Asia. This method of utilizing large tracts of otherwise agriculturally marginal areas for producing food crops as well as cash crops such as spices, a concession to steep terrain and seasonal rainfall, 'however primitive and inadequate, demonstrated survival inertia as a way of life' (Cady 1964, p. 10) in a hot, humid environment.

Besides the activities based on agricultural resources, throughout its history much of the region, due to its cross-roads situation, participated in sea-borne trade and commerce as important aspects of socio-economic life. From Roman times Southeast Asia was a source of valuable commodities like spices and ivory, and formed a corridor in the trade between China, South Asia and the Middle East as the sea arm of the celebrates 'silk route'. The seasonal reversal of the dirction of the monsoon wind systems and the pattern of storm incidence were very much factors in the direction of the sea-borne voyages before the era of steam boats. Most maritime or riparian communities in Southeast Asia developed sea-faring traditions partly as a response to the wind regimes in the island arcs or in the ocean expanses. The spread of progressively more effective means of dealing with the variability of water supply and making better use of the climatic environment would have been facilitated by the sea-going tradition of the people of Southeast Asia.

The geography of the region, the nature of soils and the aforesaid climatic conditions, particularly the rainfall pattern and seasonality of wind systems, apparently dictated the dispersion of the centres of population. The socio-economic organization was advanced in favourable areas. It is said that the 'disciplined character and economic vitality characteristic of the Vietnamese people, for example, were derived in large measure from their successful age-long struggle to curb the turbulent Tongking delta of the flooding Red River' (Cady 1964, p. 8). It appears that the very early inhabitants tended to avoid floodprone areas as well as areas of dry season water scarcity until the technology and the socio-economic and political organization necessary for water control as required by wet rice cultivation were available. Only an assured supply of water which was a *sine qua non* for the production of adequate food supplies ensured a large sedentary population which in turn was a necessary condition for evolving an integrated social structure and political stability needed for building and maintaining water storage, distribution and control works.

Throughout history those areas which had socially and politically stable conditions for the development and management of adaptive technological means to hedge against the uncertainties of the rainfall conditions had an edge over less well organized areas in terms of further advancement in the social and economic frontiers. It is said that the remains of great civilizations as evident in parts of central Myanmar, central Thailand, Angkor in Kampuchea, Borobudur in Java are testimony to such strong socio-political organization, although there is a lot of controversy as to the role of centralized, strong political power which may have a hand in the development of 'hydraulic civilization' of that nature (see Wittfogel 1957) on the concept of 'oriental despotism'). Referring to the Angkor, Brooks (1928) (cited in Lamb 1982) put forward the hypothesis that a major climatic change led to the disruption of the Angkor civilization, although there is no definitive evidence to take a stand one way or the other. The earliest organized civilizations of Southeast Asia were all based to an important degree on irrigation rather than rain-fed agriculture, and, consequently, were insulated to a certain degree from climatic vagaries. However, there are suggestions that subsequent longer term climatic desiccation became a major problem especially where water and land resources were limited relative to population.

Fisher (1964, p. 79) had traced the antiquities of the early inhabitants of Southeast Asia as a 'series of southerly migrations' and has attempted a reconstruction of the sequence of cultural evolution in the region during the pre-historic times. The earliest inhabitants are referred to as 'sub-men whose successors were the first truly human inhabitants of the region' who evolved a form of Paleolithic hunting and gathering culture not strikingly different from that of the present day Negrito or other primitive peoples in the interior of the upland regions of Southeast Asia. These were followed in Mesolithic times by the appearance of a more advanced culture in the form of 'Bascon Hoabinhian' culture and it has been inferred that their economy, though based on hunting, collecting and, especially, fishing, was also supplemented by rudimentary agriculture. Although there is no concrete evidence, it can be suggested that these people lived in hot, humid conditions with a minimum adjustment to their climatic environment but lived in harmony with it through various adaptive responses.

According to Carl Sauer (1952) (cited in Fisher 1964, p. 79) these Mesolithic peoples in Southeast Asia were responsible for developing the 'earliest form of agriculture ever practised by man' under what Fisher calls 'the first fishing-planting culture'. This is attributed to the region's natural advantages of 'high physical and organic diversity . . . reversed monsoons giving abundant rainy and dry periods, of many waters inviting to fishing, or location at the hub of the Old World for communication by water of land' (Sauer 1952, p. 24). According to Fisher (1964), there is convergence of opinion that South Asia developed an ancient agricultural centre in the Neolithic era but it is not clear whether Southeast Asia had parallel developments about the same time.

The Neolithic age in Southeast Asia was introduced by Nesiot[1] people and seems to have had a longer duration compared to that in South or East Asia where in the last centuries B.C. Bronze Age civilizations had reached a very high level. The use of metals had just been introduced to the region about

[1] Nesiots or Proto-Malays who have come to Southeast Asia around 2000 B.C. overland from the north mainly through the Assam/Myanmar border.

this time by Paraoeans.[2] The Nesiot characteristics are very marked among the upland or interior peoples whereas Paraoneans predominate elsewhere, particularly the coastal and riverine lowlands. The Paraoeans had navigational skills which did help the diffusion of the iron and bronze age culture in Southeast Asia.

The last centuries B.C. saw the advance of human migrations from India and China to the south and southeast as a series of southward movements. Besides the major groups mentioned above many minor groups were involved in the currents of southward migrations, the present day representatives of whom are found mainly in the hill communities of the mainland.

By the dawn of the Christian era, Southeast Asia appears in its most characteristic role as the 'meeting ground of cultural influences from India and China' (Fisher 1964, p. 81). Fisher goes on to say that the above statement does not mean that the Southeast Asian people were 'untutored savages incapable of initiative, but . . . in view of the severe restraints imposed by the humid tropical environment, some measure of cultural retardation *vis-à-vis* India and China is sure to be expected, and the surprising thing is not that Southeast Asia progressed more slowly than its neighbours, but rather that a region in such latitudes has attained so high a level of civilization at so early a date' (Fisher 1964, p. 81).

Rice cultivation was practised mostly as a form of shifting cultivation before the great period of Indianization and Sinicization. For instance, when Annam was made part of the Chinese empire in the 2nd century B.C., a system of shifting cultivation was still the main mode of rice cultivation, and the same was true of the Mekong Delta prior to the Indianization during the early centuries A.D. Also, some of the interior parts of the Indo-Chinese peninsular and some of the coastal areas may have been practising irrigated rice cultivation. In the archipelago region too wet rice culture may have been carried out in east-central Java. The terraced rice fields of Luzon are of great antiquity. However, in most other regions wet rice cultivation was a much more recent phenomenon which did not occur until after the 14th century (Fisher 1964). Fisher (1964, p. 82) thinks that the limited spread of wet rice cultivation is also due to its exacting requirements in terms skills, water control technology and the greater human effort required for it, and says that 'in these circumstances both the generally enervating climate and the lethargy induced by recurrent malaria and other diseases may have been potent forces making for conservatism'.

The influence of India and China on Southeast Asian culture is clearer with the beginning of recorded history. In a sense the whole region formed a great frontier zone in which the cultures of both these neighbouring lands met and interpenetrated and fused. As far as the influence of India on Southeast Asian socio-economic fabric is concerned, Fisher says that 'whether or not

---

[2] Mongoloid peoples who came from the north through the main valleys of the peninsular around 1500 B.C.

the introduction of wet rice cultivation there is to be attributed to Indian initiative, its further development within historic times owes much to that source' (Fisher 1964, p. 82). While the Indian influence was on the western two thirds of Southeast Asia along the sea lanes, the influence of China in the main was exerted overland, In the Annamite lands the imprint was 'deeper and enduring'. The primitive hoe culture of the Annamites was superseded by the much more advanced agricultural methods of the elaborate system of water control and dike building.

In the Annamite and Angkor regions large hydraulic works were established in pre-European times. The Khmer civilization of Angkor adopted the elaborate Indo-style irrigation system which relied heavily upon artificial storage lakes and tanks. Most, if not all, of the ancient irrigation installations known in the region – notably in the interior valleys of Burma and Thailand, East-central Java, Bali, parts of Sumatra and Luzon – depended on smaller scale works which do not call for elaborate, centralized political control. In all these regions the seasonality of rainfall enabled the development of irrigation technologies required for wet rice cultivation. The above implies that the establishment of civilization in Southeast Asia has more to do with circumstances borne of external origin rather than within the region and, that, in spite of the enervating tropical environment (Gourou 1974, 1975, 1982; Fisher 1964).

In any case, it is of importance to note that the elaborate hierarchical ordering of society centred on the great capital cities of the former times had already begun to give place to a simpler and more localized way of life in all parts of Southeast Asia, the Annamite lands excepted, before the first traces of European influence was felt anywhere in the region.

8. Climate and human comfort

Generally, intense heat and high humidity bringing about enervating, sultry conditions in Southeast Asia result in very high bioclimatic discomfort. Besancenot (1984) has developed a comparative index of annual discomfort (Figure 6) according to which the equatorial lowland climates have the highest discomfort (24–36 within a scale of 0–36) and the equatorial highland climates the least (0–6). Much of Southeast Asia experiences more than 20, the degree of discomfort decreasing in general with increasing distance from the Equator, but the predominantly zonal pattern is modified by the greater comfort experienced in mountain and hill areas.

At the Equator, extreme daytime discomfort is felt uniformly throughout the year except for some slightly desirable conditions immediately after the two solstices. Equinoctial periods with relatively calm conditions are the 'most trying' (Besancenot 1984). As one gets away from the Equator a seasonality in bioclimatic discomfort/comfort is evident, with the season of relative comfort occurring in the winter of the hemisphere. It is also found that, generally, the rainfall periods happen to be the seasons of strongest discomfort although there are exceptions. Diurnal variations of comfort/discomfort are noteworthy, nights being significantly cooler and afternoons and early evenings being uncomfortable.

Figure 3 shows the bioclimatic regionalization of Southeast Asia into three major types. Obviously within these there will be variations brought about by local conditions, particularly the nature of the immediate environment (rural or urban, open or forested, coastal or interior), altitudinal differences and physiography. The three main types are: (a) Equatorial type: one of 'continuous or almost continuous discomfort' (index of over 34), caused by heat, high humidity and weak ventilation, e.g. Singapore; (b) tropical climate of partial discomfort, due mainly to 'humid heat' and occurs mainly in the summer of the hemisphere; the summer can be more tiresome than in the equatorial climates; pleasant conditions are experienced during the winter; Manila and Jakarta are typical stations; and (c) highland areas having a climate similar to that in warm temperate regions, e.g. Cameron Highlands in Malaysia and such other hill resorts.

Most urban centres in Southeast Asia are regions of extreme bioclimatic discomfort when exposed to the elements (Ilyas et al. 1981; Nieuwolt 1966), necessitating artificial ventilation or air-conditioning, the creation of green areas and parks and the intake by people of large quantities of cooling substances such as water, cool drinks, ice cream, etc. It seems, however, paradoxical that a region of such bioclimatic discomfort should be densely populated. The traditional way of life of the Southeast Asian people, not only in terms of the daily routines in economic life but also in social life, clothing, housing and food habits, etc. show efficient ways of alleviating such discomfort. In more recent times, of course, the process of modernization has helped some selected communities, especially the more affluent, to reduce the bioclimatic discomfort through air-conditioning.

The traditional farmers and manual workers in Southeast Asia normally avoid the mid-day sun if they can help it and work during the relatively cooler mornings and evenings, resting during the peak hot hours. Moreover, the traditional housing arrangements in most rural societies in Southeast Asia are long on measures to ameliorate hot humid conditions and seasonal downpours. The use of such natural thatching material as dry palm fronds, straw or tall grass and structural media as wood are designed to shelter from the intense tropical heat and downpours as well as minimize heat absorption and maximize free ventilation and cooling.

The traditional rural settlements are a lesson in how to achieve bioclimatic comfort under steamy, enervating conditions, whereas what are seen in most urban agglomerations and so-called 'modern' or 'futuristic' buildings in Southeast Asia are indicative of how not to build for the environment. Of course, the advances of air-conditioning technology affordable by the relatively more affluent among the Southeast Asians is sometimes cited as a triumph of humans over nature. Air-conditioned houses, cars, offices, shopping complexes, are commonplace in most Southeast Asian urban areas. However, still, to the vast majority of the population, such facilities are very much luxuries beyond their reach, especially in low-income, high-cost-energy situations.

Architects and town planners in Southeast Asia, perhaps in common with their counterparts in other tropical areas, seem to ignore climatic conditions altogether or give only marginal attention to them in the face of economic and political realities, although this is not necessarily an inevitable approach (David and Lim 1990). The recent haze episodes in Southeast Asia arising out of the extensive forest fires in Kalimantan and Sumatra have been particularly serious in their effects on high density, urban built-up areas, for instance in the Kelang Valley region of Malaysia, due most probably to the supplementary sources of pollution arising out of unscientific planning of urban and industrial development.

A number of 'hill stations' have been developed in the highland and mountain areas of Southeast Asia mostly during the colonial times as refuges for the Europeans from the sultry conditions in the humid tropical lowlands (Spencer and Thomas 1948; Sirinanda 1983). Although these were originally used exclusively by the Europeans, since the middle of this century they have been patronized by everybody who would like to have a respite from hot, steamy conditions and can afford the expenses of such a 'get-away', so much so that they have become fast developing holiday resorts. Over thirty such stations exist, many of them in Java, Sumatra, Peninsular Malaysia, Thailand and in the Borneo island. Most of them are within the easy access of dense population centres, and the better known and popular ones are strung at elevations ranging from about 500 to 1500 m. Besides being 'summer resorts' they are also well known for their low temperature agricultural produce, especially tea and temperate vegetables and flowers. During the earlier stages, these resorts had mostly cottage or bungalow type holiday accommodation but now many international class as well as less expensive hotels abound. They tend to be particularly crowded during school holidays.

9. Climatic modification

From time immemorial humans in Southeast Asia have, to some extent, made deliberate attempts towards modifying the weather and climate, especially to hedge against hazardous water availability conditions due to rainfall variability. Religious ceremonies, rites and rituals to seek protection from inclement weather and to propitiate the 'rain-gods' have been common among the tribal as well as the rural agricultural communities; this is part and parcel of the traditional culture of Southeast Asia irrespective of the ethnic or religious backgrounds of the people. In later years, of course, efforts were made to apply more modern 'rain-making' techniques in many Southeast Asian countries both to enhance rainfall as well as control its distribution, but not much credence is now placed on the claims by the latter day rainmakers, and the programmes of rain-making which are still being continued are very much low key affairs compared to the unbounded enthusiasm of the late 1970s (Sirinanda 1980).

Modification of the micro-climate for agriculture through measures to control the heat and water balances is something which has been done with increasing effectiveness in Southeast Asia. Traditional farmers adopted numerous time-tested techniques in soil management, crop combinations, irrigation and drainage and general land use practices to minimize the adverse effects of excesses of heat and moisture or seasonal moisture deficits, the impact of high rainfall intensities or strong winds. Many agrometeorological research centres spread throughout Southeast Asia have addressed the problems of making optimum use of the agroecological resources of Southeast Asia, (e.g. IRRI 1974; Nieuwolt 1981a, 1982; Oldeman 1982, 1984a and b) particularly the abundant supply of sunshine and high but variable rainfall conditions.

The earliest agricultural settlements in Southeast Asia were centred on regions with a seasonality of rainfall favouring rain-fed shifting cultivation or sedentary wet-rice cultivation. These settlements outgrew their capacity to support increasing population numbers necessitating, over the centuries, the development of water control strategies both for protective as well as productive purposes. The Southeast Asian landscape is littered with agricultural water resource development schemes of different scales, whether it be terraced rice fields of Java, or the vast rice plantations of Burma and Thailand, the modern irrigation schemes such as MADA or MUDA in Malaysia, or the wet rice areas of the Mekong or Red River basins. Direct diversions from rivers or streams, impounding in reservoirs, groundwater tapping and rain-water harvesting are done to different degrees. Water resource development strategies have favoured the promotion of multi-purpose schemes to meet the increasing demand for agricultural requirements as well as the non-agricultural needs such as potable water supplies, industrial water requirements and recreational and environmental considerations.

Another aspect of climate alteration due to human activities is accidental modification arising out of the varying degree to which the land surface has been changed, ranging from the conversion of natural surfaces to agricultural areas to the complete transformations into urban centres. Much of the local climatic modification due to such changes cannot be easily ascertained or have not been scientifically studied, the few attempts at studying the urban centres of Singapore (Nieuwolt 1966) and Kuala Lumpur and smaller urban centres in Malaysia (Sham 1979, 1980, 1985 and 1988) notwithstanding. The presence of a large number of high density urban clusters with intensively built-up environments have created urban 'heat-islands' and air pollution domes in all the city and industrial areas in Southeast Asia leading to many changes in the local climates of the areas concerned, although quantitative estimates of the extent of the changes are not available. A significant development lately towards preventing large scale inadvertent modification of climate leading to adverse impact is the joint effort by ASEAN countries to promote greater understanding and awareness of the importance of including the climatic environment as an input in development related decision making (ASEAN 1982a and b, 1991).

10. Deforestation and climatic alteration

As has been pointed out in numerous discussions by environmental scientists (Hill and Bray 1978; MacAndrews and Chia 1979; Myers 1990; Sharp and Sharp 1982), the destruction of the tropical rainforests due to various factors (Table 3) such as large-scale land development schemes in Malaysia, transmigration projects in Indonesia, or logging which is found extensively throughout the region, human settlement development, industrial and transport development, slash and burn agriculture and fuel-wood extraction, are bound to lead to land degradation which in turn could bring about climatic alteration (Sirinanda 1984).

Where the conversion of forest is governed by the dictates of industrialization and urbanization, the alteration of the climatic characteristics is less in doubt or are better documented. The phenomenon of urban heat islands is a case in point (Nieuwolt 1966; Sham 1980). Climatic modification resulting from the conversion of forest into agricultural areas is less known. Nevertheless, numerous reports of increasing frequency of floods and droughts and their intensification in normally relatively dry as well as humid parts of Southeast Asia have raised the problem of the culpability of human activities in changing the land surfaces and thereby triggering climatic alteration. It is also known that certain areas even deep in the tropical jungle may, due to the above activities, turn into 'red deserts' or into waste-lands of degraded soil and/or of alang-alang grass.

Most people agree that the clearance of tropical forests on a massive scale can have far reaching consequences on the global as well as regional and local climates (Potter et al. 1975). One result of slash-and-burn agriculture, which is rampant in Southeast Asia, and the open burning of cleared vegetation and rubble during the dry seasons in the areas of settled agriculture is the addition of aerosols and carbon in large quantities (Table 4). On several occasions, for example in 1982/1983 (Anon 1984) and in early 1992, another dimension has been added due to the extensive fires over Kalimantan involving nearly 3.7 million hectares in 1982/1983 and Kalimantan and Sumatra, about 150,000 hectares 1992. The resulting clouds of dust and smoke have created extensive haze over Southeast Asia especially during relatively stable atmospheric conditions in the intermonsoonal months of March, April and October causing further exacerbation of dry, sultry conditions.

The impact of forest clearance on climate is particularly felt through the alteration of moisture regimes of the areas concerned by mainly affecting the evapotranspiration, runoff, and storage terms of the water balance equation. Also, although concrete evidence is lacking, it has been suggested that certain critical elements of the rainfall process may also be altered. Claims that vegetation changes alter the rainfall patterns are given credence by frequent reports of apparent changes in the amounts, incidence and frequency of rainfall, shifts in the rainy seasons, longer dry seasons, longer dry weather flow in streams and greater flood peaks. However, the effects of vegetation on rainfall remains a subject full of conjecture, conflicting claims and speculation.

11. Global warming, climatic change and social reaction

With the present lively international debate on the impact of the greenhouse effect, global warming and the various climate change scenarios along with the possible rise in sea level, the scientific community, policy makers and planners in the region have taken some of the warnings seriously especially in terms of the anthropogenic influence on the climate-society interaction. The causes and consequences of possible greenhouse warming, particularly the likely impact of rainforest destruction, the burning of fossil fuels and the depletion of wetlands on the carbon cycle in terms of global and regional climatic change are receiving attention. The role of tropical rainforests as carbon sinks or sequestering agents, burning of tropical forests as sources of atmospheric carbon, and the wet-lands as stores of carbon are the focus of study by several groups under the patronage of the respective government and international agencies in the region.

Sea-level change scenarios which envisage a meter rise in sea-level in the next century have not received the attention it should in many of the countries in the region, although the warnings are taken notice of. Southeast Asia's geography makes it an area most likely to suffer extensive adverse effects should such a rise in sea-level eventuate. Many of the coastal communities and those living along low-lying river valleys both in rural and urban settlements are the most vulnerable.

Research on the above aspects is now actively being undertaken in most of the countries of the region by local as well as outside research groups under the auspices of numerous funding bodies both national and international. Since the preparation of the IPCC Report and the Ministerial Declaration at the Second World Climate Conference in 1990 and the Earth Summit in 1992 and the signing of the FCCC an intensification of work on climatic change and possible impact on society in Southeast Asia can clearly be seen. During the last few years many meetings of scientific groups have been held with the blessings of the governments of the region. Some of the notable ones are: ASEAN Workshop on Scientific Policy and Legal Aspects of Global Climatic Change, Bangkok, September, 1990; Asia-Pacific Seminar on Climatic Change, Nagoya, January, 1991; Conference on Climatic Impact on Environment and Society, Tsukuba, January, 1991; and Workshop on Carbon Cycle and Global Climatic Change, Kuala Lumpur, October, 1991. However, still there is very little common agreement as to the causes, the magnitudes and the impacts of climatic change and of the spatial patterns of such impacts in Southeast Asia.

Some possible ways in which the Southeast Asian region may be affected by global warming have been summarised following Jauregui (1991), Maunder (1991), Parry (1991) and Tabucanon (1991). Along with the other tropical regions, Southeast Asia is likely to experience 'relatively' small changes in temperature resulting from greenhouse warming. However, even the increases in temperature envisaged are likely to affect crop calendars in countries like Thailand and decrease soil moisture conditions in Southeast Asia in general in December, January and February. An increase of rainfall of between 5–20% may also affect agriculture and its calenders and the exacerbation of the heat island effect in large urban areas may result in increased morbidity and loss of productivity despite the acclimatisation to heat by the inhabitants of the region. Demand for air-conditioning will increase as buildings move away from 'traditional passive means of cooling' and this in turn, will make the air pollution and carbon dioxide emission problems worse. However, even though present General Circulation Models (GCMs) are in poor agreement concerning regional and local changes expected from enhanced greenhouse warming, there is some consensus that in many areas the impact of climate change will be 'determined more by the magnitude and frequency of extreme events, or of events exceeding given critical thresholds than by changes in average conditions'. The sum total of the above is that we are still a long way from a reasonable understanding of what directions the climate of Southeast Asia will take in future and how changes of any form will impact its physical environment and society.

Almost all the countries in Southeast Asia via both government and non-governmental agencies have recently participated in the World Environment Day with the theme 'Climate change – the need for global partnership'. A wide range of activities was conducted with a view to promoting particularly, greater awareness in all strata of society about climate change and its possible consequences and to seek their co-operation in implementing various preventive and protective measures to hedge against any undesirable impacts. People's attitudes, perceptions and habits are at the root of the environmental problems which now threaten humanity and the success of measures taken to overcome them depend on a change in human conduct. In Southeast Asia, due perhaps to the relative abundance of the key resources of the atmospheric environment most people take weather and climate for granted, but the warnings now being given of future climatic changes and consequent ecological impact have to a certain extent shaken some knowledgeable people out of their equanimity. However, much more effort is required to create greater societal sensitivity and sensibility in the area of present as well as potential climate-society interaction issues.

References

Anon (1984): Fire in the earth: A quiet catastrophe changes the world. *Asiaweek*, July 13, 1974, 32–55.

ASEAN Sub-Committee on Climatology (1982a): The ASEAN economic atlas. ASEAN Secretariat, Jakarta.

ASEAN Sub-Committee on Climatology (1982b): The ASEAN compendium on climate statistics. ASEAN Secretariat, Jakarta.

ASEAN Sub-Committee on Climatology (1991): User's manual for the ASEAN climatic atlas and compendium of statistics. ASEAN Secretariat, Jakarta.

Asian Development Bank (1990): Asian development outlook 1990, Manila, Asian Development Bank.

Besancenot, J.P. (1984): Bioclimatological background and agricultural human settlements in Monsoon Asia. In: *Climate and agricultural land use in Monsoon Asia*, ed. by M.M. Yoshino. Tokyo University Press, Tokyo, 109–150.

Biswas, A.K. (1979): Climate, agriculture and economic development. *Ecologist*, 6, 188–195.

Biswas, A.K. (ed.) (1984): *Climate and development*. Tycooly International Publishing, Dublin.

Biswas, M.R. and Biswas, A.K. (eds.) (1979): *Food, climate and man*. John Wiley, New York.

Brooks, C.E.P. (1928): *Climate through the ages*. Yale University Press, New Haven.

Brown, L.R. (1970): *Seeds of change: The green revolution and development in the 1970s*. Methuen, London.

Bryson, R.A. and Murray, T.J. (1977): *Climates of hunger*. Wisconsin University Press, Madison.

Cady, J.F. (1964): *Southeast Asia: Its historical developments*. McGraw Hill, New York.

Carter, G.F. (1968): *Man and the land: A cultural geography*. Holt, Rinehart and Winston, New York.

Chang, Jen-hu (1968): The agricultural potential of the humid tropics. *Geographical Review*, 58, 347–373.

Chen, R.S., Boulding, E. and Schneider, S.H. (eds.) (1983): *Social sciences research and climatic change*. D. Reidel, Dordrecht.

Cho, L.J. and Bauer, J. (1987): Population growth and urbanization. In: *Urbanization and urban policies in Pacific Asia*, ed. by R.J. Fuchs, G.W. Jones and E.M. Perina Westview Press, Boulder, 15–37.

David, A. and Lim, J.T. (1990): Climatic data applications in the ASEAN region. In: *Using meteorological information and products*, ed. by A. Price-Budgen. Ellis Horwood, London, 325–341.

Dobby, E.H.G. (1950): *Southeast Asia*. University of London Press, London

Dobby, E.H.G. (1953): Rice in Southeast Asia. *Malayan Journal of Tropical Geography*, 1, 57–58.

Fisher, C.A. (1964): *Southeast Asia: A social, economic and political geography*. Methuen, London.

Fisher, C.A. (1969): Southeast Asia: The Balkans of the Orient? A study in continuity and change. In: *Man, state and society in contemporary Southeast Asia*, ed. by R.P. Tilman. Pall Mall Press, London.

Fleure, H.J. (1919): Human regions. *Scottish Geographical Magazine*, 35, 94–105.
Garbell, M.A. (1947): *Tropical and equatorial meteorology*. Pitman: London.
Gourou, P. (1960): *The tropical world* (translated by E.D. Laborde). Longmans, London.
Gourou, P. (1974): Man and the monsoon in Southern Asia. In: *UNESCO natural resources of humid tropical Asia*. UNESCO, Paris, 449–456.
Gourou, P. (1975): *Man and land in the Far East* (translated by S.H. Beaver). Longmans, London.
Gourou, P. (1982): *Terres de bonne espérance. Le monde tropicale*. Pion, Paris.
Hall, D.G.E. (1964): *A history of Southeast Asia*.
Harrison, B. (1986): *Southeast Asia: A short history*. Methuen, London.
Hill, R.D. and Bray, J.M. (eds.) (1978): *Geography and the environment in Southeast Asia*. University of Hong Kong Press, Hong Kong.
Huntington, E. (1915): *Civilization and climate*. Yale University Press, New Haven.
Huntington, E. (1924): *The principles of human geography*. John Wiley, New York.
Huntington, E. (1926): *The pulse of progress*. Charles Scribners & Sons, New York.
Huntington, E. (1927): *The human habitat*. Chapman & Hall, London.
Huntington, E. (1945): *Mainsprings of civilization*. John Wiley, New York.
Huntington, E. and Cushing, S.W. (1934): *Principles of human geography*. John Wiley, New York.
Huntington, E. and Visher, S.S. (1922): *Climatic changes: Their nature and causes*. Yale University Press, New Haven.
Huntington, E., William, F.E. and Valkenburg, S. (1933): *Economic and social geography*. John Wiley, New York.
Ichimura, S. (1988): The pattern and prospects of Asian economic development. In: *Challenge of Asian developing countries: Issues and analyses*, ed. by S. Ichimura. Asian Productivity Association, Tokyo, 7–64.
Ilyas, M., Pand, C.Y. and Chan, A.M. (1981): Effective comfort indices for some Malaysian towns. *Singapore Journal of Tropical Geography*, 2, 27–31.
IRRI (1974): *An agro-climatic classification for evaluating croping systems potentials in Southeast Asia*. IRRI, Los Banos.
Jackson, I.J. (1977): *Climate, water and agriculture in the tropics*. Longmans, London.
Jauregui, E. (1991): Aspects of monitoring local/regional climate change in a tropical region. Paper presented at the Conference of Climatic Impact on Environment and Society, Tsukuba, January, 1991.
Kamarck, A.M. (1973): Climate and economic development. *Finance and Development*, 10, 2–8.
Kates, R.W. (1985): Interaction of climate and society. In: *Climate impact assessment: Studies of the interaction of climate and society*, ed. by R.W. Kates et al. John Wiley, Chichester, 4–36.
Kates, R.W., Ausubel, J.H. and Berberian, M. (eds.) (1985): *Climate impact assessment: Studies of the interaction of climate and society*. John Wiley, Chichester.
Kayane, I. (1971): Hydrological regions in Monsoon Asia. In: *Water balance of Monsoon Asia*, ed. by M.M. Yoshino. University of Tokyo Press, Tokyo, 287–300.
Kellogg, W.W. and Schware, R. (1982): *Climate change and society. Consequences of increasing carbon dioxide*. Westview Press, Boulder.
Koteswaram, P. (1974): Climate and meteorology of humid tropical Asia. In: *UNESCO, Natural resources of tropical humid Asia*. UNESCO, Paris, 27–84.
Lamb, H.H. (1982): *Climate, history and the modern world*. Methuen, London.
Lamb, H.H. (1988): *Weather, climate and human affairs*. Routledge, London.
Lee, D.H.K. (1957): *Climate and economic development in the Tropics*. Harper Brothers, New York.
Leggett, J. (ed.) (1990): *Global warming. The Greenpeace Report*. O.U.P., New York.

Low, K.S. and Balamurugan, G. (1991): Urbanization and urban water problems in Southeast Asia: A case of sustainable development. *Journal of Environmental Management*, 32, 195–209.

MacAndrews, C. and Chia, L.S. (eds.) (1979): *Developing economies and the environment. The Southeast Asian experience*. McGraw Hill, Singapore.

Maunder, W.J. (1970): *The value of the weather*. Methuen, London.

Maunder, W.J. (1974): National economic models. In: *Climatic resources and economic activity: A symposium*, ed. by J.A. Taylor. David Charles, Newton Abbot, London, 237–257.

Maunder, W.J. (1986): *The uncertainty business: Risks and opportunities in weather and climate*. Methuen, London.

Maunder, W.J. (1989): *The human impact of climatic uncertainty*. Routledge, London.

Maunder, W.J. (1991): Regional and national responses to climate change: Implications, risks and opportunities and what each nation needs to do. Paper presented at the Conference on Climatic Impact on Environment and Society, Tsukuba, January, 1991.

Monteiro, R. (1962): Climate and man along the east coast of Malaya. Unpublished M.A. thesis, University of Malaya, Singapore.

Myers, N. (1990): Tropical forests. In: *Global warming. The Greenpeace Report*, ed. by J. Leggett. O.U.P., Oxford.

Nestmann, L. (1974): Human development in its relation to ecological conditions. *Geoforum*, 18, 7–17.

Nieuwolt, S. (1966): The urban microclimate of Singapore. *Journal of Tropical Geography*, 22, 30–37.

Nieuwolt, S. (1977): *Tropical climatology: An introduction to the climates of the low latitudes*. John Wiley, London.

Nieuwolt, S. (1981a): Agricultural drought in Peninsular Malaysia. Serdang, Malaysian Agricultural Research and Development Institute.

Nieuwolt, S. (1981b): Climates of continental Southeast Asia. In: *World survey of climatology: Southern and Western Asia*. Elsevier, Amsterdam, 1–66.

Nieuwolt, S. (1982): Climate and agricultural land use in Peninsular Malaysia. Special Report, Serdang: Malaysian Agricultural Research and Development Institute.

Oldeman, L.R. (1984a): Climate and agricultural land use in Indonesia. In: *Climate and agricultural land use in Monsoon Asia*, ed. by M.M. Yoshino. University of Tokyo Press, Tokyo, 275–296.

Oldeman, L.R. (1984b): Climate and agricultural land use in the Philippines. In: *Climate and agricultural land use in Monsoon Asia*, ed. by M.M. Yoshino. University of Tokyo Press, Tokyo, 333–352.

Oldeman, L.R. and Frere, M. (1982): A study of the agroclimatology of the humid tropics of Southeast Asia. Technical Report, FAO/ UNESCO/ WMO Interagency Project on Agroclimatology, Rome.

Ooi, J.B. (1959): Economic development in rural areas with special reference to Malaya. Special Issue, *Malayan Journal of Tropical Geography*, 12, 1–222.

Osborne, M. (1979): *Southeast Asia: An introductory history*. George Allen and Unwin, Sydney.

Pagney, P. (1984): Climatic problems of agricultural land use in Monsoon Asia, tropical Africa, and Latin America: A comparative analysis. In: *Climate and agricultural land use in Monsoon Asia*, ed. by M.M. Yoshino. University of Tokyo Press, Tokyo, 355–385.

Parry, M.L. (1985): Impact of the climatic variations on agricultural margins. In: *Climate impact assessment: Studies of the interaction of climate and society*, ed. by R.W. Kates. John Wiley, Chichester, 351–367.

Parry, M.L. (1991): Potential impact of climate on agriculture. Paper presented at the Conference on Climatic Impact on Environment and Society, Tsukuba, January, 1991.

Pelzer, K.J. (1968): Man's role in changing the landscape in Southeast Asia. *The Journal of Asian Studies*, 27, 259–270.
Pittock, A.B., Frakes, L.A., Jenssen, D., Peterson, J.A. and Zillman, J.W. (eds.) (1978): *Climatic change and variability: A southern perspective*. Cambridge University Press, London.
Potter, G.L., Elsesser, H.W., McCracken, M.C. and Luther, F.M. (1975): Possible climatic impact of tropical deforestation. *Nature*, 258, 697–698.
Rigg, J.D. (1991): *Southeast Asia, a region in transition: A thematic human geography of the Asian Region*. Unwin Hyman, London.
Sauer, C.O. (1952): *Agricultural origins and dispersals*. New York.
Sham, S. (1979): *Aspects of air pollution climatology in a tropical city: A case of Kuala Lumpur – Petaling Java, Malaysia*. National University of Malaysia Press, Bangi.
Sham, S. (1985): Air Pollution and air quality management in Malaysia. *Environmental Professional*, 7, 168–177.
Sham, S. (1988): Inadvertent climatic change through urbanization: A challenge to planners. In: *Environmental monitoring and assessment. Tropical applications*, ed. by S. Sham and A. Badri. National University of Malaysia Press, Bangi.
Sharp, D.M. and Sharp, T. (1982): The desertification in Asia. *Asia 2000*, 1, 40–42.
Sirinanda, K.U. (1976): Variability of water supply and paddy production: A case study of Sri Lanka with application to Asia at large. *Ilmu Alam*, 5, 31–47.
Sirinanda, K.U. (1980): Precipitation management efforts in some South and Southeast Asian countries. *Malaysian Geographers*, 2, 21–36.
Sirinanda, K.U. (1983): Climate land use and human settlement in the hill stations in Peninsular Malaysia. Paper presented at the Meeting of the IGU Working Group on Tropical Climatology and Human Settlements, July 25–August 2, Honolulu, Hawaii.
Sirinanda, K.U. (1984c): Land degradation and its impact on climatic alteration in Monsoon Asia. In: *Climate and agricultural land use in Monsoon Asia*, ed. by M.M. Yoshino. University of Tokyo Press, Tokyo, 151–179.
Spencer, J.E. (1954): *Asia east by south: A cultural geography*. John Wiley, New York.
Spencer, J.E. (1966): *Shifting cultivation in Southeast Asia*. University of California Press, Berkeley.
Spencer, J.E. and Thomas, W.L. (1948): The hill stations and summer resorts of the Orient. *Geographical Review*, 38, 637–651.
Tabucanon, M.S. (1991): Global environmental congress. Paper presented at the Conference on Climatic Impact of Environment and Society, Tsukuba, January, 1991.
Tarling, N. (1966): *A concise history of Southeast Asia*. Praeger, New York.
Tate, D.J.M. (1971): *The making of modern Southeast Asia*, Vol. 1. *European Conquest*. O.U.P., Kuala Lumpur.
Taylor, G. (1957): *Geography in the twentieth century*. Methuen, London.
Tilman, R.O. (ed.) (1969): *Man, state and society in contemporary Southeast Asia*. Pall Mall Press, London.
Watts, I.E.M. (1955): *Equatorial weather – with special reference to Southeast Asia*. University of London Press, London.
Wigley, T.M.L., Huckstep, N.J., Ogilvie, A.E.G., Farmer, G., Mortimer, R. and Ingram, M.J. (1985): Historical climate impact assessments. In: *Climate impact assessment: Studies of interaction of climate and society*, ed. by R.W. Kates et al. John Wiley, Chichester, 526–563.
Wittfogel, K.A. (1957): *Oriental despotism*. Yale University Press, New Haven.
World Bank (1987): World development report. O.U.P., New York.
World Bank (1989): World development report. O.U.P., New York.
World Bank (1990): 1989 World Bank Atlas: Gross national product, population and growth rates. Washington, D.C. World.

Yoshino, M.M. (ed.) (1971): *Water balance of Monsoon Asia*. University of Tokyo Press, Tokyo.

Yoshino, M.M. (1984a): *Climate and agricultural land use in Monsoon Asia*. University of Tokyo Press, Tokyo.

Yoshino, M.M. (1984b): Water balance of Monsoon Asia. In: *Climate and agricultural land use in Monsoon Asia*, ed. by M.M. Yoshino. University of Tokyo Press, Tokyo, 57–77.

Yoshino, M.M. (1984c): Ecoclimatic systems and agricultural land use in Monsoon Asia. In: *Climate and agricultural land use in Monsoon Asia*, ed. by M.M. Yoshino. University of Tokyo Press, Tokyo, 81–108.

K.U. Sirinanda
Department of Geography
University of Brunei Darussalam
Gadon 3186
Brunei Darussalam

3.4. CLIMATIC AND PATHOLOGICAL RHYTHMS IN A HUMID TROPICAL AREA, THE CASE OF THE PHILIPPINES

JOCELYNE PÉRARD AND JEAN-PIERRE BESANCENOT

1. Introduction

The fact that weather significantly affects yearly, monthly and daily morbidity in human populations has been documented for a long time by a number of individuals. They have illustrated very well some of the basic structure and underlying mechanisms of seasonality in morbidity. But until now almost all these efforts were carried out in countries with temperate mid-latitude climates. Furthermore, meteoropathological investigations have for the most part concentrated upon mortality statistics for the usual reasons of data availability and comparability. However, an important question that arises when considering the effects of weather on diseases is how the relationship varies under different climatic regimes. On the other hand, the highest number of deaths for a given disease does not always coincide with the highest number of cases. Thus, the present investigation asks whether or not a warm and humid low latitude climate does in fact influence health. Notifiable disease cases in the Philippines as a whole have been analyzed over a six-year period ending in 1986, and especially for the years 1985 and 1986, to detect and characterize some rhythms of risk in various illnesses. Insofar as a seasonal pattern is clearly established, and as weather seems a main determinant of morbidity, such documentation could lead to a better understanding of disease patterns in the population and could identify more effective public health intervention strategies.

M. Yoshino et al. (eds.), Climates and Societies – A Climatological Perspective, 235–254.

2. Materials and methods used

Notified cases and deaths from various communicable and non-communicable diseases were kindly supplied on a weekly basis, Sunday to Saturday, by the Health Intelligence Service, Ministry of Health, Manila. The morbidity reported in the *Weekly Health Intelligence Bulletin* covers the main government medical centres and all public hospitals, but private medical establishments are not included. The available data chiefly concern the lower socio-economic strata, since the upper-classes do not much use the government facilities covered by the study, temporal variation of morbidity is not greatly compromised. Twenty-four nosologic entities are regularly distinguished in weekly reports; however in several cases, as chickenpox, filariasis, leprosy or syphilis, it was found that the numbers of reported cases were so small as to make interpretation difficult. Therefore, the analysis focused on respiratory and enteric infections, measles and heart failures. Some other diseases were however briefly discussed. In the same way, only data on reported "cases" were considered since the number of reported "deaths" was often too small for satisfactory comparison. The relationship between weather and weekly mortality in the Philippines remains a subject for future biometeorological research.

In other respects, morbidity data are only presented for all 67 provinces and 61 chartered cities combined, i.e. on a national level. Thus it was impossible to detect whether or not the degree of effect of a climatological or meteorological stimulus may vary markedly according to the geographical location of individuals. Thereupon, as Manila is at or near the weighted centre of population of Philippines and itself contains the highest proportion of the total population of the archipelago, Manila International Airport was chosen as the main, but not exclusively, the reference climate station. Meteorological data were obtained from published weather records (*Annual Tropical Cyclone Report, Daily and Monthly Summaries of Meteorological Observation*, National Weather Office, PAGASA, Manila).

The availability of suitable detailed statistics prompted an examination of weekly variations in the morbidity rates over two successive years (1985 and 1986). Only 0.17% of the values were missing and these were filled in by linear interpolation. Besides, the mean archival data for years 1981–1986 were used as reference and in some cases use was also made of a longer period (1965–1986). Yet at that time the data set was less trustworthy and less detailed, covering only a few communicable diseases.

Various statistical techniques have been proposed for the detection of cyclical trends, and for the evaluation of biometeorological correlations. But the clinical material collected for the present study was not very extensive as a rule and the methods were not sufficiently standardized to allow of a clear-cut statistical analysis. In addition, seasonal changes in the incidence of diseases as well as weather-disease relationships were often obvious on

Table 1. Maximum and minimum temperature (°C), relative humidity (%), rainfall normals (mm) for selected Philippines stations.

Station		J	F	M	A	M	J	J	A	S	O	N	D
Manila	T. Max	30.3	31.3	32.9	34.3	34.3	32.6	31.4	30.8	31.0	31.2	30.8	30.2
Airport	T. Min	20.7	20.8	22.0	23.7	24.7	24.4	24.0	24.0	23.8	23.3	22.4	21.3
	RH	75	70	67	65	69	78	82	83	84	82	80	79
	R	13	4	16	16	124	243	343	438	323	177	127	59
Legaspi	T. Max	28.7	29.1	29.9	31.1	32.2	32.5	31.9	31.7	31.7	31.3	30.3	29.2
	T. Min	22.3	22.3	22.8	23.6	24.2	24.0	23.7	23.6	23.5	23.1	23.1	22.8
	RH	84	82	82	82	82	82	84	85	85	85	85	85
	R	302	176	208	173	162	205	230	283	247	307	478	446
Masbate	T. Max	29.5	30.0	31.3	32.6	33.4	33.1	32.4	32.1	32.0	31.7	30.9	29.9
	T. Min	23.4	23.1	23.7	24.8	25.4	25.4	25.1	25.1	25.0	24.8	24.7	23.9
	RH	83	82	81	79	78	79	82	83	83	83	84	85
	R	171	75	64	43	106	141	179	205	181	225	239	228
Surigao	T. Max	28.6	28.7	29.8	30.8	31.7	32.0	31.4	31.8	31.8	31.2	30.1	29.2
	T. Min	22.5	22.4	22.6	23.1	23.7	23.6	23.5	23.9	23.9	23.4	23.1	22.9
	RH	84	82	82	82	82	82	84	85	85	85	85	85
	R	606	479	369	247	188	134	178	156	171	268	411	607

Manila Airport: 14°31′N, 121°01′E
Legaspi: 13°08′N, 123°44′E
Masbate: 12°22′N, 123°31′E
Surigao: 9°48′N, 125°30′E

simple visual inspection of disease calendars. Accordingly, only morbidity curves have been built, one for real years, one for a period average. In this latter case, data were systematically smoothed with 5-week running means.

3. The climatic and socio-economic context of the Philippines

3.1. *The climatic environment*

In the Western Pacific, east Asia, the Philippines is made up to 7107 islands, located between almost 4 and 21°N and between 116 and 127°E. The country forms a true puzzle of often mountainous lands surrounded by permanently hot water on all sides (Pérard 1984). As a consequence, a tropical humid climate prevails throughout the year over the archipelago (Table 1). Mean temperature fluctuates around 27°C, with a weak seasonal range because of the flow of latent heat. Relative humidity is almost always high, mean values of Manila, for instance, being at or below 70% only during February, March, April and May. Annual mean rainfall is about 2000–3000 mm in the western parts of the archipelago and 3000–4000 mm along the eastern coasts. Fortunately this hot and humid climate, that includes many excessive features and demands continuous regulation efforts by the human body (Besancenot 1984, 1986b), is moderated, at least at a local scale, by many breezes due to the relief or proximity to the coast.

Furthermore, the Philippines is located in the world region most frequently visited by tropical perturbations. On average, about twenty cyclones per year are reported, about 80% of which reach the tropical storm or typhoon stage. Nearly every second cyclone crosses the islands. Usually, the main cyclonic season lasts from July to November, but some cyclones may occur at any time of the year with strong winds and accompanying destructive storm surges (Pérard 1988). Around 15% of the tropical disturbances cause extensive losses to life, property and crops. Traffic is disrupted by torrential rains, the plains of Luzon and the lower districts of Manila are flooded for several days at a time; some flood events last several weeks following the passage of a typhoon. In the end though, the cyclones do also have beneficial aspects. By bringing needed rain to drought-affected agricultural areas, they represent a major component of daily life in the Philippines (Vèlimirovic and Subramanian 1972).

If, on the whole, the climate of the Philippines shows a very simple pattern with only a small but monotonous fluctuation around a hot and humid theme, there are some marked regional variations. A year-round rainy climate prevails in the eastern parts of the archipelago under the permanent influence of eastern circulations (cf. Table 1; Legaspi, Masbate, Surigao), and in Mindanao which is nearly free of cyclonic disasters. The western façade and most of Luzon experience a tropical climatic regime, with two pronounced seasons, one wet during summer, the other less humid, even dry from November to April. The length of the dry season gradually increases from south to north, as well as from coastal areas to the interior (cf. Table 1; Manila).

Broadly speaking, the year is arranged as follows:

(a) From the *winter* solstice to March, during clear and dry nights with strong northeastern winds, the plains and ranges of Luzon and occasionally the southern islands experience cold surges; the recorded temperatures are largely less than 20°C over that period (for instance, 11°C at Laoag and 7°C at Baguio in January 1971, both stations located in Luzon, the latter at an elevation of 1500 m). These cold surges which give a true feeling of cold to under protected and often vulnerable populations, can be expected to increase suddenly, the incidence of some meteorosensible diseases (Besancenot 1986a).

(b) The *spring* begins another painful season, because of the very high temperatures associated with a relatively dry air, that favours some other diseases, namely measles. During this season, at least in Luzon, May is well

Table 2. Main mortality causes in the Philippines per 100,000 inhabitants.

Pneumonia	89.3
Heart diseases	61.0
Tuberculosis	52.9
Circulatory diseases	39.6
Cancer	30.2
Diarrheas	27.8
Accidents	16.8
Malnutrition	13.4

known as being the most uncomfortable month, coming just before the rainfall increase. At that time, the population is confronted not only with great food and energy deficiency (this is the time of long brownouts), but also great discomfort caused by a combination of high temperatures and rapidly rising humidities. People are adversely affected by the sultriness of an oppressive weather condition.

(c) Generally in early June, the monsoon which was previously confined to the southern islands, reaches the whole of Luzon. Normally, the passage of a violent disturbance over the archipelago establishes the southwesterly air current at the beginning of the season. The airstream is unstable and is characterized by frequent convective activity. The *summer* is marked with: (i) extremely heavy rainfall following the passage of a cyclone; this also intensifies the southwest flow; (ii) a very high hygrometry (in some years the atmosphere remains persistently saturated for several weeks at time); and, (iii) falling but very sultry and unpleasant temperature.

(d) At the end of the summer the monsoon retreats southwards. *Autumn* rainfall is noticeably lower, but once more rhythmed on tropical cyclones which are still very numerous in the area.

3.2. *The socio-economic context*

The impact of climate on health cannot be studied outside the social, economic and political context. It is known that the Philippines has lately experienced a series of shocks with unavoidable consequences to life. As many economic or demographic indices certify, the archipelago is a developing country: agriculture provides 70% of both employment and income; the percentage of the total population living in rural areas is as high as 59; the rate of natural growth is 25.6 per thousand; the population is very young with life expectancies for males and females under 62 years of age. Up to the 1970s the Philippines maintained a moderate economic growth. But in the early eighties the trend was reversed and the economic crisis grew worse day after day. Economic activity decreased (by as much as 12% between 1984 and 1987), inflation soared (37% during the only year 1985), employment remained scarce (20% of the active population being unemployed, and perhaps 50% underemployed), real wages dropped nearly 50%, external debt climbed to unprecedented heights, rural to urban migration grew rapidly (mainly towards the Greater Manila Area), and insecurity spread in many parts of the country. All these problems took a leading part in the political turmoils of 1985. However, a more democratic government did not reverse the trend; poverty keeps worsening, chiefly among the rural population and in the slums of Manila. A recent survey (Burg 1988) has pointed out that about three out of five families are now living on less than US$ 117 per month. According to the FAO, a large part of the population is today suffering from nutritional deficiencies and protein malnutrition. In spite of some improvements, the health services remain very precarious; in 1982 there was only one physician for every 1090 persons, one hospital bed for every 756 people and the mortality rate remained very high, especially among children. These findings fully confirm an impression of generally unfavourable conditions, that undoubtedly form the backdrop against which any consideration of the health-disease relationships must take place. In fact, the direct climatosensible diseases occupy a dominant place among the mortality causes (Table 2). Together, of the first eight causes of death, pneumonia emerges as the most frequent underlying cause while heart diseases rank second (Hansluwka and Ruzicka 1982).

4. The mean pathological rhythms: The seasonal disease calendars

Climate may contribute directly or indirectly to the appearance of a disease, aggravate pain and suffering or accelerate death. Nevertheless, in the tropics weather and climate are seldom the sole cause of disease as in cases of heat stroke due to extreme temperatures (Weihe 1986). More important is the contributory stress or aggravating effect of climatic factors on humans with some existing pathology and weakened adaptive capacity. A climatic stress may be harmless or even invigorating for individuals when they are young and adaptable, but fatal in their vulnerable old age, particularly if a predisposing disease exists (Weihe 1979). As the majority of diseases is of multifactorial origin including various failures of homeostatic regulation, the importance of climate in the chain of nosologic factors which leads to the final breakdown of the system is not always easy to ascertain. However, as the infectious agents and the climatological factors which stimulate or trigger pathological events fluctuate in number and intensity during the course of the year, it is evident that most diseases visually appear to have a seasonal or fluctuating occurrence. These seasonal or pseudo-seasonal phenomena and some other long-periodical relationships can be summarized for a number of meteorotropic, both infectious and non infectious, diseases. For communicable illnesses, climate may affect the virulence of the causative agent (and/or the vector), the development and survival of the pathogen during the periods outside the homeothermic human host, or/and the resistance of the host. For non-communicable diseases, climate may directly perturb the normal working of the human organism.

The most remarkable feature of this study is that a different pattern emerged for almost each disease. Here the discussion is limited to a small number of the most important human illnesses (Figure 1). A comparison was made between morbidity patterns for the six year period 1981–1986 inclusive and patterns for a longer period, starting in 1965. It must be emphasized, in all cases coefficients of correlation (Pearson Product Moment) between both time periods are in excess of 0.7.

A few diseases revealed no definite seasonal trend. For instance, malignant neoplasms and schistosomiasis did not vary significantly from week to week or from month to month. If such a result was expected for cancers, it might be considered to be in contradiction of the usual pattern for schistosomiasis. Indeed, this disease is due to parasites which spend parts of their life cycle in water in one or more non-mammalian hosts and which have one or more free-living stages; therefore, in many tropical countries, and even in equatorial climates, the prevalence of schistosomiasis rises steeply during the wettest season. Perhaps the Philippines is permanently wet enough, so that schistosomiasis occurs year round; but it is still difficult to interpret these reports.

In any case, striking seasonal changes occurred for almost all other illnesses:

(a) The respiratory diseases, notably influenza, acute bronchitis and pneumonia, showed predominant concentration in September. Needless to say, the influenza curve ran parallel to the curve for bronchitis and pneumonia; the rise, however, began earlier in the curve for influenza. The application of Edwards test for seasonality showed that there is a very clear seasonal variation. For instance, bronchitis peaked up in September, waned rapidly soon in October, then formed a seven-months plateau without any marked peak from November to June, and gradually curved up from June to September, with a small trough in early August. Incidentally, the morbidity of pneumonia-bronchitis in winter has decreased in recent years due to the successful introduction of artificial climate. The precise weather features that favour the occurrence of influenza are not known. Clearly, the death rate of influenza virus is high at 50–90% relative humidity, but low at 15–40%, and therefore in the northern hemisphere mid-latitude areas the maximum number of cases from influenza infections occur between October and March, and in the tropics the maximum incidence is generally observed in the dry season and in dry areas. However, in the Philippines the virus is surprisingly endemic, with sporadic flurries. But why do these outbreaks chiefly occur in September, i.e. at a time when

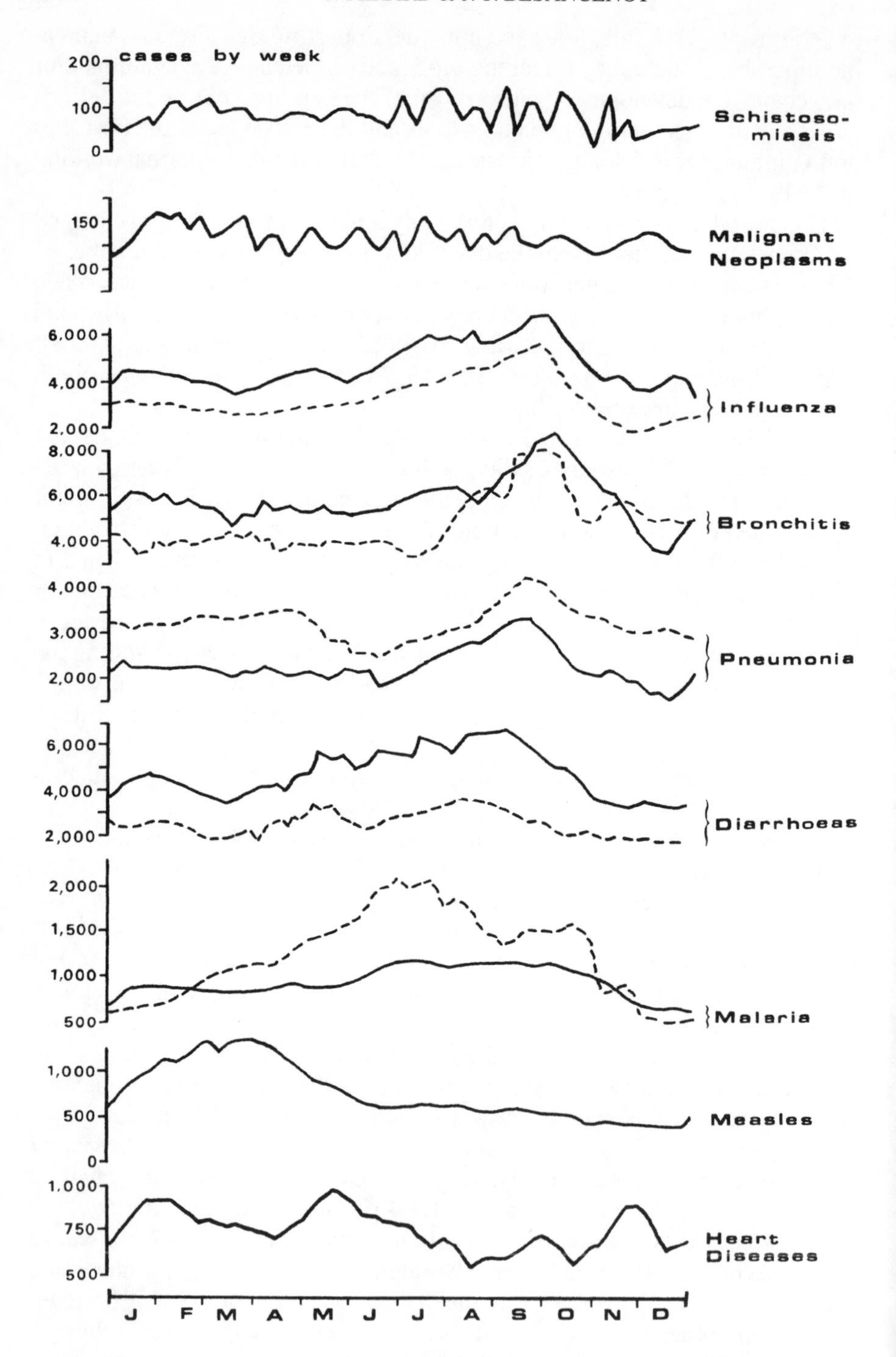

Figure 1. Averaged diseases calendars in the Philippines. Five-week running means. Continuous line; 1981–1986. Dashed line; 1965–1986.

the relative humidity reaches its yearly maximum (82 to 85%)? Much more research is required to clarify the exact significance of weather in relation to the development of influenza epidemics.

(b) Incidence of measles were highest from February through April, i.e. during the dry months. It may be remembered that measles virus keeps dry. Moreover, there is a world-wide pattern; a single spike of measles appears almost everywhere in the spring, but not every year, the seasonal pattern recurring in 2 or 3 year cycles.

(c) As expected, diarrheas and other gastrointestinal diseases, as dysentery and cholera, i.e. infections from several types of enteroviruses, were least common during the winter months and reached their highest peak during the hottest months of the year, i.e. between May and August; this coincides with the maximum incidence of *shigella* dysentery.

(d) Lastly, heart diseases peaked several times in the year, without any well-established pattern. This result is rather surprising because studies by various authors indicate a clearly seasonal incidence of coronary thrombosis, myocardial infarction and angina pectoris. Indeed, both cold and heat (mainly humid heat) may bring about considerable disfunction of the thermoregulation mechanism of the human body and of all the endocrinal and nervous functions in the body controlled by the hypothalamus. As a result, profound circulatory alterations occur, which may cause an over- or under-stimulation of the heart-muscles. This may be detrimental if the functioning of the heart is a labile mechanism, as in the case of potential heart patients, e.g. persons with an impaired circulatory system or those suffering from severe arteriosclerosis. Moreover, the climatic stress becomes much greater if physical activity is carried out, and heart failure during storm events is not uncommon. But probably, in the Philippines' case, averaged data are obscuring this influence. Besides the same phenomenon was previously established for a few other tropical climates.

To summarize, the seasonal prevalence in morbidity from infectious disease as well as the systematic deficiency diseases in some seasons are specific medical-biological problems of human-climate relationships. While seasonal variations may not prevail in several disease types, such variations were found in many of the common illnesses in the Philippines. Obviously, the presence of such a significant seasonal variation in a particular disease indicates that environmental factors are involved. However, the day-to-day course of the weather is also very important.

5. The "true" pathological rhythms

Table 3 and Figures 2 to 7 show the annual values and the variations of respiratory infections, diarrheas, measles and heart diseases reported in the Philippines in 1985 and 1986. They provoke a few remarks or interrogations:

Table 3. Diverse diseases reported in the Philippines in 1985 and 1986 (weekly data).

Period	Influenza	Bronchitis	Pneumonia	Heart diseases	Diarrheas	Measles
1985						
Cases*	7,544	9,734	3,323	853	8,901	1,072
Standard deviation	2,518	2,613	921	233	2,320	525
1986						
Cases*	5,216	6,366	2,084	638	6,648	578
Standard deviation	2,435	2,622	747	277	2,126	192
1981–1986 mean						
Cases*	4,622	5,888	2,310	746	4,700	768
Standard deviation	977	1,220	414	185	1,136	330

* Number of reported cases.

(a) While the reported case number in 1986 was near the average value (a little higher for influenza, bronchitis and diarrheas, a little lower for heart diseases, pneumonia and measles), in many respects 1985 was an abnormal pathological year. The morbidity rates were as much as twice the 1981–1986 mean.

(b) Both years, and especially 1986, were characterized by very marked variations from one week to another, particularly for heart diseases. Thus, graphs of the time series of morbidity by cause, on a weekly basis, revealed a big departure from the mean pattern, with some very sudden and sometimes very sharp peaks in the number of diseases. However, as can be seen, the recorded data tended to corroborate the assumption that there was a certain seasonal coherence.

(c) For most illnesses, perhaps except major diseases and measles, both 1985 and 1986 differed not only from the mean pattern, but also one from the other. Smoothed curves (not presented here due to space limitation) highlight this difference, with 1985 displaying a rather bi-cyclic curve. The main diseases (bronchitis, pneumonia, diarrheas) had a major peak in summer or in early summer (influenza) and a deep trough at the end of autumn (November–December). For instance, the influenza peak in May was three to four times the mean rate for November–March. A minor maximum occurred in winter (January–March) and an unremarkable minimum appeared in April, i.e. at the height of the dry and torrid season for most areas of the archipelago. On the other hand, the 1986 curves were rather plurimodal, pointing to a sharp decrease of many illnesses in August–September in contrast with a strong

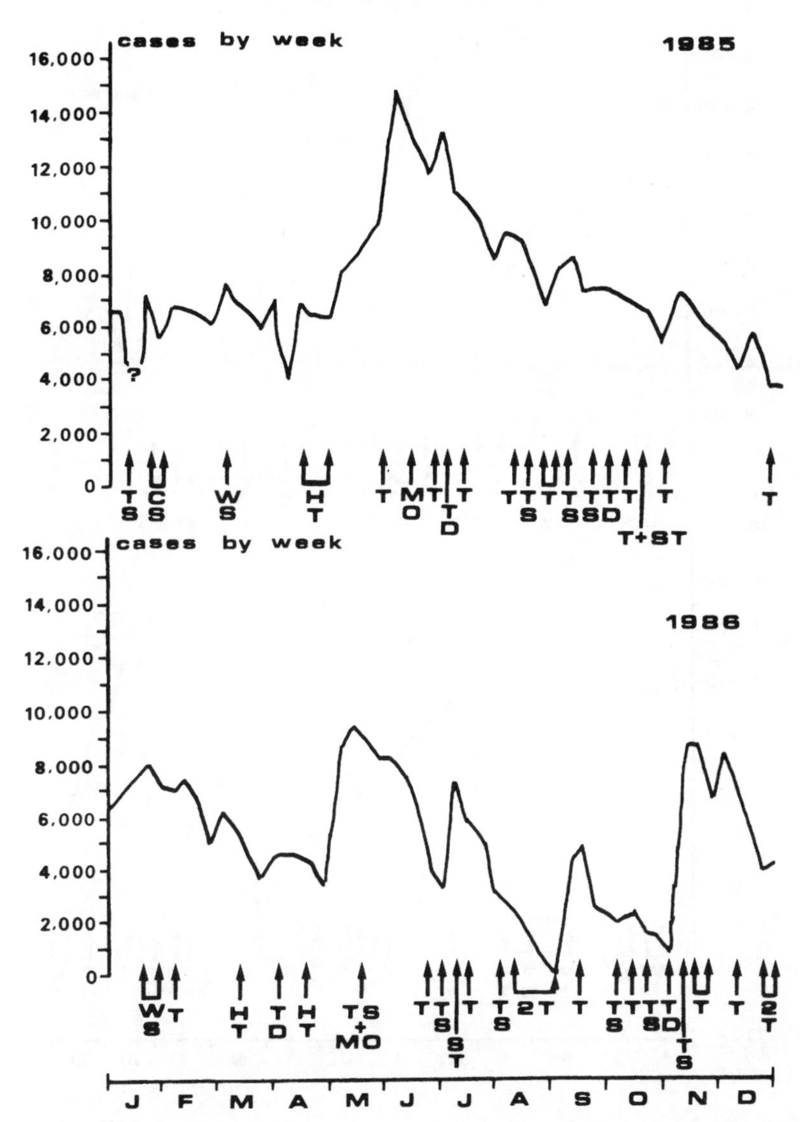

Figure 2. Reported influenza cases by week and major weather events in the Philippines (1985–1986). cs = cold surges; ws = trade wind surges; HT = Highest Temperatures; MO = southwest Monsoon Onset over Luzon; ST = Super Typhoon; T = Typhoon; TS = Tropical Storm; TD = Tropical Depression.

morbidity at both the beginning and end of the year (cf. pneumonias and diarrheas).

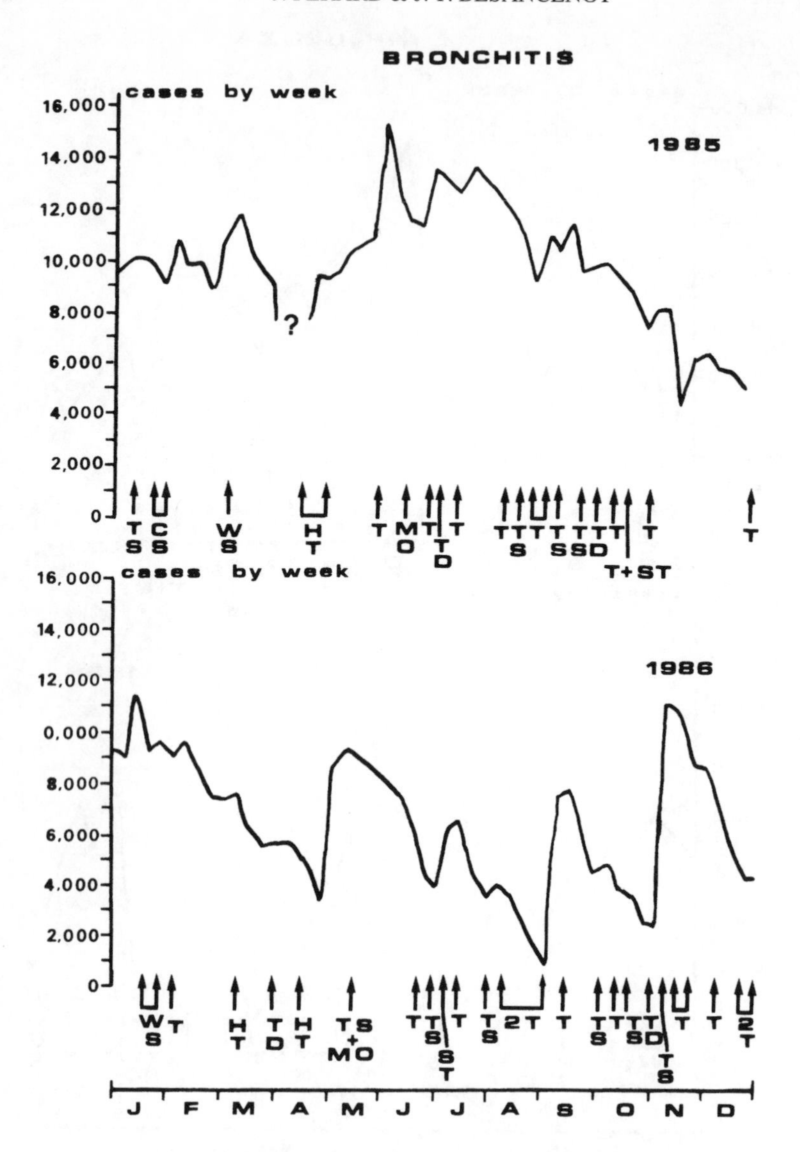

Figure 3. Reported cases of bronchitis by week and major weather events in the Philippines (1985–1986). See legend to Figure 2.

(d) However, for a given year, many diseases had a rather similar seasonal development. Indeed, it is well established that drastic weather changes are associated with increases in morbidity. Yet the reasons for this have not always been found. The most pronounced upsurges of morbidity could not be causally related to a single, separate climatic element. Even the use of complex climatic factors, such as bioclimatic indices, was only marginally

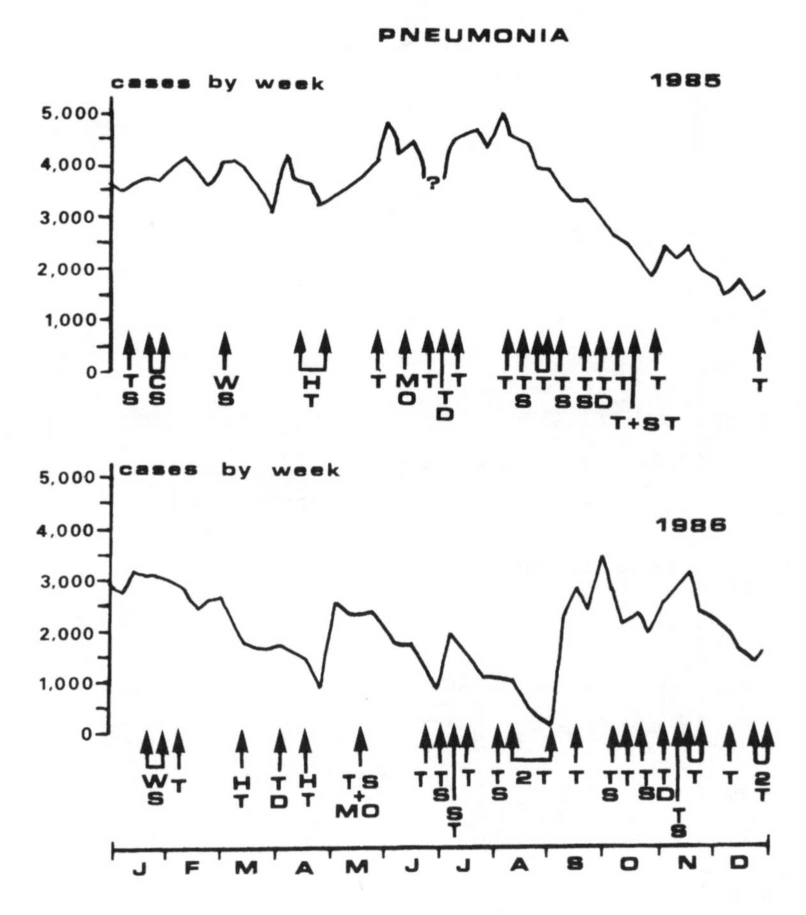

Figure 4. Reported cases of pneumonia by week and major weather events in the Philippines (1985–1986). See legend to Figure 2.

successful. By contrast, a simple visual inspection of the figures suggests, at least in part, a parallelism between certain major weather events and upswings or downswings in morbidity.

5.1. *1985*

The beginning of January 1985 experienced some vigorous cold surges of the northeast 'monsoon', with day temperatures higher than 30°C and night temperatures of around 17–18°C. As usual these cold surges followed the passage of a vigorous typhoon east of the region, a fair distance from the islands over the Philippine Sea. The short-term variations of temperature are stressful in an environment with continuous humid heat. This relatively cold weather was accompanied by increases in diarrhea and respiratory diseases

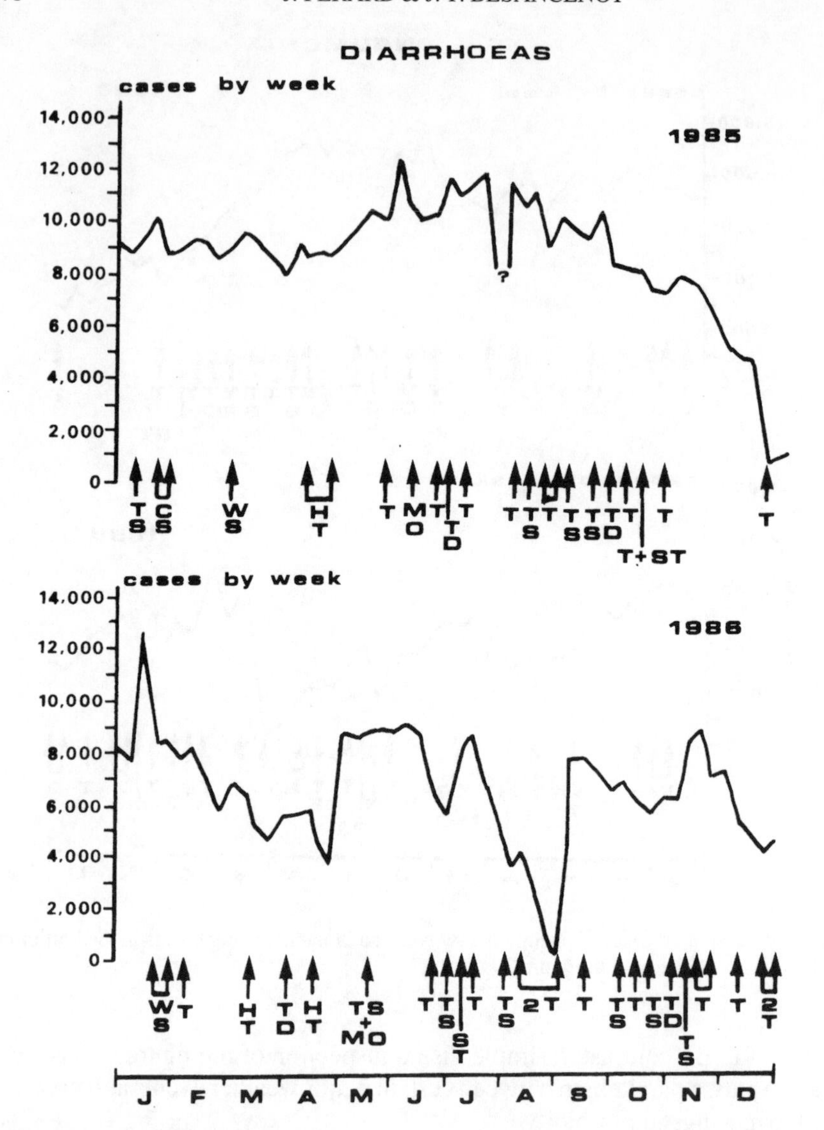

Figure 5. Reported cases of diarrheas by week and major weather events in the Philippines (1985–1986). See legend to Figure 2.

(influenza and bronchitis, mostly). Young children, in particular, paid a heavy toll during these periods, which of course is not surprising. Nevertheless, the mechanisms involved are not well understood. On one hand it may be assumed that, under certain humidity condition sudden drops in temperature favour the development and virulence of pathogenetic agents, while on the other, short-term variations of temperature may cause a weakening of the

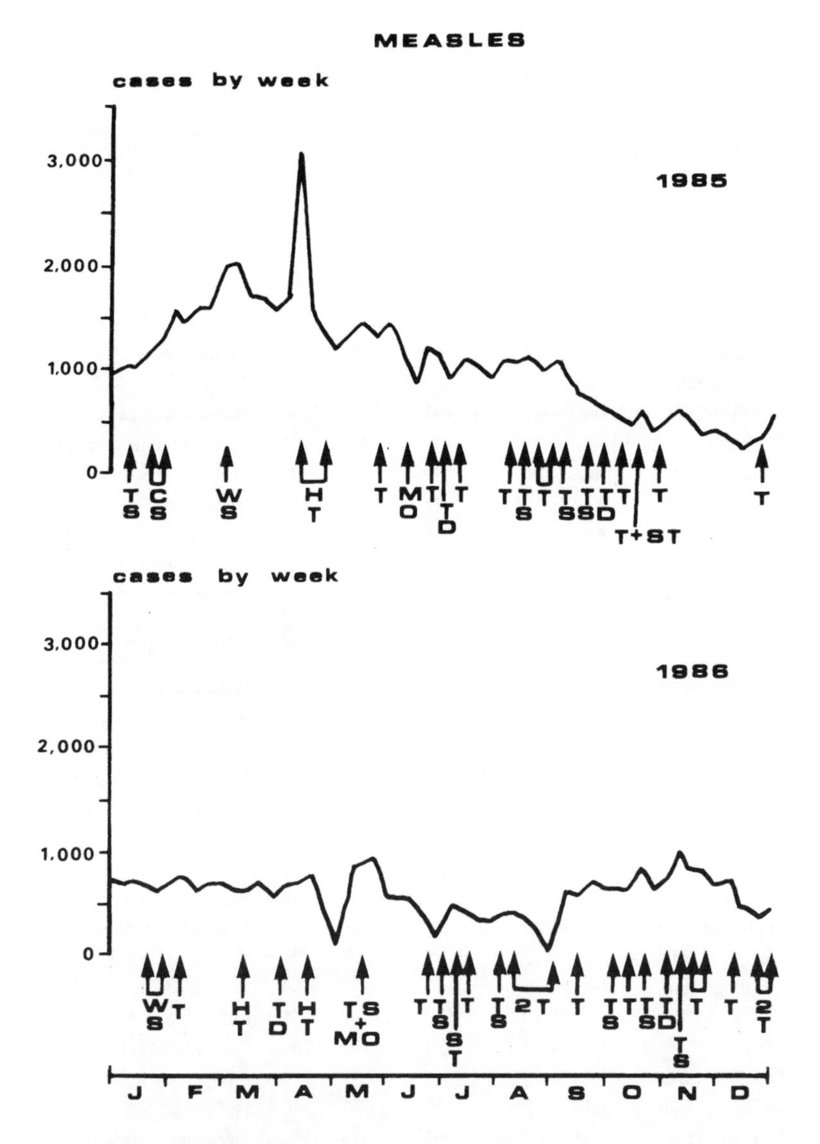

Figure 6. Reported cases of measles by week and major weather events in the Philippines (1985–1986). See legend to Figure 2.

organism, making it more sensitive to infections. Moreover, in any case, the cooling may well aggravate the development of infectious diseases and consequently worsen the prognosis.

After the hot weather of April, characterized by a very sharp peaking of measles epidemic, southwest monsoon and massive rainfall burst on Luzon as early as the first days of June, just after the incursion of typhoon Gay

Table 4. Tropical cyclones in the Philippine area of responsibility (ST = Super Typhoon; T = Typhoon; TS = Tropical Storm; TD = Tropical Depression).

	Classification-Name	Period	Max. wind (kph)	Min. pressure (hPa)	Max. 24h rainfall (mm)	Cyclone track
1985						
1.	TS Atring/Fabian	Jan 5–7	110	990	156	Philippines Sea
2.	T Bining/Gay	May 21–25	185	957	39.4	Philippines Sea
3.	T Kuring/Hal	Jun 20–24	165	961	304.6	Balintang Channel
4.	T Daling/Irma	Jun 25–29	165	958	344.5	Philippines Sea
5.	TD Elang	Jul 4–7	35	999.2	281.4	Crossed Samar and S. Luzon
6.	T Goring/Jeff	Jul 27–29	130	968	28.5	Philippines Sea
7.	TS Huling/Lee	Aug 10–12	85	990	134.4	Philippines Sea near Taiwan
8.	T Ibiang/Nelson	Aug 19–22	120	963	173.8	Philippines Sea near Taiwan
9.	T Luming/Pat	Aug 26–30	110	971	175.3	Taiwan Strait (SW–NE)
10.	TS Miling/Tess	Sept 1–4	140	983	197.7	Crossed N. Luzon
11.	TS Narsing/Val	Sept 15–17	85	1004	–	Crossed Batanes Islands
12.	TD Openg	Sept 24–26	–	–	52.8	Crossed N. Luzon
13.	T Pining/Brenda	Sept 29–4 Oct.	165	964	84.3	Philippines Sea near Taiwan
14.	TS Rubing/Cecil	Oct 12–14	140	974	67.6	S. China Sea-Palawan
15.	ST Saling/Dot	Oct 15–20	240	893	262.4	Crossed Central Luzon
16.	T Tasing/Faye	Oct 23–31	165	963	227.7	Crossed Luzon Recurvature
17.	T Unsing/Hope	Dec 20–24	185	948	77.8	Philippines Sea Recurvature
1986						
1.	T Akang/Judy	Feb 2–5	150	972	46.8	Philippines Sea
2.	TD Bising	Apr 5–6	55	1006	219.7	Landfall over Samar
3.	TS Klaring/Mac	May 24–29	75	998	221.5	Balintang Channel
4.	T Deling/Nancy	Jun 22–24	140	955	99.5	Philippines Sea
5.	TS Emang/Owen	Jun 28–2 Jul	100	987	151.2	Philippines Sea
6.	ST Gading/Peggy	Jul 6–10	220	894	709.6	Crossed North Luzon
7.	T Heling/Roger	Jul 13–15	140	955	116.6	Philippines Sea
8.	TS Iliang/Sarah	Jul 30–4 Aug	85	983	345.4	Philippines Sea – Recurvature
9.	T Loleng/Vera	Aug 13–25	140	939	159.8	Philippines Sea – Erratic Track
10.	T Miding/Wayne	Aug 17–4 Sept	175	950	313.9	S. China Sea – Erratic Track
11.	T Norming/Abby	Sept 15–19	185	943	135.9	Philippines Sea-Crossed Taiwan
12.	TS Oyang/Dom	Oct 5–9	95	1002	195.6	Crossed Central Luzon
13.	T Pasing/Ellen	Oct 10–14	150	975	150.8	Crossed Samar and Visayas
14.	TS Ruping/Georgia	Oct 17–20	110	989	170.6	Crossed Samar and Visayas
15.	TD Susang	Oct 30–1 Nov	45	1004.6	136.6	Crossed North Luzon
16.	TS Tering/Herbert	Nov 7–10	55	987	157.0	Crossed S. Luzon and Visayas
17.	T Uding/Ida	Nov 11–14	130	986	150.8	Crossed Visayas
18.	T Weling/Joe	Nov 18–24	150	940	85.5	Philippines Sea – Recurvature
19.	T Yaning/Kim	Dec 7–11	150	951	12.6	Philippines Sea
20.	T Aning/Marge	Dec 20–24	110	947	74.1	Crossed N. Mindanao-Visayas
21.	T Bidang/Norris	Dec 30–1 Jan	100	958	41.6	Crossed Visayas

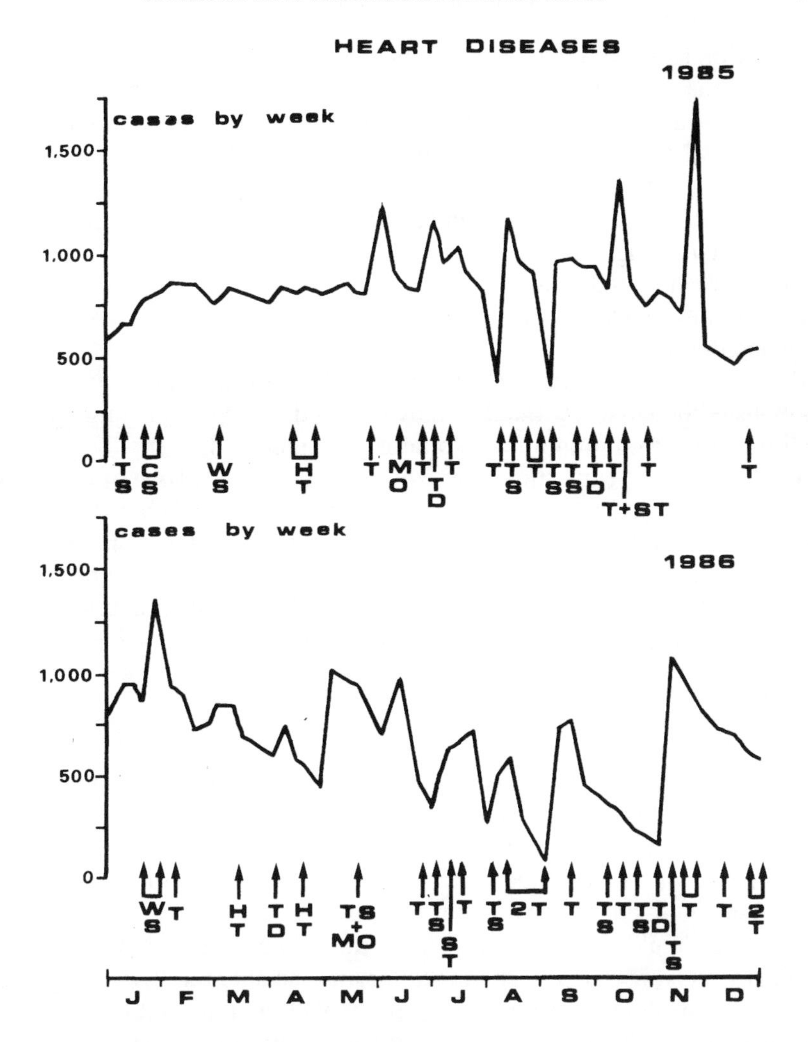

Figure 7. Reported cases of heart diseases by week and major weather events in the Philippines (1985–1986). See legend to Figure 2.

(Table 4). So a recrudescence of the main diseases suddenly occurred. June and July experienced a monsoon which intensified and expanded northwards whenever a vigorous typhoon passed over the Philippine Sea. Rainfall was heavy and widespread, except in Mindanao. The main upswing of respiratory diseases occurred in this very disturbed early summer; at the weekly scale, all curves showed a good correspondence between climatic paroxysms and epidemic bursts (namely, at the end of May and at the beginning of June, at the time of passage of typhoons and storms). From August to October many strong disturbances were still crossing the islands. However, the rainfall was generally deficient and in spite of a few peaks coinciding with some of these typhoons, the morbidity rates tended to decrease for most illnesses. It must be said, that most typhoons of this period crossed the islands or veered to the north towards Taiwan (Table 4). Because of that, the considerable rainfall they brought with them was restricted to a small area in the archipelago (José 1978; Pérard 1986). It is noteworthy that when a typhoon crosses to the north of Luzon and brings an absolute deluge over a part of the great island, the other areas come under the influence of the subsidence sector of the storm track. Such areas experience a rather stable and dry, and in most respects, more healthy weather.

During the last months of 1985 the Philippines hardly experienced typhoons, although a few intense rainy events did occur. This might explain the timing of the minimum for respiratory and enteric diseases.

There remains the question as to why 1986 experienced one of the highest levels of morbidity ever recorded in the Philippines. Weather data cannot constitute a complete explanation. It can be hypothesized that the hypermorbidity was heavily conditioned by the poor living conditions outlined earlier, namely by the various crises that culminated in 1985. But, of course, such an answer must be approached with considerable caution.

5.2. *1986*

1986 was characterized in the Philippines by very abundant rainfall and a very intense cyclogenesis. From the beginning of February to the last week of December there were 2 cyclones, 13 typhoons and 6 tropical storms (Table 4). Half of these storms occurred between the middle of September and the end of December. A good many had a long duration due to erratic tracks (22 days, for example, for *Wayne*). Accordingly, intense monsoon and heavy rainfall events repeated and continued far into the autumn. It was interesting to relate these climatic events to the weekly pattern of morbidity. The curves lend credence to the idea that a relatively high number of diseases has been favoured by very disturbed weather-types (influenza, bronchitis, diarrheas). Additionally, the large number of sharp peaks for various illnesses appeared to closely follow the main climatic stresses (cf. the typhoon *Akang* in February, followed by a trade wind surge, the sudden onset of the monsoon and the storm *Klaring* in May, the super typhoon *Gading* in July, etc.). Note, too, that the major maximum for respiratory infections occurred in autumn coinciding with a mid November culmination of cylconic activity when three cyclones on adjacent tracks passed through the area in close sequence. Lastly, we turn to the rather conspicuous minimum of August 1986. The low number of cases is surprising because typhoon *Wayne* rambled for several weeks over the China Sea at this time and brought torrential rains over the most heavily populated regions of Luzon. A possible explanation for the minium is that several cases were missed. This is supported by considerable verifiable evidence. The difficult weather conditions considerably delayed or obstructed the transmission of morbidity data, so that cases from about twenty provinces and ten cities were not included in the *Weekly Bulletin*. Fortunately, that is the only case when the data introduced such a bias.

Finally, incidence of heart diseases for 1985 and 1986 are examined. The curves for both years were deseasonalized. Well defined coherence between sharp peaks in heart disease and extreme weather was lacking due perhaps to the specific characteristics of the diseases. However, it is noteworthy that the pronounced maximum of October 1985 agreed fairly well with the super-typhoon *Dot* crossing over Luzon when the atmospheric pressure fell to 893 hPa within a few hours while winds blew at 240 km/h. And equally, in 1986 the major upsurges of heart diseases were always coincident with strong typhoons (for instance, in February).

6. Conclusion

Despite its limitations, this study has demonstrated a good association between morbidity-rates and the major weather-types experienced in the Philippines as a whole. It seems that such a high association may only be a causal one. In other words, we suggest that the inter-week variation in some morbidity-rates is closely dependent on a large number of climatic factors and may be largely due to the impact of adverse climatic phenomena on the human body. However, in a number of areas, the study has done no more than suggest lines for more in-depth research which would enable many hypotheses to be confirmed or the weight of the various factors to be assessed. Peculiarly, limitations due to data preclude the use of visual comparison as a principal means of evaluating the existence of associations between peaks in weekly morbidity and weather. This, as well as the possibility that time lags exist between related events, will further require the use of unbiased and more rigorous statistical tools, such as lagged correlation and variance spectrum analysis.

The way is thus open for further research, but even now the broad lines of a health policy can be drawn up on the basis of the results of this study. The findings presented here appear to have significance for the clinician charged with the care of patients. Specifically, the association of increased risk for disease following unusual weather typical of typhoons during the summer, and drops in air temperature during the winter, indicates that uncontrolled exposure to such weather situations could cause or advance illness. Thus, properly managed exposure and activity during weather situations such as those described in this chapter is warranted, especially for people for which weather-sensitivity is particularly obvious, a group that includes new borns, babies and the elderly.

References

Besancenot, J.P. (1984): Bioclimatological background and agricultural human settlements in Monsoon Asia. In: *Climate and agricultural land use in Monsoon Asia*, ed. by M.M. Yoshino. University of Tokyo Press, Tokyo, 109–128.

Besancenot, J.P. (1986a): Froid et santé en zone intertropicale. In: *Climatologie tropicale et établissements humains: Tropical climatology and human settlements*, Université, Dijon, 159–171.

Besancenot, J.P. (1986b): Recherches de bioclimatologie humaine en zone intertropicale. In: *Etudes de climatologie tropicale*. Masson, Paris, 11–28.

Burg, P. (1988): Philippines. In: *Encyclopaedia Universalis*. Universalia, Paris, 331–333.

Hansluwka, H. and Ruzicka, L.T. (1982): The health spectrum in South and East Asia: An overview. In: *Mortality in South and East Asia. A review of changing trends and patterns 1950–1975*. Geneva, World Health Organization, 49–82.

Jose, A.M. (1978): *Rainfall pattern in the Philippines during occurrence of tropical cyclones*. PAGASA, Manila, 63.

Pérard, J. (1984): Recherches sur les climates de l'archipel malais; les Philippines. *Dijon, Centre de Recherches de Climatologie*, 4, 922.

Pérard, J. (1986): Actions et rétroactions de la cyclogenèse tropicale sur la mousson Philippine. In: *Climatologie tropicale et établissements humains; Tropical climatology and human settlements*. Dijon, Université, 103–119.

Pérard, J. (1988): Typhons sur les Philippines: fléauou providence? In: *Climats et risques naturels*. Association Française de Géographie Physique, Paris, 45–49.

Vèlimirovic, B. and Subramanian, M. (1972): The pattern of morbidity after typhoons in a tropical country. *International Journal of Biometeorology*, 6, 343–360.

Weihe, W.H. (1979): Climate, health and disease. In: *Proceedings of the World Climate Conference*. World Meteorological Organization, Geneva, 313–368.

Weihe, W.H. (1986): Life expectancy in tropical climates and urbanization. In: *Urban climatology and its applications with special regard to tropical areas*. World Meteorological Organization, Geneva, 313–353.

Jocelyne Pérard
Centre de Climatologie
Université de Bourgogne
6, boulevard Gabriel
B.P. 138
21004 Dijon Cédex
France

Jean-Pierre Besancenot
G.R.D. "Climat et Santé"
Faculté de Médecine, 7
boulevard Jeanne d'Arc
21033 Dijon Cédex
France

3.5. POSSIBLE IMPACT ON AGRICULTURE DUE TO CLIMATIC CHANGE AND VARIABILITY IN SOUTH AMERICA

FERNANDO SANTIBÁÑEZ

Climatic fluctuations are a risk factor in agriculture. They hinder agricultural development in extensive arid and semi-arid tropical areas of South America. The main uncertainty factors in the tropics are the amount and distribution of rainfall in dryland-farming areas and, in the subtropics, possible spring frost where temperate and subtropical species coexist. Economic losses, caused by these climatic phenomena, should not be underestimated. In Chile, annual losses in agriculture, due to frost, are estimated from 12 to 20 million US$, and due to droughts from 40 to 120 million US$ (Santibáñez 1983). To appreciate the magnitude of this problem at a South American level, these values would have to be multiplied by a factor of at least 8 to 10. Drought also affects irrigated areas, because streams and rivers depend on runoff water and/or the accumulation of snow in the upper Andes. The Brazilian northeastern arid zone, one of the agricultural areas with most variable precipitation in South America, is almost entirely dependent on rain for crop production. However, in the last few years underground water has been tapped for irrigation. Livestock production is especially uncertain in areas with less than 300 mm rainfall per year. In spite of this, there is an important segment of the population who depend on the rain to survive in the Chilean mediterranean climates and arid steppes of Tierra del Fuego.

In the last few years, these risk factors have been compounded by possible climatic change. Climatic change would increase interannual fluctuations, shift the arid zone limits further South and shift the Equatorial rain regimes toward the tropics. On the other hand, these climatic modifications could induce a rise in temperature, hence deeply affecting agriculture also in large areas of South America.

At the southern edge of the Atacama desert, in the arid steppes and semi-arid zones of Chile, the historic series of annual precipitation indicates a

M. Yoshino et al. (eds.), Climates and Societies – A Climatological Perspective, 255–277.

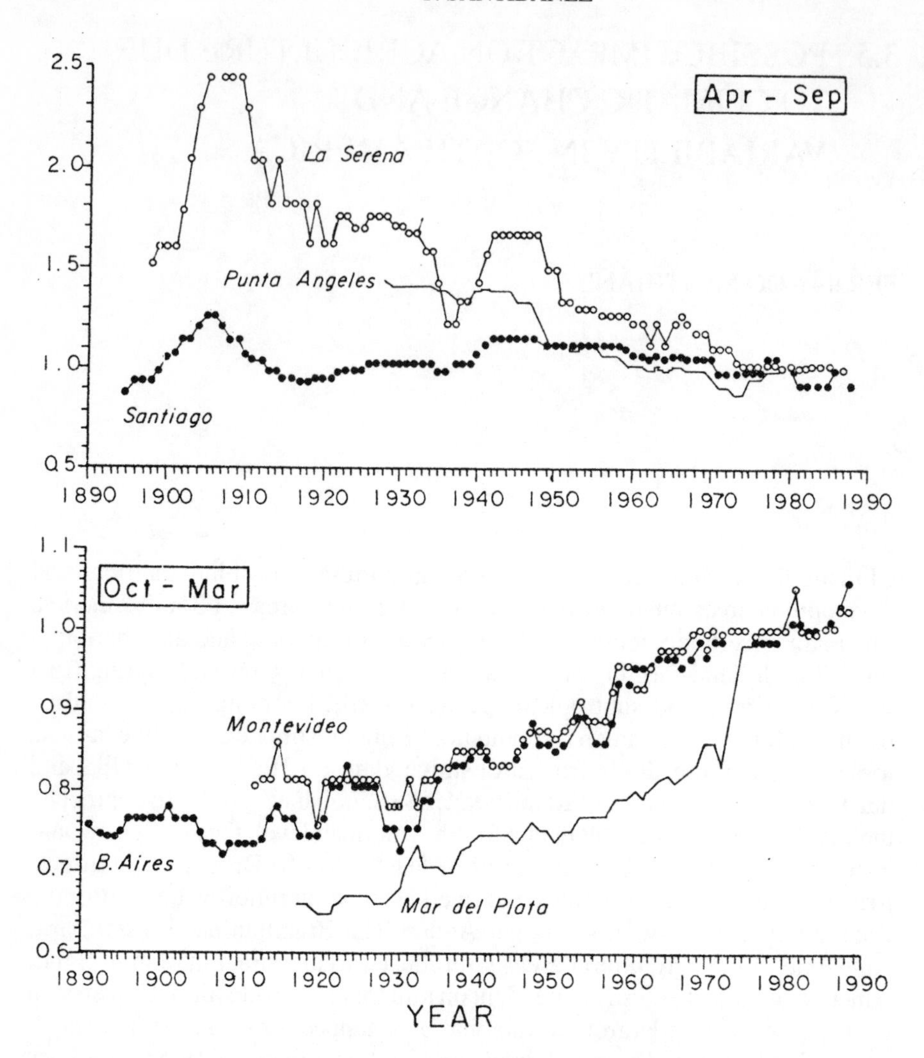

Figure 1. Moving average for 30-year period, taking 1951 to 1980 as reference. Courtesy of Dr. P. Aceituno, Facultad de Ciencias Físicas y Matemáticas. Universidad de Chile. Beaucheff 850, Santiago (unpublished data).

clear temporal negative slope tendency. This would suggest a certain advance of aridity in areas of South Temperate climates. Precipitation decline has occurred at a large number of stations and this information is consistent enough to conclude that a real climatic phenomenon is taking place (Santibáñez et al. 1988). In spite of this, there is no evidence to suggest that this tendency will result in a climatic change. This is probably the negative phase of a long-term cycle. Whatever the origin of a temporal variation of this kind may be, it affects agriculture and therefore, its impact should be evaluated together with possible mitigating strategies. In contrast, to the desertification occurring on the South Pacific shoreline, there is a hint of an increase of precipitation of similar proportions on the Atlantic side. The climatological linkage of these two processes has not been established. In spite of this, we could infer that we are probably facing a long-term interconnected situation due to a systematic behaviour of the general circulation of the atmosphere in the southern Cone of South America (Figure 1).

Should these trends become a permanent feature of the climate, we would have to project the possible changes to the environment that would ensue from them to enable us to restructure the productive infrastructure, protect watersheds, and train farmers to adapt their production systems to the new conditions. For farmers this could eventually include a change in their crupping systems (Conicyt 1989).

The impact of fluctuations and climatic change on cultivated crops is difficult to quantify, because of the complex interactions of the plant with climate, which also vary with specie ecophysiology. The change of a single climatic factor, such as a temperature rise, would favour productivity as growth rate would increase, but it could also accelerate development, shortening crop life cycles and decreasing the final yields. The complexity of these responses requires the use of a model which by integrating present-day knowledge through realistic algorithms enables a reasonable projection of cultivated plant behaviour under a new climate.

1. Modelling of crop behaviour to climatic variations: SimProc model

To evaluate the possible responses of crops to climatic stimuli, we have developed a dynamic simulation model. This simulator of crop productivity (SIMPROC) integrates the main ecophysiological relations that operate dynamically during the biological cycle of several cultivated species, from the seedling stage to harvest maturity (Santibáñez 1986, 1990).

Growth simulation is based on levels of solar radiation, leaf interception, temperature and relative satisfaction of evapotranspiration needs. The crop develops from one phenological stage to the next based on thermal or development units accumulation.

Starting at emergence, dry matter accumulation is simulated at every stage of crop development based on absorbed photosynthetic active solar radiation (PARa), which depends on the incoming solar radiation (PARo), exposed leaf area index (LAI), leaf albedo (AL) and amount of radiation transmitted to the soil surface.

$$\mathrm{PARa} = (1 - \mathrm{AL}) * \mathrm{PARo} * (1 - \exp(\mathrm{k} * \mathrm{LAI}))$$

The biochemical fixed light or gross photosynthesis (GFOT) depends on: PARa; the photosynthetic efficiency (FE), which varies according to light intensity (PARo); species type (Jm) C3 or C4; and atmospheric CO_2.

$$\begin{aligned} \text{GFOT} &= \text{PARa} * \text{FE} \\ \text{FE} &= \text{A1} * \text{Dc}/(\text{A2} + \text{A3} * \text{PARo}) \\ \text{Dc} &= 1 + \text{Jm} * (1 - \text{EXP}(\text{Jo} * (330 - CO_2))) \end{aligned}$$

A1, A2 and A3 are empirical parameters.

Dry matter production (DMP) depends on total maintenance respiration (MAINT) and the biochemical growth efficiency (GE) of each species (Penning de Vries and van Laar 1982).

$$\text{DMP} = (\text{GFOT} - \text{MAINT}) * \text{GE}$$

Maintenance respiration is proportional to total plant biomass (BIOM) and temperature (T).

$$\text{MAINT} = \text{BIOM} * \text{km} * f(T)$$

Km is a proportional factor which depends on the concentration of protoplasm in each organ (leaves, stems, roots); f(T) is a thermal control of respiration.

Growth is related to temperature through a function (T_c) specified by the following thermal parameters: minimum temperature, f_u (no growth); optimum temperature, f_o (maximum growth); and maximum temperature, f_x (no growth).

$$\text{Tc} = (1/f) * X^b * e^{-X}$$

$$X = (f_x - f_i)/(t * f_x)$$

f and b are parameters depending on f_o and f_u.

Phenological stage is proportional to relative thermal growth unit accumulation. Because of its simplicity we have taken the degree-day concept as a thermo-chronological counter. Relative phenological stage (Pa_i) at time 'i' is calculated as:

$$Pa_i = (1/DDo) * dd_i,$$

where DDo is the total degree-days at harvest maturity, and dd_i is the partial degree-day accumulation at time 'i', starting at emergence.

Crop is at phenological stage 'x' when $Pa_x < Pa_i < Pa_{x+1}$, where Pa_x and Pa_{x+1} are the minimum and maximum phenological age respectively, for the 'x' stage. Phenology modulates photosynthetic partition of total dry matter production (DMP) into different organs in such a way as to cause a harmonic plant growth architecture. The sum of all partition coefficients (u_1: leaves;

u_2: stems; u_3: roots and u_4: fruits) at any moment of the phenological cycle is always equal to 1:

$$(u_1 + u_2 + u_3 + u_4) = 1$$

Growth rate of an individual organ is equal to the product: U1·DMP.

Phenology also modulates cardinal temperatures, crop sensitivity to frost, and water deficit. The model contains complex algorithms to simulate frost injury, frost hardiness and effect of water shortage on production.

Growth can be hindered by frost, destroying exposed organs, leaves and growth apices. Frost, water stress sensitivity and temperature thresholds are simulated by a phenological submodel that gives each phase a different sensitivity.

Leaf area index (LAI) is generated as a balance between the dry matter partition to leaves (a) and the rate of leaf senescence rate (SEN).

$$\begin{aligned} \mathrm{LAI} &= \mathrm{S}\{\mathrm{a\ DMP} - \mathrm{SEN}\} \\ \mathrm{S} &= \mathrm{surface/mass\ ratio\ of\ leaves} \\ \mathrm{LAI} &= \mathrm{S}\ \{\ \mathrm{ul.\ DPM} - \mathrm{SEN}\} \end{aligned}$$

Natural leaf senescence is controlled by phenological age, Pa.

$$\mathrm{SEN} = \mathrm{S1} + \mathrm{S2} * \mathrm{Pa\hat{}2} + \mathrm{S3} * \mathrm{Pa\hat{}3},$$

where S1, S2 and S3 are empirical constants

Towards the end of the plant life cycle, the rate of leaf senescence is greater than the generation of new leaf area. This reduces the leaf area index or photosynthetic surface.

There is a subroutine to simulate the water balance of the soil-plant system. Evapotranspiration, or water consumption, by the crop increases gradually as the crop covers the soil. Simultaneously, direct evaporation from the soil decreases. Water in the soil is replenished by precipitation or irrigation. In the latter case, the user of the model can fix a criterion for watering or change the efficiency of the water applied. The water stress is represented by the quotient ETc/ETMAX which requires, through a production function, on the one hand the growth phase, and on the other, the process of senescence, when soil water content drops below a certain critical specific threshold.

Temporary excesses of water in the soil, caused by precipitation, are detrimental to plant growth. The model has a system that operates every time the soil is sasturated, and precipitation (mm/day) exceeds evapotranspiration over a certain threshold. This simulates the effect of water-drenched soils on plant growth when puddles form.

In order to calibrate the model, many years of experimental work was needed. Each tested crop was sown at monthly intervals throughout the year. A record of the fluctuation of dry matter, leaf area index, growth of aerial parts, senescence of leaves, and phenological observations was kept throughout

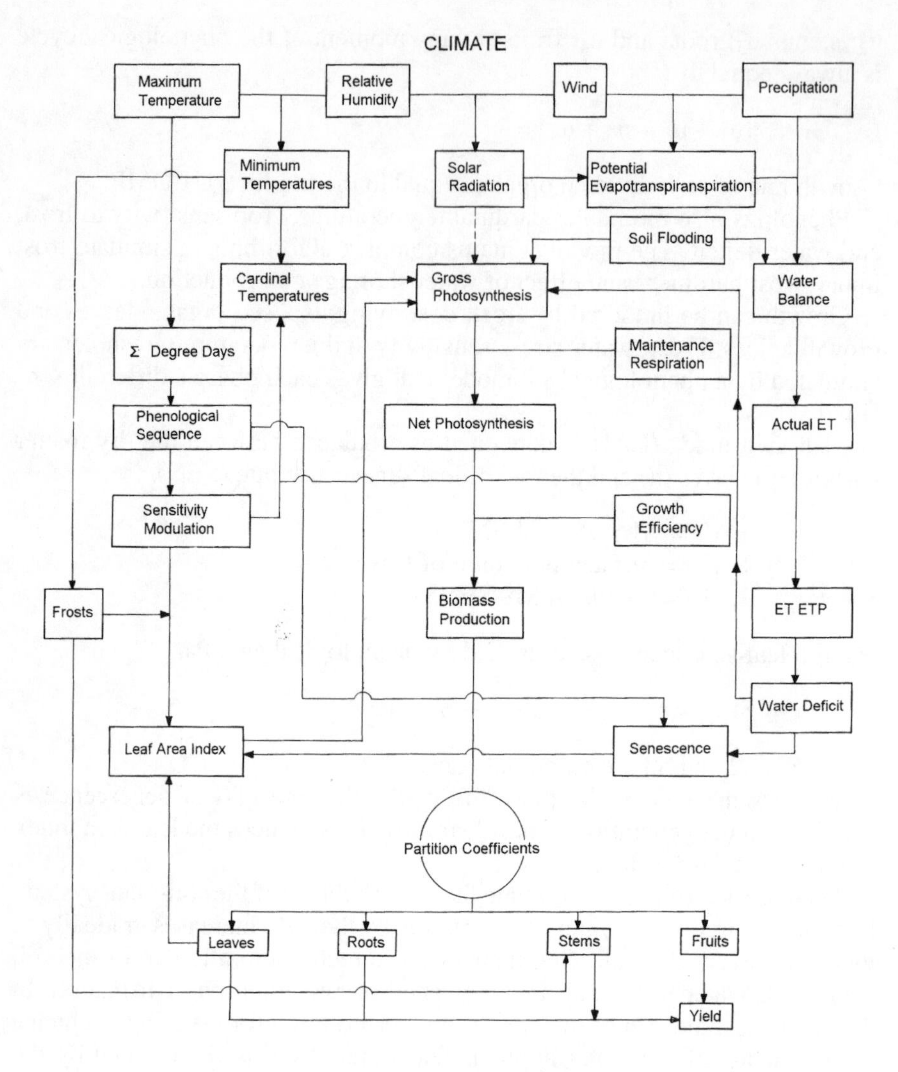

Figure 2. Basic structure of the SIMPROC model.

the growing cycle for each trial. Figure 2 shows the basic structure of the SIMPROC model.

2. Climatic change: Thermal and hydric modifications induced by climatic change

The agroclimatic parameters which determine the thermal and hydric regimes of a given place form an interdependent group, where the modification of an element means the readjustment of a whole chain of variables. From a crop production standpoint the most important variables which define the thermal system, are the extreme temperatures, frost intensity and frequency, effective temperature accumulation, length of vegetative rest period, chilling hour accumulation, and the number of hot days. A variation in the mean or extreme temperatures forces a modification in all the variables. In a similar way, a variation in precipitation modifies the rain-evaporation equilibrium, altering the water deficit and surplus, the duration of the dry and rainy season, and much of the irrigation required for each crop. To evaluate the degree of variation in growth parameters due to temperature or precipitation changes, a model (VARICLIM) was formulated. This model recalculates all parameters every time temperature or precipitation input is changed. The model scans the South American continent, using a data base obtained from about 250 stations. For every station it calculates some 20 variables every time a fluctuation in temperature or precipitation is simulated. Figure 3 shows the climatic relations established by the VARICLIM model.

Depending on the latitude and longitude of each station a thermal and pluviometric variation input is created through an algorithm that assumes an increase of up to 2°C at the equator, and up to 8°C at the poles (Schlesinger 1984; Manabe and Wetherald 1980; Gates et al. 1981). The pluviometric variations are more difficult to simulate using a simple algorithm. As a working hypothesis, we assumed that the heating of the atmosphere would increase precipitation in the regions under the equatorial or circumpolar cyclonic influence, and decrease it in those under the tropical or polar anticyclonic influence. The area with greatest pluviometric increase (equator, circumpolar front) and decrease (tropics and poles) may vary depending on the earth's meridian, considering the general circulation of the atmosphere (Hansen and Takahashi 1984; Schlesinger and Mitchell 1987; WMO-UNEP 1988).

3. Simulation of the impact of climatic changes on agricultural yields

Crop phenology and productivity under climatic change is simulated by joining the VARICLIM-SIMPROC models. In a first stage the VARICLIM model generates a scene of climatic change which may vary according to external inputs. In the second stage the SIMPROC model is fed with the modified temperature data of the hydric regime, and evaluates the effect of the new climatic condition on crop productivity and phenology.

4. Generation of climatic change scenarios in South America

The climate of South America varies from tropical rain forests in the Amazon Basin, where annual precipitation exceeds 3000 mm, to the driest desert in the world (the Atacama) with large areas, under 10 mm rainfall. In the Amazon

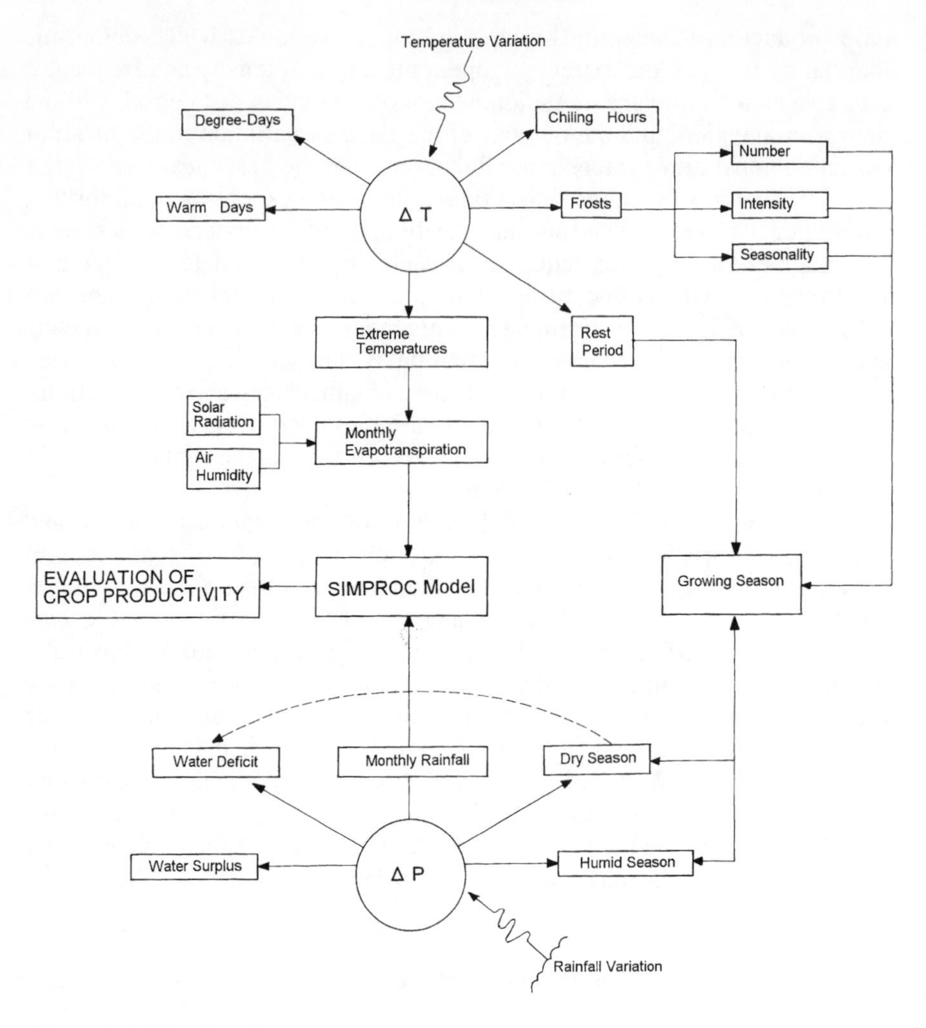

Figure 3. Climatic relations established by the VARICLIM model, for the evaluation of climatic modification induced by a thermal or pluviometric change.

Basin and northern coastal area tropical Af and Aw climates according to Koeppen, dominate, reaching as far as 20°S. Towards the southern Cone, climates evolve into temperate Cw and Cf types in the eastern sector (Southern Brazil, Uruguay and the central North of Argentina). The desertifying effect of the southeastern Pacific anticyclone creates a typical east coast desert climate (BW) in the south Peruvian and north Chilean coastline, and an arid steppe in the central Chilean zone (BS). Towards the southwestern coastline, temperate mediterranean climate Cs develops, changing to Cf in the southern coastline due to the proximity of the circumpolar cyclone. The central and southern parts of Argentina have a dry steppe BS climate, that are separated from the

Atacama desert by the Cordillera de Los Andes whose elevation ranges from 3000 to 5000 m.a.s.l. This generates polar climates (E) at high elevations over the length of this mountain chain.

On the one hand, global warming of the atmosphere could change the limits of the A climates slightly, moving them further south to the eastern regions where C temperate climates now prevail. On the other hand, as an effect of the activity of the Southeastern Pacific anticyclone, the southwestern limits of the temperate climates could be displaced by subtropical arid B climates, pushing them towards the Atlantic shoreline. In consequence, the Atacama desert could extend south a few degrees, covering part of the Chilean central zone and pushing the temperate climates of the Pacific coast towards the south. The altitudinal limit of the polar climates of the Cordillera de los Andes would rise a few hundred metres, driving the E climates to even higher grounds.

Rainfall increase in the equatorial Amazon would force this basin to release to the ocean significantly greater volumes of water than at present, increasing the soil erosion hazard in the areas that are now being deforested. The accelerated embankment rate of estuaries could cause flooding of large areas near its outlet to the sea. The present hydric surplus of 1000–2000 mm could increase to 2000–3000 mm. In the arid and desert areas of Chile and Argentina the water surplus would fall sharply, placing vast areas on the zelo isoline. This could affect the replenishment of the underground water table in large areas of these two countries. The expansion of the arid zone in the southern Cone could be very large, especially in central Chile and northern Argentina, based on evidence from a past dry period.

A rise in temperature would extend the Amazon equatorial thermal condition to all of Paraguay and the eastern coastline of Brazil, as far as Rio de Janeiro. The 3000 degree-day isoline, which marks the lower limit of the subtropical climates, could move, covering the Chilean and Argentinean central zone, and making the now temperate climates below 30 °S subtropical. The frost regime would undergo considerable changes and a1l areas of Southern Brazil, Uruguay, the Buenos Aires area of Argentina, and all of the western coastline of the Chilean territory would be frost free. This system would continue to exist in a reduced form, on high ground, over the Andean mountain ranges and the southern Argentine coastline. The climate in the Andean Altiplano would become less severe, with temperature ranges that would allow for a greater crop diversity. The decrease in the chilling hour regime in the temperate climates of the southern Cone could present certain problems to temperate trees production fruit, especially to chill demanding species (apples, pears), displacing their crop areas further south. Figures 4, 5, and 6 show the climatic change pattern of several climatic variables calculated on the basis of previously described algorithms.

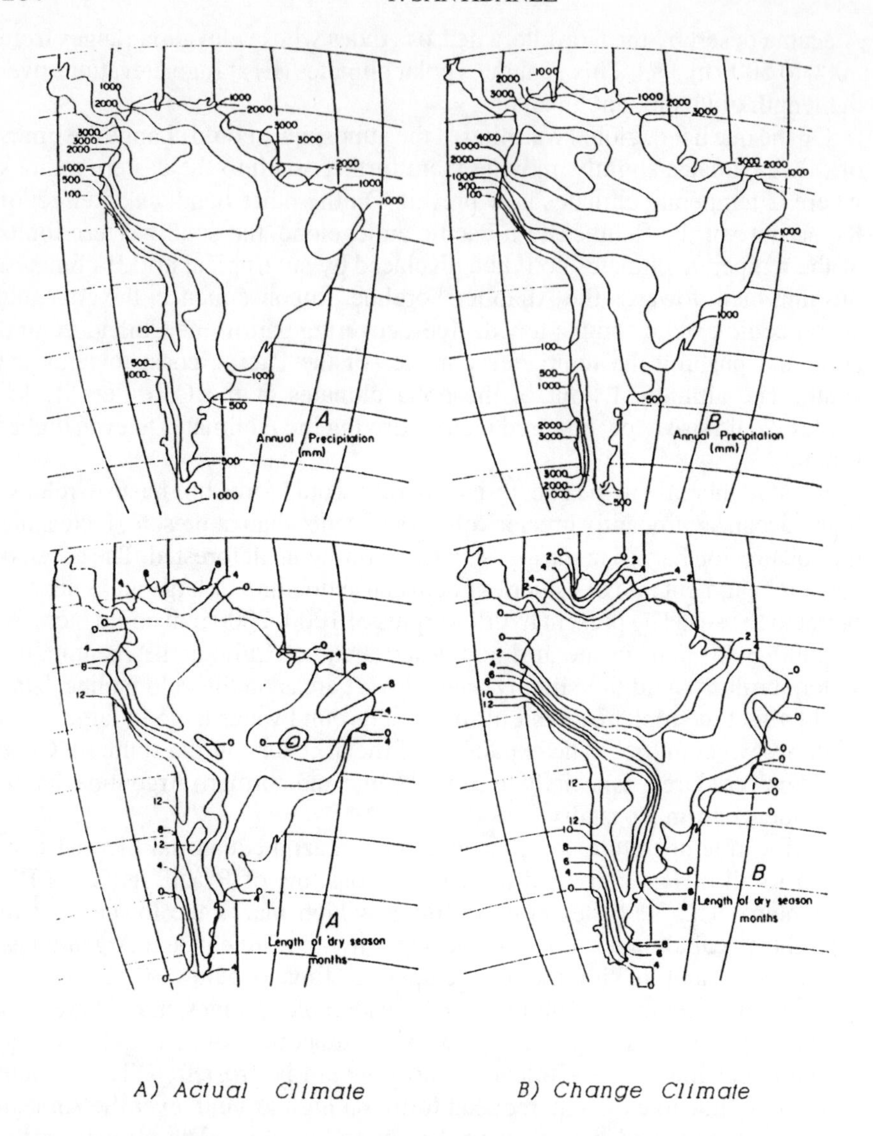

Figure 4. Climate change pattern of annual precipitation and length of dry season in South America.

5. The effect of climatic change on crop behaviour

From the results obtained, temperature variation on crop productivity is difficult to predict. Because temperature regulates considerable parts of the growth and development processes, some of these can be antagonistic to one another in determining final yield. In the colder climates of South America (southern part and Andean areas) growing conditions could improve considerably,

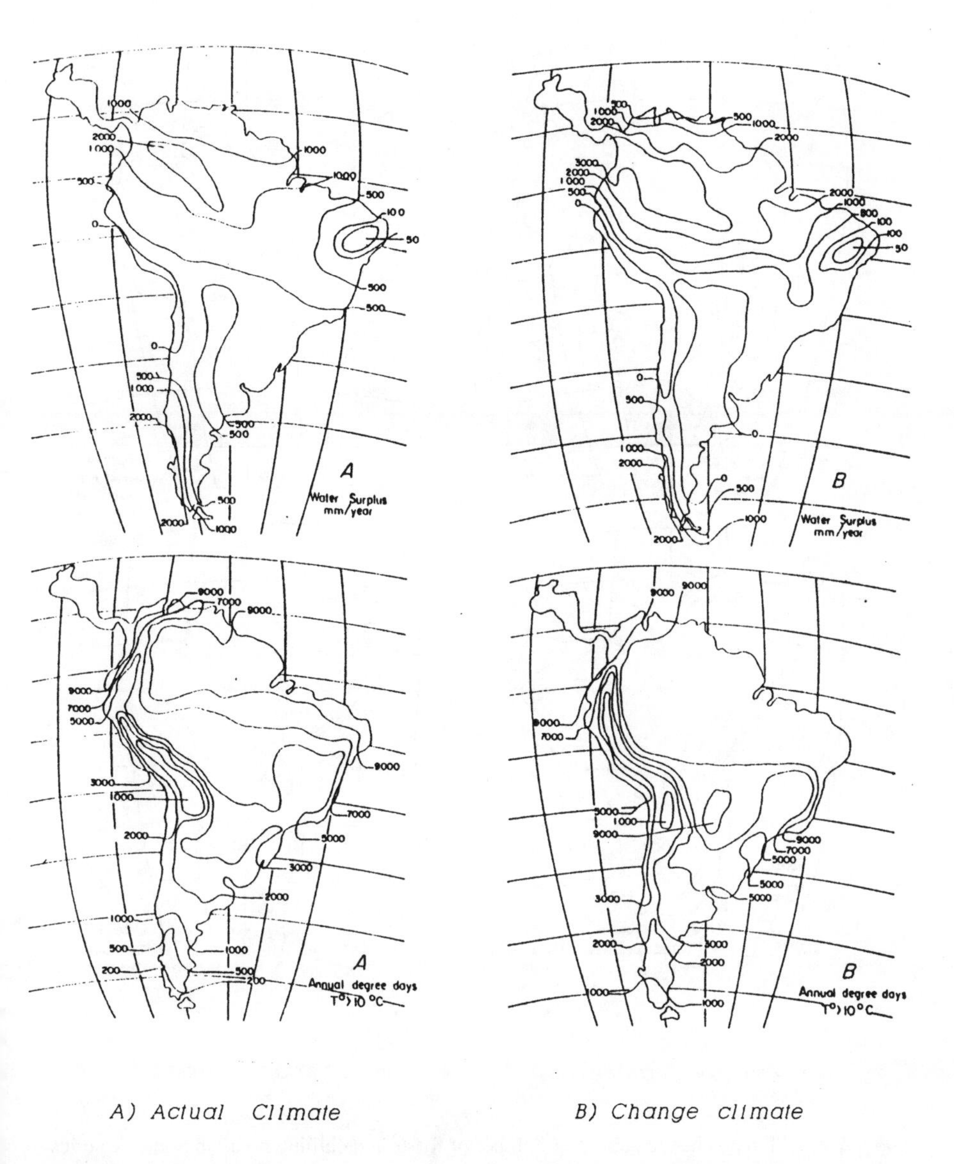

Figure 5. Climate change pattern of water surplus (annual precipitation – annual potential evapotranspiration), and annual degree days in South America.

allowing the introduction of species which cannot be grown there at present. In the warmer parts of the continent (the Amazon Basin and the intertropical system adjoining it), an increase of temperature would be rather unfavourable, because the climate of this region already limits crop diversity due to high temperature. Overall, atmospheric warming would accelerate the speed of

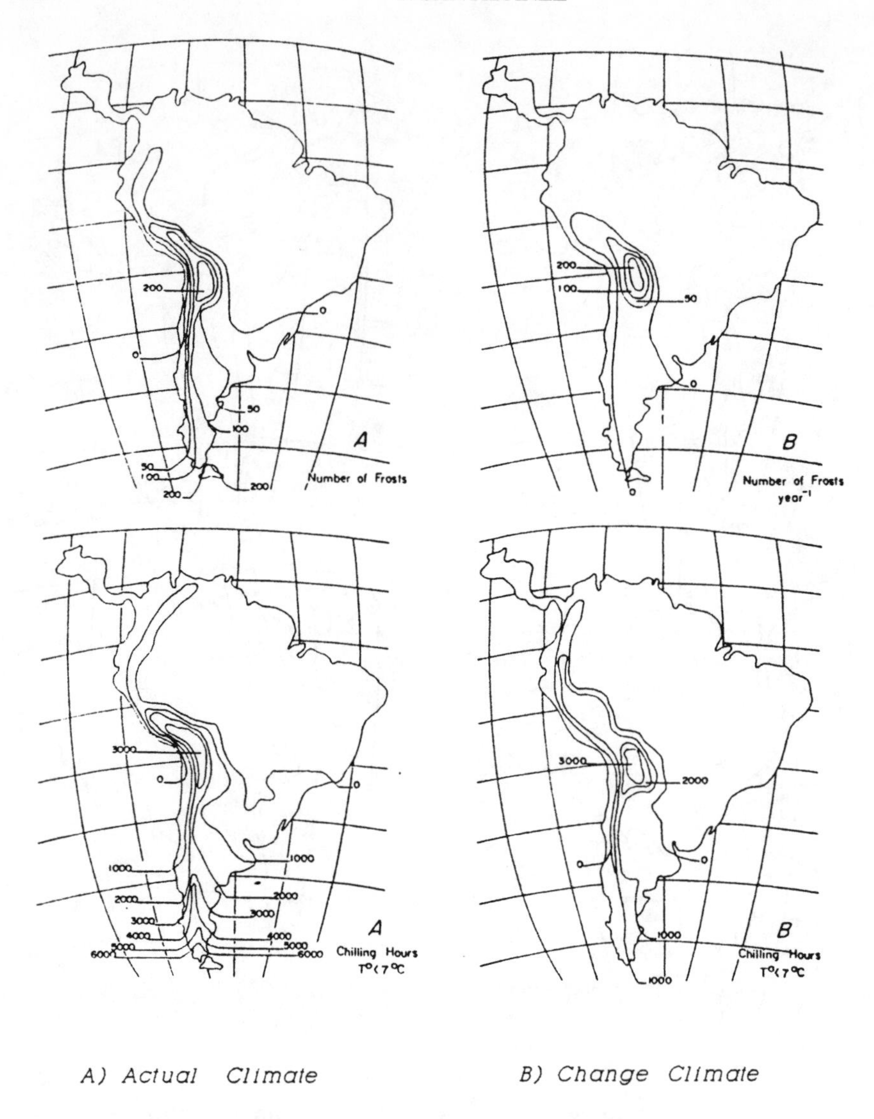

Figure 6. Climate change pattern of number of frosts and chilling hours in South America.

plant growth, shortening the phenological cycle, and reducing the time of crop growth. This would decrease crop yield in temperate zones. On the other hand, the increase of CO_2 in the atmosphere from 300 to 500 ppm should increase the photosynthetic rate by 40 to 50% in C_3 plants (wheat, potatoes, beans, sugar-beet, alfalfa, vegetables, and other plants). This phenomenon would tend to compensate for the shorter life cycle.

An increase in rainfall in the equatorial region would make annual crop growing more hazardous, reducing sowing possibilities and increasing seed loss risk due to soil flooding. In the subtropical areas where precipitation would decrease, crops, and especially cereals, would suffer a drop in yield, because higher temperatures would increase evapotranspiration and therefore crop water requirement and the risk of droughts.

In the cold temperate climates of the southern Cone, crop yield tends to increase with climatic change. This also happens in high altitude Andean climates (Cuzco, La Paz). The southern wheat cropping area limit would be pushed as far as the extreme south of South America (Usuahia), provided it is kept under irrigation. In warm temperate climates of south-central Chile, crop yield would drop because of the acceleration of the vegetative cycle. In these climates crops sown in autumn would have higher yields, because the greater part of the life cycle would occur in winter. As a consequence of this, water requirements of cereals and potatoes could, as an apparent paradox, decrease with global climate change; on the other hand, its irrigation requirements would increase. The growing area of this crop would tend to be displaced towards the centre of the southern Cone, where it could even be grown in autumn, because there would be no risk of winter frost (Santiago, Mendoza, Buenos Aires). Potato yield would tend to increase in Andean climates, because winters would become mild (La Paz, Cuzco). In temperate climates, the time of year for the highest yield would be autumn, because the warmer summers would cause thermal stress to these crops.

On a whole, the crop vegetative cycle would shorten. Hence, crop yield would decrease but, in warm temperate and tropical climates, two to three crops might be grown a year. Two strategies could be implemented: The first would be to seek longer vegetative cycle varieties within the same species, so as to compensate for the natural life span shortening caused by heating; the second would be to favour the use of short cycle varieties, so as to grow two or more crops a year, of the same or different species. Figures 7, 8, and 9 show the change in wheat, maize and potato productivity induced by climatic change. As for temperate fruit crops in the southern Cone, the drop in winter chilling hours could make fruit growing difficult. One of the strategies would be to replace the varieties grown presently by others which require less chilling hours. Alternatively, the growing area could be moved further south. Irrigation required for fruit growing would increase drastically in arid and mediterranean climates of Chile and Argentina. Fruit harvests would foreseeably occur 20 to 30 days earlier than now. This would slightly modify the international fruit trade between the northern and southern hemispheres.

6. Irrigation requirement and crop yield

Global climatic change would affect crop yield basically through (a) the increase of excess water periods in the Amazon region; (b) temperature

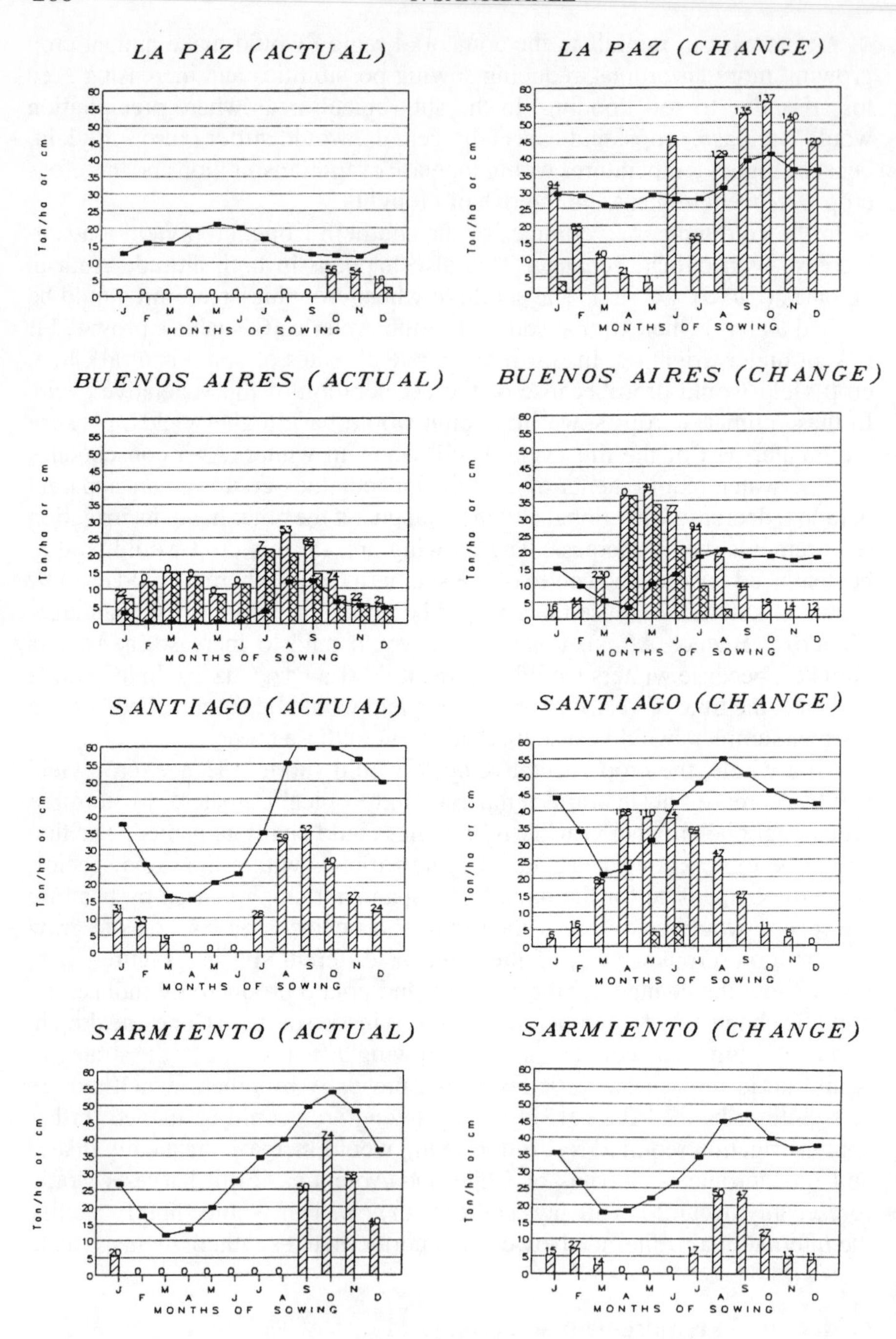

Figure 7. Rainfed (mesh) and irrigated (inclined line) monthly crop yield of sown potato (t/ha) in selected localities; on the left, under the present climatic situation; on the right, the same quantities, modified by a climatic change. The continuous line indicates the total irrigation requirement (cm) for each sowing date. The number over the column expresses the water use efficiency (kg dry matter/mm applied water).

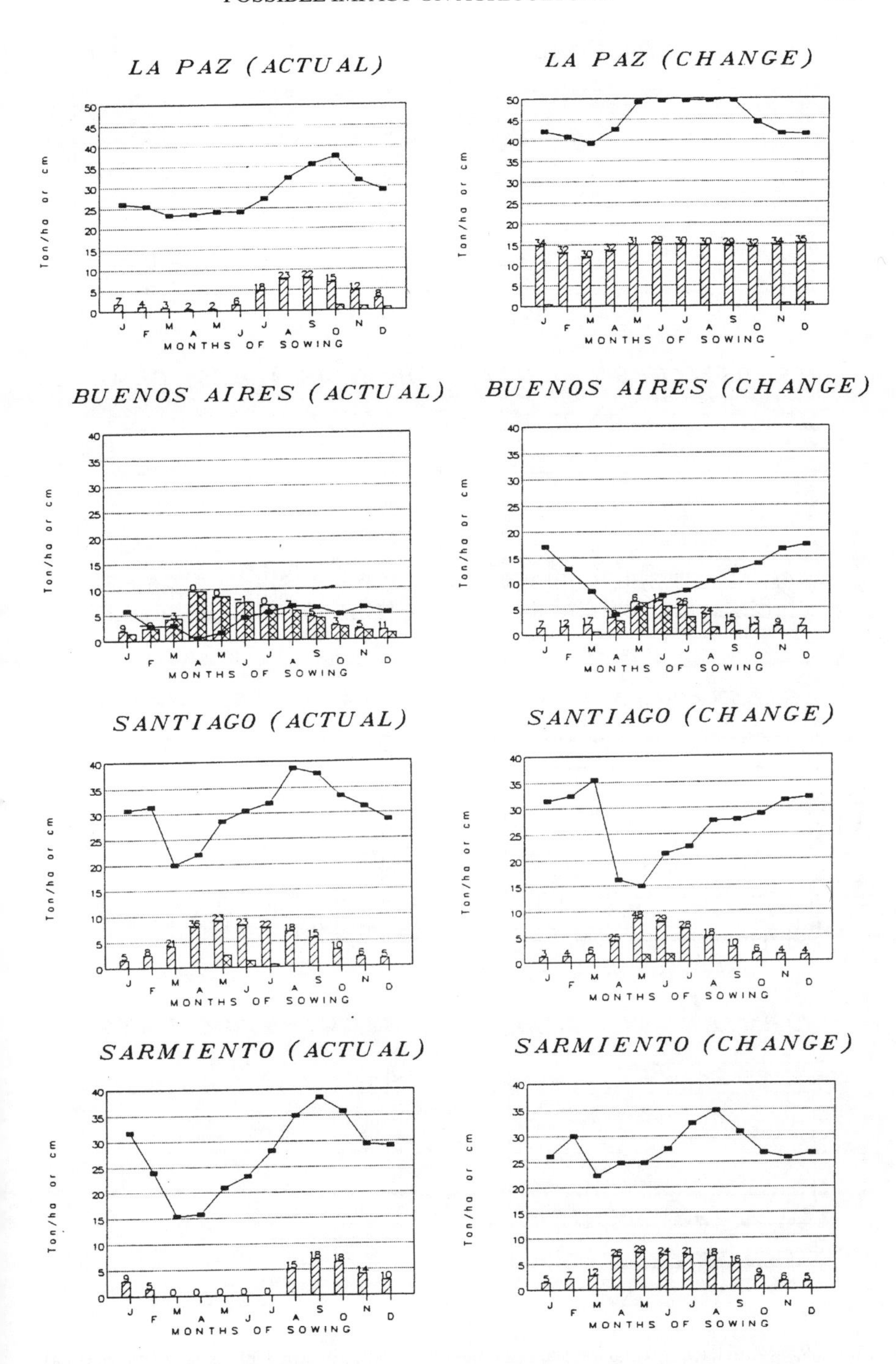

Figure 8. Rainfed (mesh) and irrigated (inclined line) monthly crop yield of sown wheat (t/ha) in selected localities; on the left, under the present climatic situation; on the right, the same variables, modified by a climatic change. The continuous line indicates the total irrigation requirement (cm) for each sowing date. The number over the column expresses the water use efficiency (kg dry matter/mm applied water).

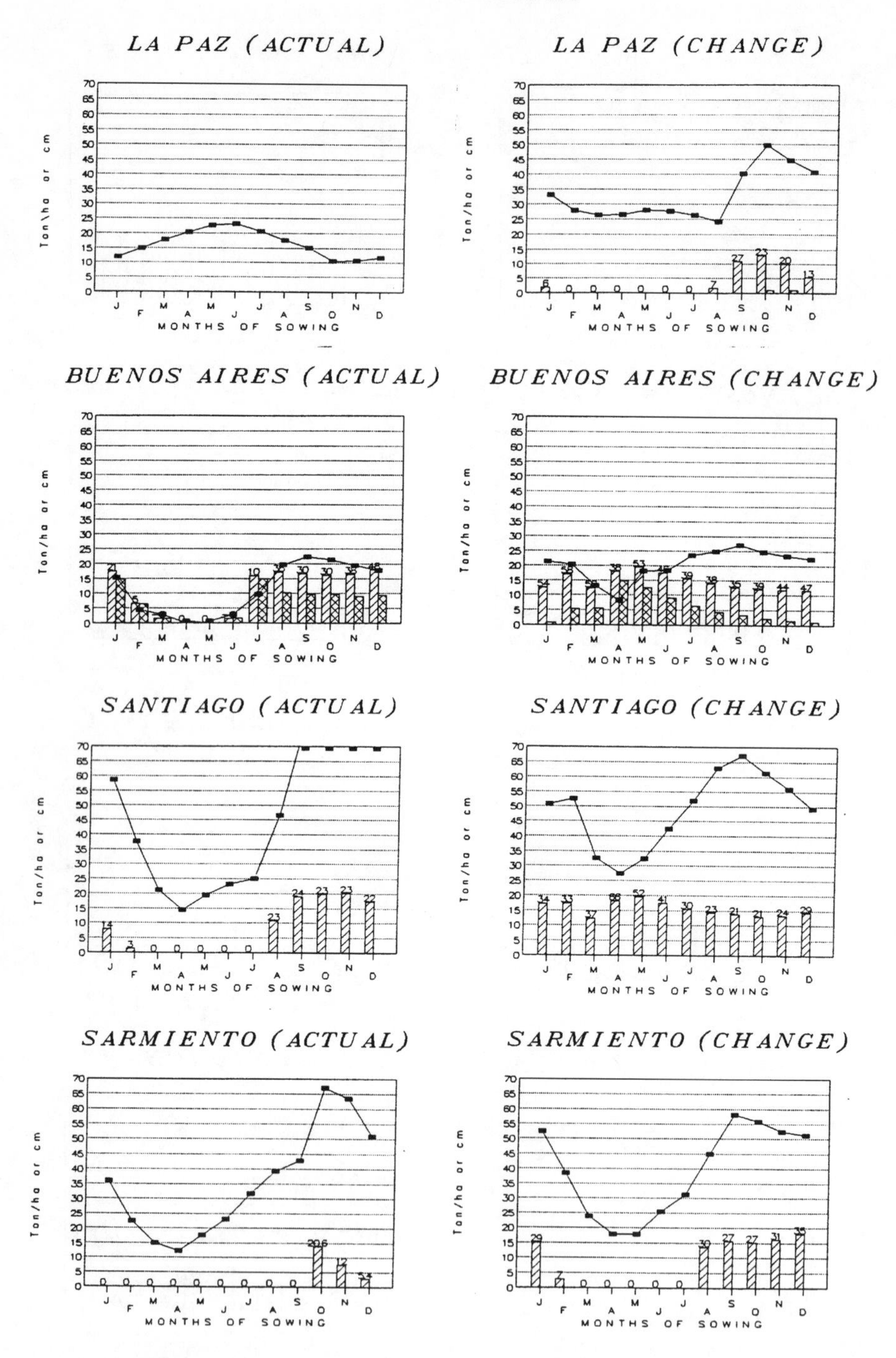

Figure 9. Rainfed (mesh) and irrigated (inclined line) monthly crop yield of sown maize (t/ha) in selected localities; on the left, under the present climatic situation; on the right, the same variables, modified by a climatic change. The continuous line indicates the total irrigation requirement (cm) for each sowing date. The number over the column expresses the water use efficiency (kg dry matter/mm applied water).

increase causing thermal stress in crops like potatoes, peas, and some temperate climate fruit trees; and (c) a drop in precipitation in subtropical areas, making dry-land farming, principally cereals, more hazardous.

The rise of evapotranspiration, due to the increase of mean temperature, would not necessarily mean greater irrigation requirements for all crops, because the increase in irrigation rates would be compensated by the shortening of their growth cycle and an increase in water use efficiency due to high CO_2 level. This phenomenon is especially noticeable for winter crops in mediterranean type climates, where a temperature increase would allow for a shift in the sowing date, so as to make crop growth coincide, in the most advantageous way possible, with natural precipitation. The greatest rise in irrigation requirements could occur in the extreme south temperate climates, and those of high altitudes (La Paz), for almost all crops. In these areas, the implementation of new irrigation projects would be especially profitable if irrigation water efficiency (kg/ha.mm) is taken into account: from 100 to 120 for potato, and from 30 to 50 for wheat and maize (Figures 8 and 9). In some subtropical and mediterranean areas, such as southern Brazil, Argentina and central Chile, irrigation projects could benefit because moderate amounts of water could result in high production efficiency.

7. Fluctuation in the precipitation regime and its economic influence on agricultural production – cereal and potato production: A case of a small farmer in a rainfed area

The interannual variability of rainfall is the factor that most affects net income fluctuation in dry-land farming. So as to grasp the magnitude of its economic effect as well as determine the probable hazard this may represent to agriculture, a model is needed to integrate unit production (farm) and yield variability (generated by the SIMPROC model), with production costs and the commercial value of the harvested crop.

For this reason a farm model was created (MODAP) to combine the following variables: several technological levels, different intensities of commodity (input) use, the land use structure, commodity costs, hand labour requirements at different levels of crop development and labour, and the market value of what is produced. As a consequence of joining the MODAP and SIMPROC models, an economic-productive value is obtained, at the level of an agricultural production unit, whenever the value of one of the variables is changed. The output of the model is: production, hand labour requirement, capital needed for operation, total production costs, and net income. When there is a variation in the climatic condition input, there is a change in yield. The model, complete with 90 variables, generated results which are shown to be close to reality for small farmers in a pilot project area located in the central zone of Chile. Figure 10 shows the coupling of the SIMPROC and MODAP models for economic impact evaluations.

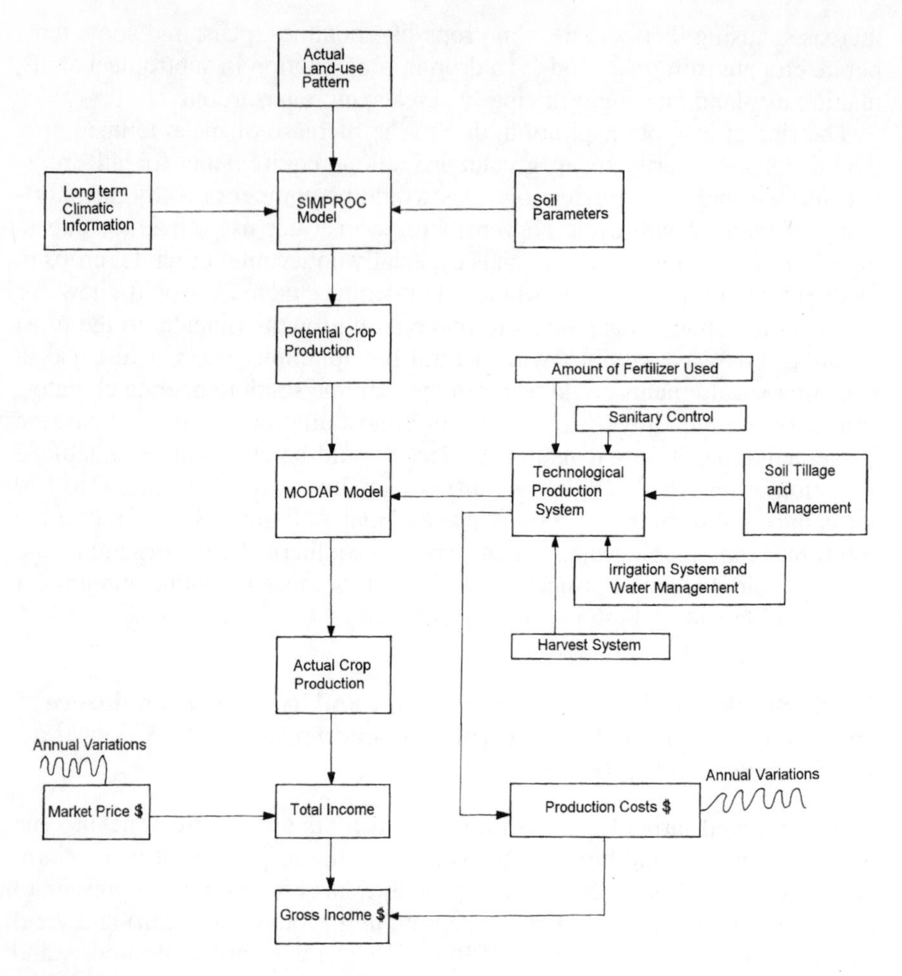

Figure 10. Coupling of the SIMPROC-MODAP models for economic impact evaluations of climatic change.

In this section, the model will be used to simulate the economic position of a small (10 ha), low technology farmer located in Traiguen (38°15′ S, 72°40′ W) southern Chile. The climate is temperate with a mediterranean precipitation regime. Using a 30-year precipitation series (1951–1980) and the SIMPROC-MODAP models, an evaluation of the variability in production and its economic impact was made, taking into account the present local production costs and market prices for wheat. A second simulation followed in which a climatic change algorithm, including a 15 to 20% decrease in current annual precipitation values, was utilized.

Under a climatic change scenario, the results clearly indicated a yield drop in wheat. For several years net income was near 0, and in four out of the

WHEAT

LONG TERM VARIATION IN GROSS INCOME

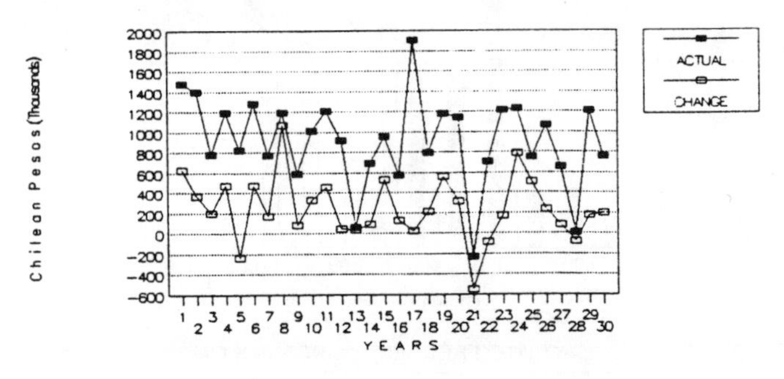

POTATO

LONG TERM VARIATION IN GROSS INCOME

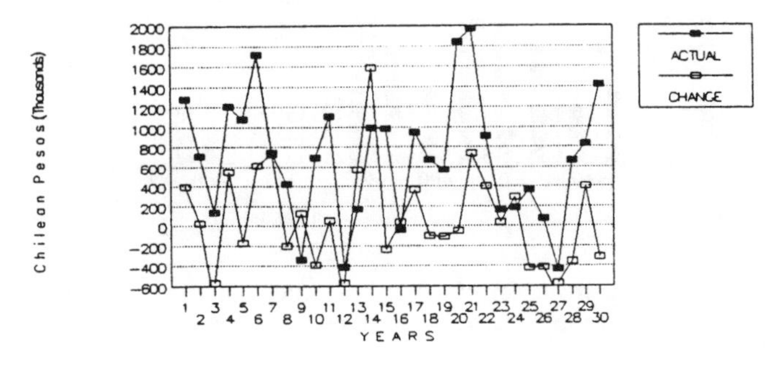

1 US$ = 350 Chilean pesos (July 1991)

Figure 11. Long-term variation in gross income of wheat and potato, based on a 30-year climatic simulation (1951–1980).

30 years under review net income was negative. The increase of economic vulnerability, due to climatic change, is clear.

In the case of potatoes its sensitivity is even greater. The damaging effect of a decrease in precipitation appears, the increase of temperature decreases yield due to the shortening of the tuberization phase, and a reduction of tuber formation in favour of aerial crop development. The variation in dry-farming potato yield is greater than for wheat, because of its greater sensitivity to water stress. With climatic change, potato yields drop and variability increases. Because of this, net income also falls, making this crop hazardous to grow without irrigation. Figures 11 and 12 show simulated interannual variations of gross income for wheat and potato crops over a long period of time.

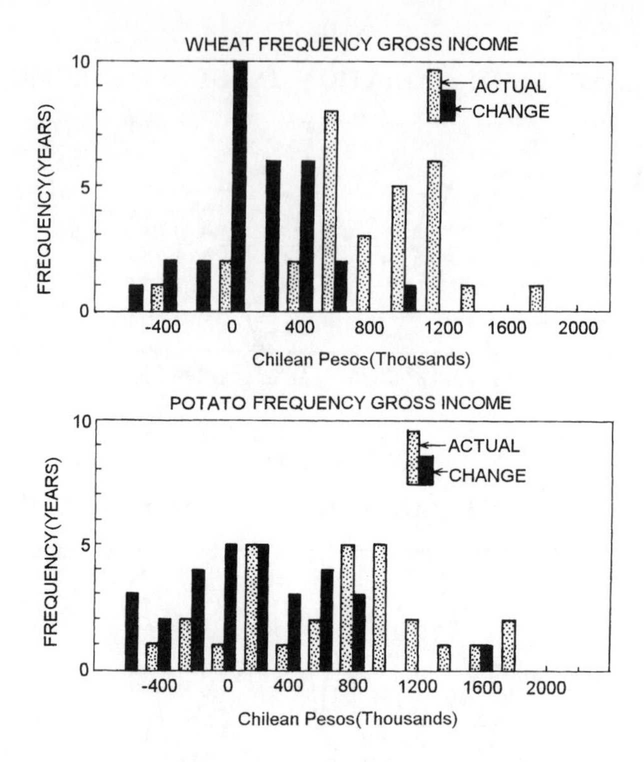

Figure 12. Gross income frequency of a 5 ha production unit of wheat or potato under the present climatic situation and, the same, modified by a climatic change (based on simulation showed in Figure 11).

8. Annual semi-arid mediterranean grassland behaviour and sheep load

In arid and semi-arid winter-precipitation climates of South America, climate discourages the use of intensive till farming. In these areas, animal production is the only activity which sustains the agricultural population economically. Precipitation variability is high and for that reason, a great interannual variation in forage and load capacity of the grassland represents a serious problem in herd management in a rational productive scheme. In dry years, excessive animal grazing load leaves the soil without protecting plant cover, leading to soil degradation and desertification, both with an important socio-economic impact.

From this point of view, it is important to evaluate the effect of interannual variability of precipitation on dry matter production of grasslands, and to quantify the animal load capacity using a probability approach. On the basis of the historical records of precipitation of the semi-arid mediterranean area of central Chile (Santiago, 33°27′ S, 70°42′ W), and the annual mean of 320 mm,

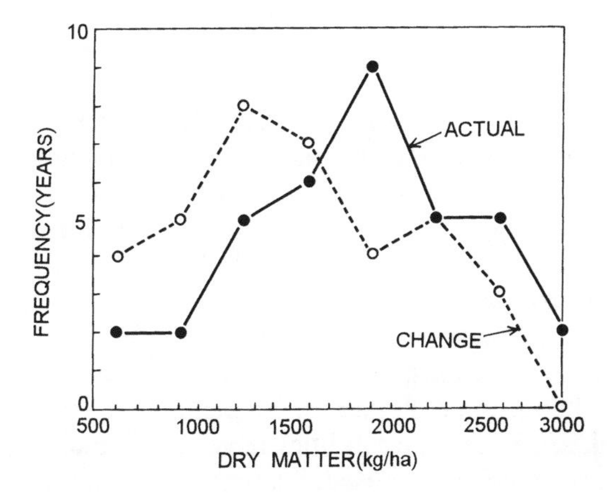

Figure 13. Simulated dry matter production frequency, under the present condition (actual) and after a climatic change, based on a 30-year climatic simulation (1951–1980).

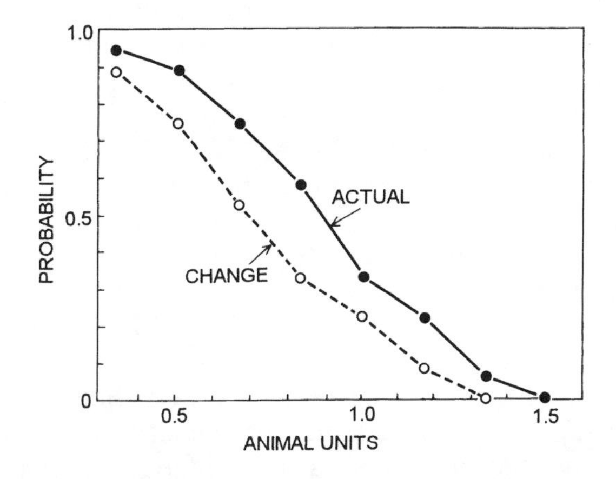

Figure 14. Variations in the animal load capacity of an annual mediterranean prairie as a consequence of the simulated climatic change. The ordinates indicate the probability of success with different animal loads (sheeps). Based on a 30-year climatic simulation (1951–1980).

the SIMPROC model has been useful in establishing the annual curves of dry matter accumulation, which have served as a base for a frequency analysis.

The model shows that mean annual biomass production would range from 1800 to 2000 kg DM ha^{-1} (that is within the same range as the production observed in the area) in well managed grasslands. Maximum values range from 650 to 2950 kg DM $ha^{-1}y^{-1}$. Following a precipitation decline of 25% induced by a climatic change, the mean annual biomass production dropped into the 1200 to 1400 kg DM ha^{-1}range, with maxima in the same order as before (Figure 13). The animal load capacity varies as shown in Figure 14.

9. Final remarks

Even though certain areas will suffer a foreseeable reduction in annual precipitation, it is difficult at present to estimate how this reduction would affect the seasonality and type of precipitation. One possibility is that seasonality does not change and that the amount of water falling every time it rains, is a fixed proportion of what falls now. Another possibility is that rainfall increases this variability, while the total amount of rainstorms per year decreases. A third possibility is the combination of the two previous ones where both the amount of rainstorms and volume of rainfall per storm decrease. If we add seasonality to the possible change in precipitation, the picture becomes even more complex. Tests of the first three possibilities have shown that the models have a certain sensitivity to the criteria used, making predictions uncertain and arbitrary. In this approach, we decreased the total amount of precipitation per year, which is a severe condition; therefore, the length of the dry period increased. This strongly limits production, because spring drought hazard increases during the flowering and grain filling stages of wheat. If this possibility were to occur, two groups of strategies would have to be implemented so as to avoid a drop in farmers' net income. One would be to improve production technology and increase yields while rationalizing the use of commodities that have the greatest effects on lowering production costs. Another group of strategies would consist of the introduction of soil management systems that would allow rain water to be efficiently used, and the implementation of back-up irrigation programes in areas that are considered irrigable. If none of these strategies were implemented, semiarid to subhumid subtropical areas will be very vulnerable. The joining of the VARICLIM-SIMPROC-MODAP models is creating a useful tool in the prospection and zonification of potential climatic change economic impact in different areas of South America. Before coming to general conclusions, a significant amount of work has to be done to determine the sensitivity of the models to the different assumptions contained in the simulation.

References

Berger, A. (1988): Milankovitch theory and climate. *Rev. of Geophysics*, 26(4), 624–657.

Conicyt (1989): El cambio global del clima y sus eventuales efectos en Chile. Comisión Nacional de Investigación Científica y Tecnológica (CONICYT), Comité Nacional del Programa Internacional de la Geófera-Biósfera IGBP), July 1989, Santiago, Chile, 26 pp.

Gates, W.L., Cook, K.H. and Schlesinger, M.E. (1981): Preliminary analysis of experiments on the climatic effect of increased CO_2 with an atmospheric general circulation model and a climatological ocean. *J. of Geophysical Research*, 86, 6385–6393.

Hansen, J. and Takahashi, T. (1984): *Climate processes and climate sensitivity*. Geophysical Monograph, 29, Maurice Ewing, Vol. 5, American Geophysical Union, Washington, D.C.

Manabe, S. and Wetherald, R.T. (1975): The effect of doubling the CO_2 concentration on the climate of a general circulation model. *J. of Atmospheric Sci.*, 32, 3–15.

Penning de Vries F.W.T. and van Laar, H.H. (1982): *Simulation of plant growth and crop production*. Centre for Agricultural Publishing and Documentation (PUDOC), Wageningen, 308 pp.
Santibáñez, Q.F. (1983): Antecedentes para la evaluación del sistema de información agrícola. Ministerio de Agricultura, Oficina de Planificación Agrícola. Chile, 62 pp.
Santibáñez, Q.F. (1986): Modelisation agroecologique appliquée à l'analise et zonification de la productivité des cultures. Thèse Doctorat d'Etat des Sciences, Université de Paris, 350 pp.
Santibáñez, Q.F. (1990): SIAM, Sistema de Información Agrometeorólogica para la evaluación de la prodcutividad y necesidades de riego de los cultivos y del impacto de las variaciones climáticas sobre estos parámetros. Edited by the author, Santiago de Chile, 86 pp.
Santibáñez, Q.F., Gajardo, R. and Denham, V. (1988): Quelques remarques sur les rapports historiques climat-végétation dans le Chili Central: cas d'une forêt laurifoliée relictuelle soumise à un déficit hydrique seculaire. In: *Time scales and water stress, Proceedings of the 5th International Conference on Mediterranean Ecosystem*. IUBS, Paris, 483–487.
Schlesinger, M.E. (1984): *Climate model simulation of CO_2-induced climatic change*. Advances in Geophysics, Vol. 26. Academic Press, New York, 141–235.
Schlesinger, M.F. and Mitchell, J.B.F. (1987): Climatic model simulations of equilibrium climatic response to increased carbon dioxide. *Rev. of Geophysics*, 25(4), 760–798.
WMO-UNEP (1988): Developing policies to respond to climatic change. WMO/TD-N, 225.

Fernando Santibáñez
Facultad de Ciencias Agrarias y Forestales
Universitad de Chila
Casilla 1004
Santiago
Chile

3.6. CLIMATE AND AGRICULTURE IN CHINA

AI-LIANG JIANG

1. Introduction

China has made great efforts in developing agriculture and achieved significant success in agricultural production during the last 40 years. The total annual production of staple food crops (grains and others) was 132 mill. t. in 1950, it increased to 435 mill. t. in 1990. Even so, China is still stressed for food due to it excessively large population (1,134 mill. in 1990).

The basic characteristics of agricultural production in China can be identified as the distinct spatial differentiation and remarkable temporal (inter-annual) variation. Both characteristics are closely related to the climate of China.

2. General view of the climate of China

2.1. *Spatial distribution of the main agro-climatic elements*

2.1.1. *Sunshine duration*

By comparing the maps of the distribution of sunshine duration and global radiation (Figures 1 and 2) it is found that the general patterns are very similar in many ways. Hence, only the distribution of sunshine duration is discussed in detail (Figure 1). It can be seen that the duration of sunshine is very large in the northern and western parts of China where annual totals are mostly between 2400 and 3200 hrs/yr, or 6.6 and 8.8 hrs/day on average. The largest value amounts to 3600 hrs/yr or almost 10 hrs/day at Lenghu (38°50′ N, 92°23′ E, 2733 m). The region of lowest sunshine duration comprises Sichuan Basin, Guizhou Province and a part of Guangxi Autonomous Region where, on average, annual totals drop to 1150–1400 hrs/yr or only 3–4 hrs/day.

Annual sunshine duration in China seems to have no significant influence on agriculture in general. For example, Sichuan Basin, although located in the

M. Yoshino et al. (eds.), Climates and Societies – A Climatological Perspective, 279–307.

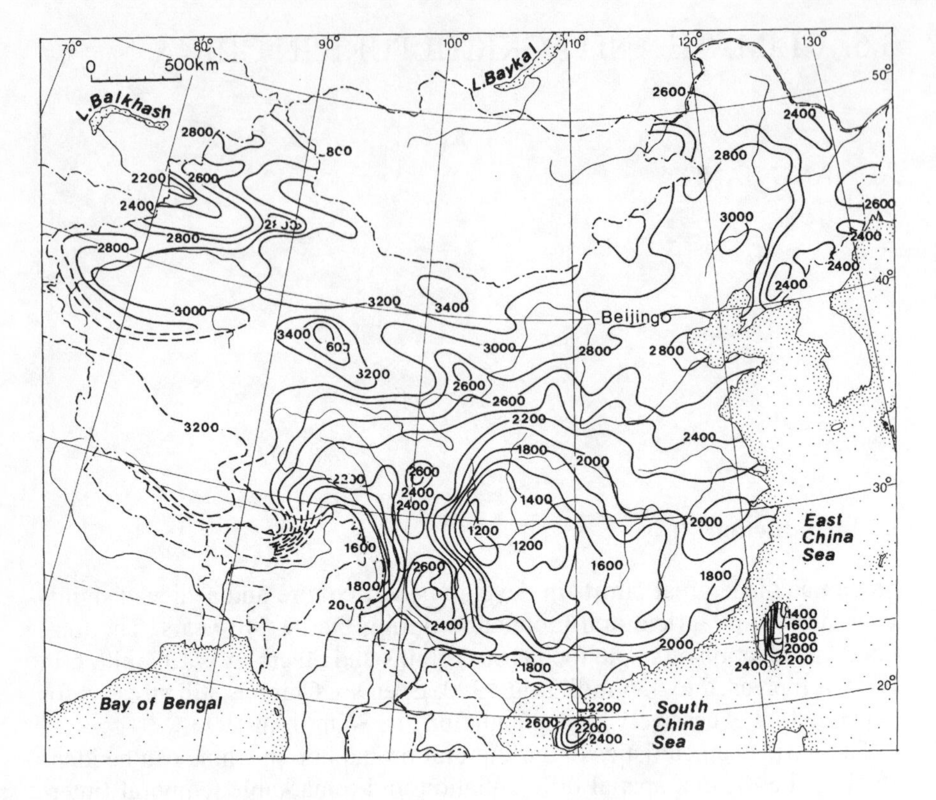

Figure 1. Distribution of mean annual sunshine duration (hrs).

region of the lowest sunshine duration, is known as the "land of abundance", due to its fertile soil, favourable temperature and ample precipitation. At the opposite end, despite very large amounts of sunshine, agricultural production is very limited, in general, in most areas in the western part of China, due to either arid or cold climatic conditions, and very limited arable lands. But in some specific cases, rich sunshine usually promotes high crop yields. For example, Lhasa (29°42′ N, 91°08′ E, 3658 m) and Chungqing (29°31′ N, 106°29′ E, 351 m), although located in the same latitude, receive different sunshine totals 3000 hrs/yr and 1540 hrs during October–March, 1200 hrs/yr and 330 hrs, respectively. The difference in winter values appears to correlate with differences in winter wheat yields which may exceed 1500 jin/mu (11.25 t/ha) in some experimental stations in Lhasa, while it is very difficult to get yields of 600 jin/mu (4.5 t/ha) in Chungqing.

2.1.2. *Temperature*

Owing to the generally warmer summers and much colder winters in China than in most localities at the same latitude and elevation, mean annual tem-

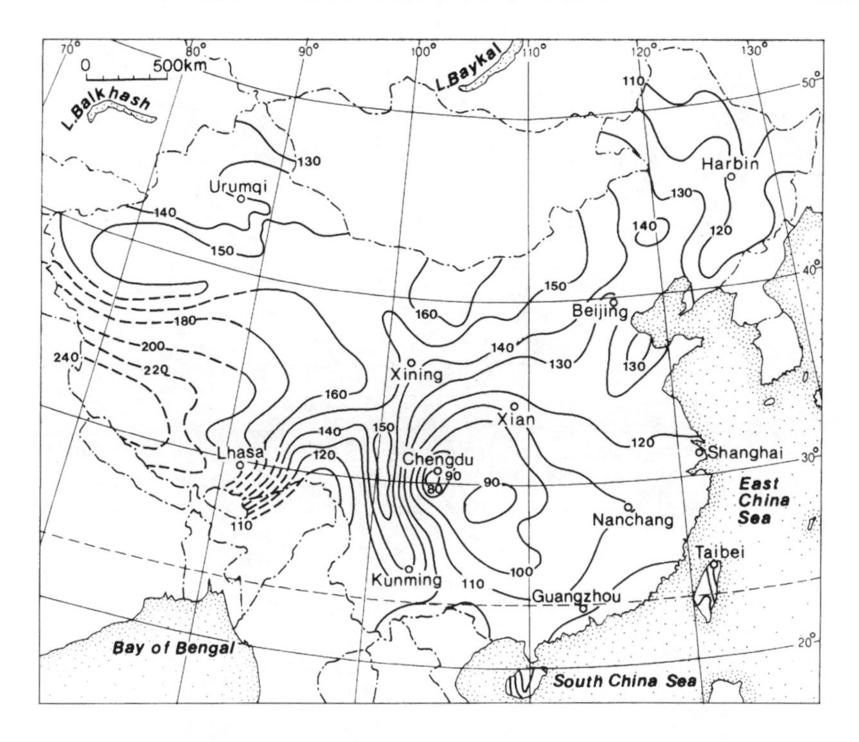

Figure 2. Distribution of annual global solar radiation (kilocal/cm^2 $*$ year) (1 kilocal/cm^2 $*$ year = 41.868 $*$ 10^6 J/m^2 $*$ yr).

peratures do not seem to have a clearly defined significance in agriculture. For example, the mean annual temperature of Nenjiang (49° 10′ N, 125° 13′ E, 222 m) is −0.4°C. Such a low temperature seems to account for no or only very limited agriculture in Nenjiang. However, because there are many crops, even thermophilic crops like rice, maize, etc. which flourish in Nenjiang in the warm summer, many Chinese agro-climatologists consider the accumulated temperature a better thermal index for crop growing potential. Two expressions of accumulated temperature are widely used for agriculture in China (Figures 3 and 4). One is the accumulated temperature > 0°C, usually used to analyze the thermal regime of temperate-and-cool-zone crops, such as wheat, barley, etc.; the other is the accumulated temperature > 10°C which is considered more appropriate for analyzing the thermal regime of thermophilic crops.

The mean monthly temperatures of the coldest and warmest months and the number of frost-free days are considered to be useful as agroclimatological indices.

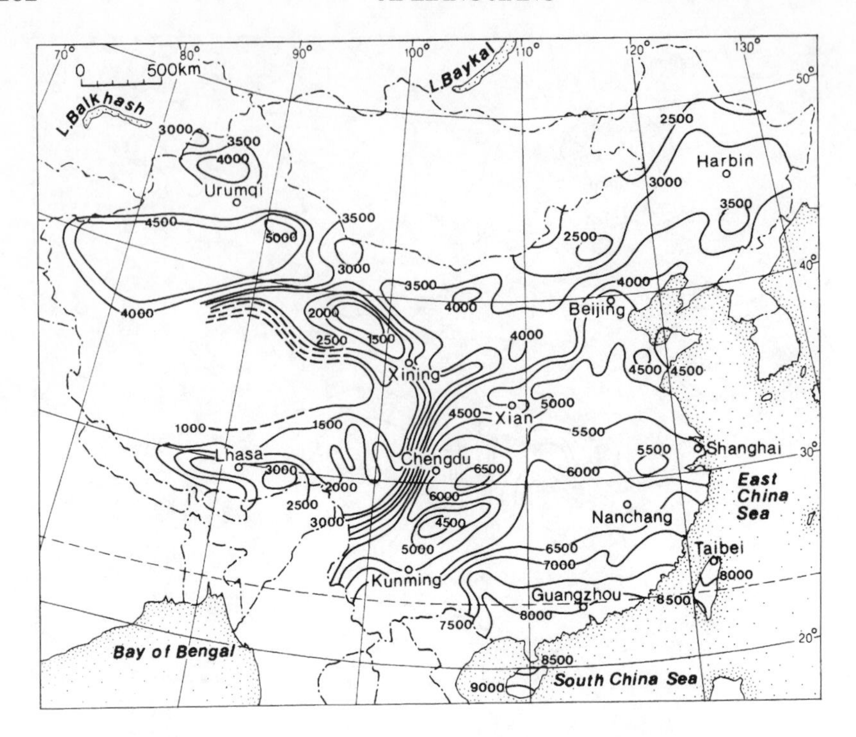

Figure 3. Distribution of annual totals of the accumulated temperature GT > 0°C (degree days).

2.1.3. *Precipitation and evaporation*

Annual precipitation is generally considered the most important element for agriculture; its distribution is shown in Figure 5. Yet, to study the water balance in a crop field, evapotranspiration must also be considered. Figure 6 shows the distribution of annual potential evapotranspiration in China calculated on the basis of Penman's formula. Figure 7 shows the water balance in terms of a surplus or a deficit, which is obtained by subtracting potential evapotranspiration from precipitation. A positive value indicates water surplus, a negative one expresses water deficit.

2.2. *Warm summer – cold winter*

The most striking feature of China's climate is the fact that latitude for latitude and elevation for elevation, summers are warmer and winters much colder in China than elsewhere on Earth. Proof is evidently given by a comparison of the monthly mean temperatures of the coldest (January) and warmest (July) months and the annual temperature range for four stations at 20, 30, 40 and 50°N in China, on the one hand, with the corresponding latitudinal mean

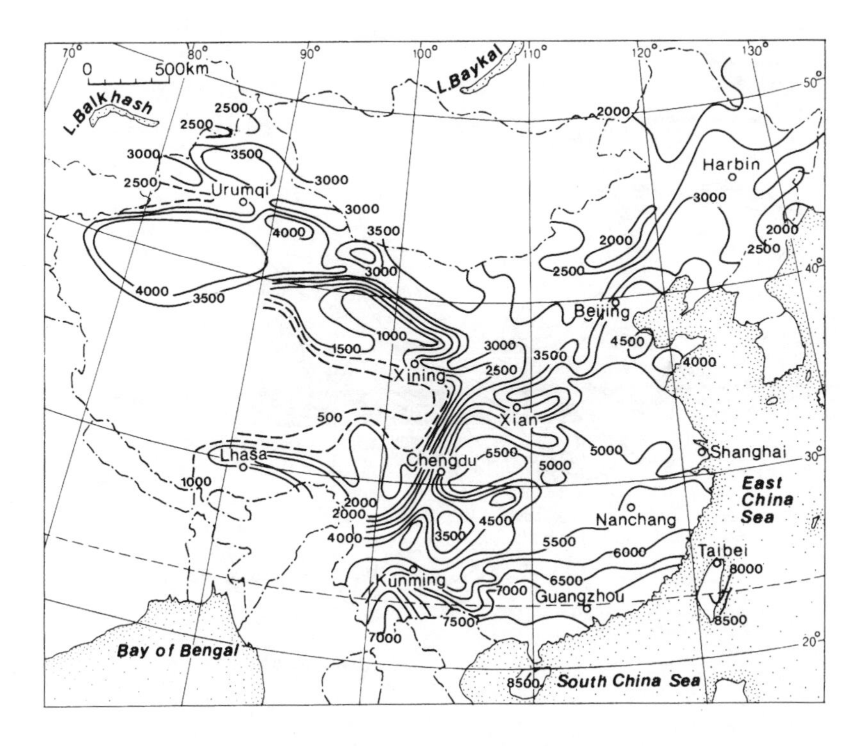

Figure 4. Distribution of annual totals of the accumulated temperature GT > 10°C (degree days) (after Zhang and Lin 1985).

Table 1. Comparison of temperatures (°C) for stations in East China with those averaged for the latitudinal circle of northern hemisphere (1961–1980, after Jiang 1984a).

Name of station and coordinates	East China			Latitudinal average		
	Jan.	July	Annual range	Jan.	July	Annual range
50 °N Nenjiang (49° 10′ N, 125° 13′ E, 222 m)	–25.0	20.7	45.7	–6.3	16.9	23.2
40 °N Beijing (39°48′ N, 116°28′ E, 32 m)	–4.5	25.8	30.3	5.7	23.2	17.5
30 °N Wuhan (30°30′ N, 114°04′ E, 23 m)	3.1	28.7	25.6	15.2	27.3	12.1
20 °N Haik°u (20°20′ N, 110°21′ E, 14 m)	17.1	28.3	11.2	22.3	28.0	5.7

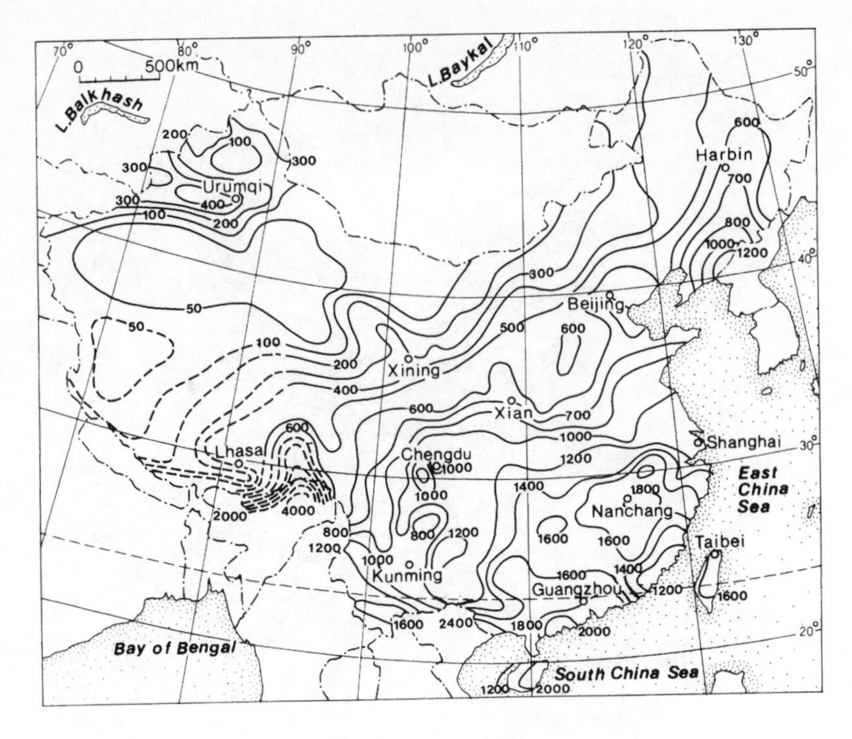

Figure 5. Distribution of annual precipitation (mm).

values in the northern hemisphere, on the other (Table 1). North China has much colder winters and larger annual temperature ranges.

The well-marked contrast between warm summers and cold winters is of direct agro-climatological significance. For example, rice thrives well in Nenjiang due to high summer temperatures and abundant sunshine; rainfall is also favourable in most years. Beijing experiences such a hot and long summer that the length of the period with temperatures > 20°C exceeds 110 days, the accumulated temperature (T > 0°C = 4544°C deg. days and T > 10°C = 4130°C deg. days) and the length of the growing season (269 and 199 days for temperature > 0°C and > 10°C) fulfill the requirements for double-cropping of winter wheat and rice. This system is, however, not widely accepted due to the high labour demand of wheat harvest and timely transplanting of rice. At Wuhan, the summer is even hotter and longer so that double-cropping of rice is widely practiced.

The cold winters in North China will occasionally cause some freeze damage to winter wheat. In some winters, drop of temperature in subtropical and tropical China is so great and the duration of cold spells so long that frost damage to subtropical plants and chilling injury to tropical plants occur.

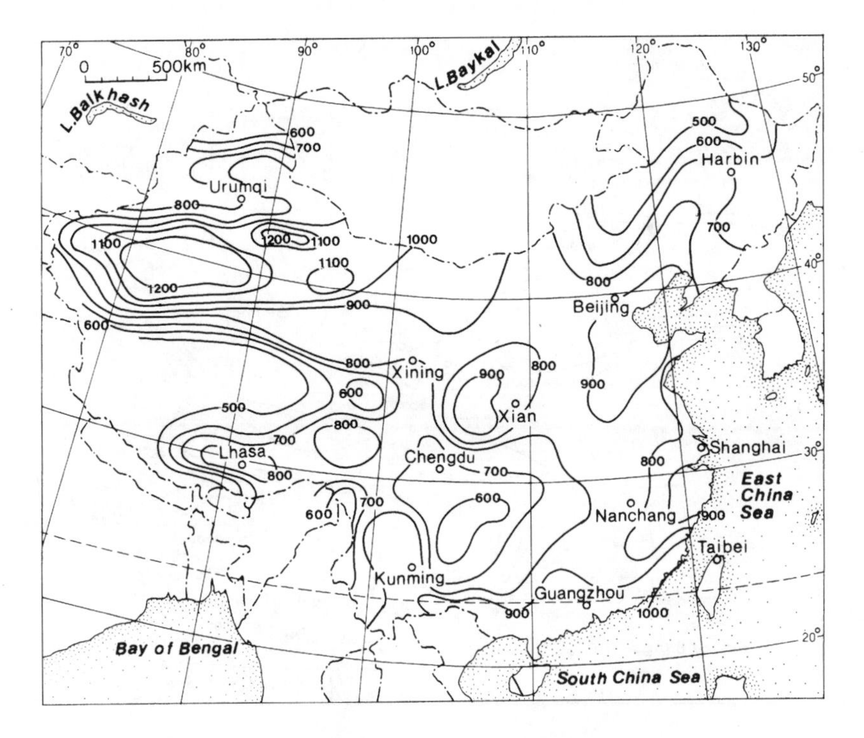

Figure 6. Distribution of annual potential evapotranspiration (mm).

2.3. *"Synchronization" of precipitation and temperature, seasonality of precipitation*

China has a typical monsoon climate. Precipitation is concentrated on the warm summer. Therefore, crop growing is greatly favoured in general; this is called by some Chinese climatologists the "seasonal synchronization of water and heat". However, sometimes excessive precipitation in a short period leads to floods and water-logging for crops. Besides, excessive precipitation also causes heavy soil erosion on slopes in case of a large-scale deforestation. Hence great attention must be paid to conservation of forests on slopes in order to prevent soil erosion.

Jiang (1984) divided the year equally into two parts: a dry half-year from October/ November to March/April, and a wet half-year from April/May to September/October. Both periods coincide with the regimes of the winter and summer monsoons. The wet half-year usually receives more than 70% of the annual total (Table 2). In Beijing, even more than 90% of the annual precipitation is experienced from May to October, corresponding to the summer monsoon.

The most striking seasonal variation of precipitation occurs in Lhasa, where in the dry half-year only 3% of the annual total is recorded, but 97% in the

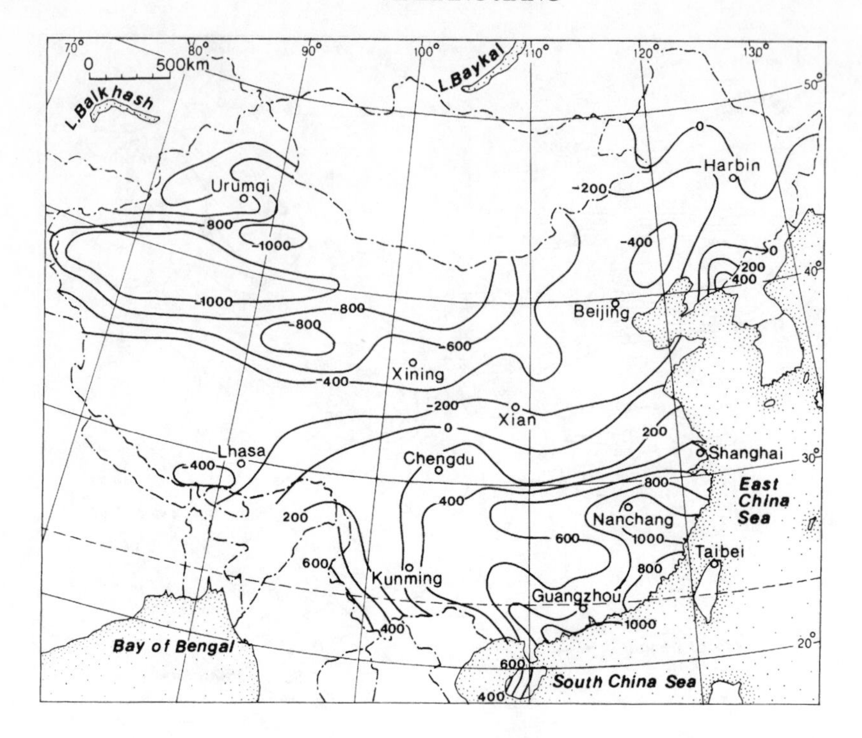

Figure 7. Distribution of annual field water balance (surplus or deficit) (mm) (−: deficit, +: surplus).

wet half-year. The reason for this is mostly the effect of the Indian southwest monsoon which usually onsets in May and retreats in September/October.

3. Regionality of climate and agriculture

3.1. *Main natural regions and agricultural production*

Dating back to the fifties, Chinese geographers divided China into three main natural regions: East Monsoon Region, Northwest Arid Region and Qinghai-Xizhang (Tibetan) Frigid Plateau (Commission on Nature Regionalization 1959). These three regions occupy 47, 26 and 27% of the total area of China. Comparing the percentages of the annual production of food crops (including grains) for the regions from 1977–1986 (Table 3) significant discrepancies can be seen between the percentage shares of space and food production for the three regions concerned.

Table 2. Seasonality of precipitation at different stations (1954–1983).

Localities	Stations' coordinates	Annual precipitation (mm)	Precipitation in the half year to the annual total (%)		Duration of the wet half year
			dry	wet	
Beijing	39°08′ N, 116°28′ E, 32m	576	9	91	May–Oct.
Shanghai	31°10′ N, 121°26′ E, 5m	1077	29	71	Apr.–Sept.
Nenjiang	49°10′ N, 125°13′ E, 222m	458	9	9	May–Oct.
Urumqi	43°47′ N, 87°37′ E, 918m	211	36	64	Apr.–Sept.
Wuhan	30°30′ N, 114°04′ E, 23m	1148	29	71	Apr.–Sept.
Chengdu	30°40′ N, 104°01′ E, 506m	921	12	88	Apr.–Oct.
Lhasa	29°42′ N, 98°01′ E, 3,658m	444	3	97	May–Oct.
Kunming	25°01′ N, 102°41′ E, 1,891m	1035	12	88	May–Oct.
Guangzhou	23°08′ N, 113°19′ E, 6m	1672	18	82	Apr.–Sept.
Haikou	20°20′ N, 110°21′ E, 14m	1639	19	81	May–Oct.

Table 3. Food production and climate of the three main natural regions in China (1977–1986).

	East Monsoon Region	Northwest Arid Region	Tibetan Plateau
Area to the total of China (%)	47	26	27
Annual food production to the total (%)	95	4.5	0.5
Annual precipitation (mm)	500–1500	50–200	50–500
Frost-free days*	100–365	100–200	<100
Annual sunshine duration (h)*	1200–2800	2600–3200	2600–3000

*For most localities.

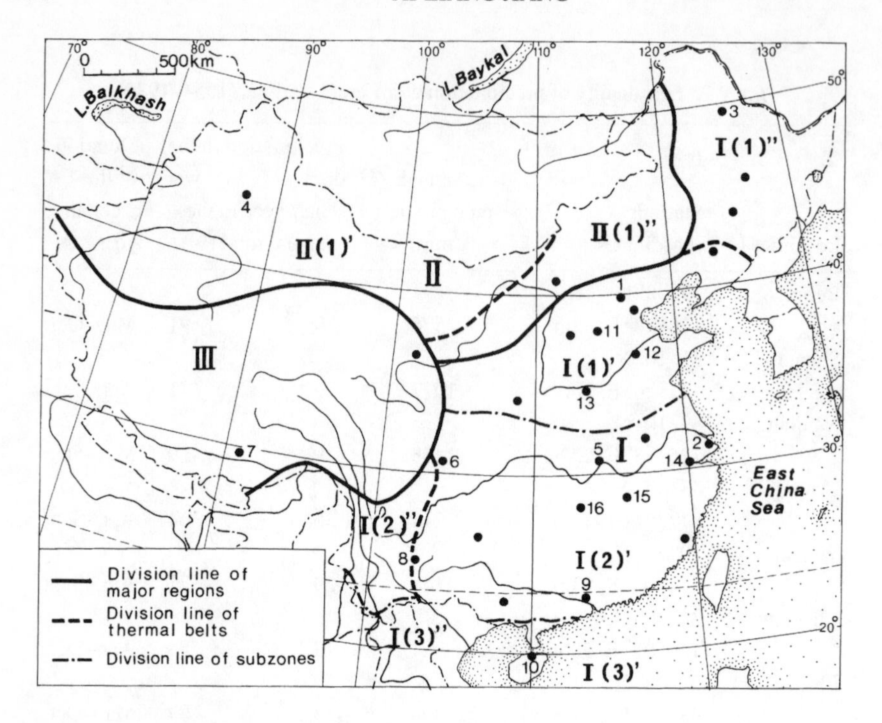

Figure 8. Regionalization of climate for agricultural land use in China (a slight modification was made to the original one) (Jiang 1984b). The difference is that the semi-arid subzone was located in the East Monsoon Region in the original regionalization, but now the semi-arid subzone is demarcated in the Northwest Arid Region.

3.2. *Climate regionalization for agricultural land use*

In a map showing the climate regionalization for agricultural land use (Jiang, 1984; see Figure 8) thermal regimes of temperate, subtropic and tropic belts are indicated by symbols (1), (2) and (3). More detailed divisions are indicated by subscript ($'$) and described in Table 4.

3.3. *Agroclimate regionalization of China*

According to a detailed regionalization of the agroclimate of China by Li et al. (1988), China is firstly divided into three regions which are similar to the main natural regions as mentioned above; secondly, the country is divided into 15 thermal belts and, finally, into 55 agroclimatic subzones. An objective appraisal to this regionalization has been carried out by a State Academic Committee of China.

Table 4. Regionalization of the climate of China for agricultural land use (Jiang 1984).

Main natural region	Thermal belt	Subdivision	Ways of land use
I: Eastern Monsoon Region	I(1): Temperate belt in East Monsoon Region	I(1): Subzone with warm temperate temperature and moderate rainfall.	Crop cultivation with a mixing system of single, and a half and double cropping.
		I(1): Subzone with lower temperature and moderate rainfall.	Crop cultivation with single cropping.
	I(2): Subtropical belt in East Monsoon Region	I(2): East part of subtropics with higher temperature in summer but lower minimum temperature in winter	Double even triple cropping; many subtropical plants can be cultivated but with frost damage occasionally in northern part.
		I(2): West part of subtropics with lower temperature in summer but higher minimum temperature in winter	Single cropping and the overwintering of subtropical plants is better below some altitudes. The upper limits of cultivation of these plants and crops are higher than in I(2)
	I(3): Tropical belt in East Monsoon Region	I(3): East part of tropics with higher temperature in summer but lower minimum temperature in winter	Multiple cropping, many tropical plants can be cultivated but with chilling injury occasionally.
		I(3): West part of tropics with lower temperature in summer but higher minimim temperature in winter	Single cropping in general owing to very little rainfall in dry half-year; many tropical plants can be cultivated without advectional chilling injury. The upper limits are higher.
II: Northwest Arid Region		II(1): Arid subzone	Small part of area single cropping only if irrigation is available, small part for grazing.
		II(1): Semi-arid subzone	Partial grazing, partial cropping with or without irrigation.
III: Qinghai-Xizhang (Tibetan) Plateau			In the southern part, the upper limit of cropping may reach 4,000 m; in the northern part, the upper limit for cropping is about 2,500–3,000 m.

4. Variability of climatic elements and agricultural production

4.1. *General description of the interannual climatic variability*

4.1.1. *Precipitation*

One of the principal features of the monsoon climate of China is the remarkable interannual variability. Since the variability of precipitation is considered most important to agriculture, great attention has been drawn to this parameter by many Chinese climatologists. Variability was studied earliest by Liu (1936) who formulated the following index of variability (Vy):

$$\mathrm{Vy} = \frac{1/n \sum_{i=1}^{n} |\mathrm{Xi} - \bar{\mathrm{X}}|}{\bar{\mathrm{X}}}$$

where Xi is the annual precipitation of this year, $\bar{X}$ the mean value of Xi for n years, i = 1, 2, ... n, and n = number of years.

Based on a larger and better data series data, some detailed studies on precipitation variability with the same definition as before were carried out by Chinese climatologists. In addition, calculation of precipitation variability, expressed by the coefficient of variation (Cn), was made by Jiang (Table 5). Cn is defined as:

$$\mathrm{Cn} = \frac{\sigma \mathrm{x}}{\bar{\chi}}$$

where σx is the standard deviation of X.

For most stations the variability of annual precipitation is smaller than the monthly variability. Secondly, in North China, variability is smallest in July, being the warmest month with the highest precipitation. There a considerable amount of precipitation with highest reliability is supplied to crops during their growing season and this combination of high moisture supply and rainfall reliability is very beneficial to crop yield. Thirdly, in the middle and lower Yangtze River Basin (Wuhan and Shanghai), the smallest variability of monthly precipitation occurs in April or May, which are rainy months. This condition is not favourable for crops because winter wheat and other crops ripen in April or May. Ample precipitation, together with long spells of rainy days, accompanied by low sunshine, often cause diseases to crops.

4.1.2. *Temperature*

The values of variability of accumulated temperature (Table 5) are much smaller than those of precipitation, but some considerable reduction at some localities in North China, such as Nenjiang and Harbin, appears noticeable.

Table 5. Interannual variability of precipitation (Coefficients of variation, Cu) and accumulated temperature (GT $>$ 0°C) for selected stations (1954–1983)

Station and geographical coordinates	Variability (Cu) of precipitation (%)													Variability ($C\nu$) of accum.temp. $\Sigma T > 0°C$ (%)
	J	F	M	A	M	J	J	A	S	O	N	D	Ann	
Nenjiang 49°10′ N,125°13′ E, 222m	0.76	1.02	0.71	0.85	0.46	0.48	0.49	0.60	0.50	0.80	0.75	0.77	0.17	0.04
Harbin 45°41′ N, 126°37′ E, 172m	0.74	0.77	0.77	0.70	0.62	0.54	0.30	0.52	0.53	0.61	0.58	1.14	0.18	0.03
Changchun 43°54′ N, 125°13′ E, 237m	0.74	0.83	0.90	0.66	0.47	0.53	0.48	0.56	0.56	0.71	0.65	1.03	0.19	0.03
Dalian 38°54′ N, 121°38′ E, 94m	1.19	1.04	0.99	0.73	0.70	0.75	0.54	0.63	0.90	0.63	0.77	1.13	0.28	0.03
Hohhot 40°49′ N, 111°41′ E, 1,063m	1.35	1.31	0.96	1.05	0.68	0.65	0.60	0.66	0.53	0.70	1.35	1.41	0.35	0.03
Beijing 39°48′ N, 116°28′ E, 31m	1.74	1.08	0.75	1.25	0.81	0.75	0.50	0.66	0.80	1.09	0.78	1.61	0.38	0.04
Lanzhou 36°03′ N, 103°53′ E, 1,517m	1.21	1.41	0.73	0.79	0.68	0.50	0.49	0.58	0.50	0.75	1.17	2.01	0.28	0.03
Altay 47°44′ N, 88°05′ E, 735m	0.81	0.82	0.94	0.73	0.81	0.81	0.66	0.71	0.71	0.61	0.84	0.64	0.28	0.05
Urumqi 43°54′ N, 87°28′ E, 654m	0.84	0.56	0.61	0.52	0.58	0.72	0.67	0.85	0.60	0.65	0.51	0.64	0.26	0.07
Yining 43°57′ N, 81°20′ E, 663m	0.92	0.59	0.65	0.52	0.66	0.54	0.72	0.72	0.64	0.73	0.54	0.72	0.27	0.04
Yinchuan 38°29′ N, 106°13′ E, 112m	1.40	1.25	1.11	1.05	0.88	0.68	0.74	0.62	0.73	0.82	1.59	1.99	0.36	0.02
Xi'an 34°18′ N,108°56′ E, 397m	0.79	0.81	0.57	0.47	0.44	0.70	0.62	0.72	0.54	0.59	0.70	1.18	0.20	0.02

Table 5. (Continued)

Station and geographical coordinates	Variability (Cu) of precipitation (%)													Variability ($C\nu$) of accum.temp. $\Sigma T > 0\,°C$ (%)
	J	F	M	A	M	J	J	A	S	O	N	D	Ann	
Zhengzhou 34°43′ N,113°3′ E, 110m	0.95	0.92	0.74	0.80	0.94	0.88	0.58	0.66	0.76	0.69	0.93	1.44	0.24	0.02
Shanghai 31°10′ N,121°26′ E, 5m	0.65	0.53	0.44	0.37	0.41	0.40	0.68	0.75	0.56	0.94	0.69	0.82	0.17	0.02
Wuhan 30°30′ N,114°04′ E, 23m	0.56	0.59	0.48	0.40	0.41	0.59	0.85	1.10	0.74	1.05	0.80	0.88	0.26	0.02
Jinan 36°41′ N, 116°59′ E, 52m	0.13	0.89	0.72	0.78	0.74	0.80	0.51	0.58	0.74	0.84	0.97	1.12	0.31	0.03
Changsha 28°12′ N, 113°04′ E, 45m	0.57	0.44	0.40	0.31	0.35	0.48	0.74	0.85	0.85	0.62	0.76	0.58	0.14	0.02
Fuzhou 26°05′ N, 119°17′ E, 84m	0.94	0.69	0.49	0.48	0.37	0.42	0.76	0.59	0.75	0.92	0.80	0.90	0.15	0.01
Guangzhou 23°08′ N, 113°19′ E, 6m	1.28	0.99	0.77	0.54	0.54	0.40	0.56	0.56	0.68	0.79	1.10	1.27	0.20	0.01
Haikou 20°02′ N, 110°21′ E, 14m	0.84	0.78	0.74	0.69	0.38	0.52	0.60	0.42	0.56	0.79	1.29	1.16	0.20	0.01
Nanning 22°49′ N, 108°21′ E, 72m	0.74	0.71	0.56	0.62	0.39	0.49	0.48	0.42	0.59	0.57	1.00	1.20	0.13	0.01
Kunming 25°01′ N, 102°41′ E, 1,891m	0.70	1.09	0.99	0.94	0.67	0.40	0.30	0.36	0.47	0.67	1.05	1.13	0.15	0.02
Chengdu 30°40′ N, 104°01′ E, 506m	0.67	0.62	0.52	0.56	0.45	0.65	0.43	0.51	0.47	0.41	0.45	0.72	0.19	0.01
Lhasa 29°42′ N, 91°08′ E, 3,658m	2.07	2.01	1.26	0.94	0.82	0.50	0.36	0.36	0.42	1.15	2.73	3.01	1.23	0.04
Nanchang 28°36′ N, 115°55′ E, 47m	0.70	0.58	0.36	0.40	0.43	0.47	0.75	0.75	0.53	0.68	0.83	0.82	0.20	0.02

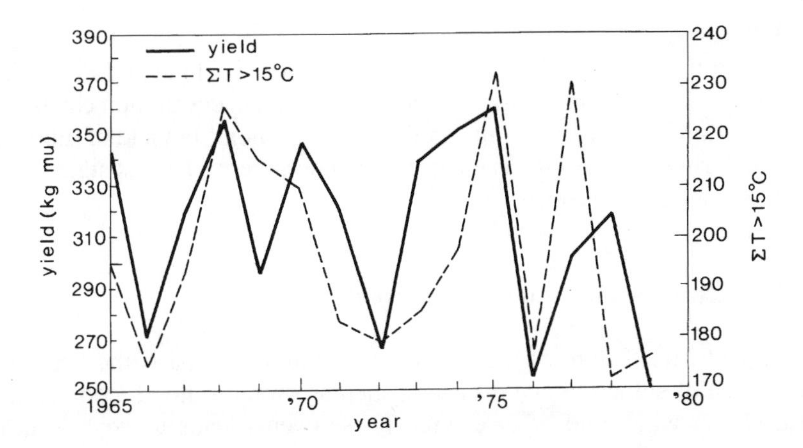

Figure 9. The relation between the annual yield of rice (jin/mu) and the accumulated temperature (GT > 15° C degree days) during the stage of flowering-grain-filling in Beijing.

4.2. *Impacts of interannual variability of climatic elements on crop yields*

4.2.1. *Precipitation*

In North China, precipitation is generally insufficient and unreliable for crops. For a non-irrigated field, a noticeable or even serious reduction of crop yields is normal. Lu (1963) and Ho (1979) who analyzed the relation between the variability of precipitation and yields of winter wheat in North China showed a positive correlation, expressing higher yields of wheat with higher precipitation. But in some southern provinces, where precipitation exceeds demand in most years, the correlation is negative. For example, a study on the relationship between the field water regime and the yield of winter wheat in Sazhou county (about 100 km west of Shanghai) showed an increasing loss of yield with a growing surplus water in the field (Shen 1984).

4.2.2. *Temperature*

Accumulated temperature in North China is usually not quite excessive for such thermophilic crops, as rice, maize and cotton. A good harvest would be expected in case of a warmer growing season, while a poor harvest would be the case when the temperature is much lower than normal. Proof is given by the analysis of observations at Shuangqiao State Farm in Beijing (Figure 9) showing the year-to-year yields of rice and the corresponding variability of accumulated temperature ($\Sigma T > 15°C$, deg. days) during the flowering and grain filling stages. As a result, a close synchronization can be seen between the fluctuation of rice yield and the accumulated temperature.

4.2.3. *Sunshine*

Analyzing the relationship between variability of sunshine and the yield of dry rubber for three State Farms (Jiang, 1988) a good correlation can be seen between monthly sunshine hours and the average monthly yield of rubber per tapping during 1981–1986. The correlation coefficient (0.45) is significant at the 99% confidence level.

5. Climatic disasters and extreme events

A characteristic feature of the monsoon climate of China is the occasional occurrence of some climate disasters, such as drought, flood, cool summer, frost, strong wind, and other events. These events leads to great losses in agricultural production in China, amounting to about 10 mill. t/yr, averaged over the last 40 years. In some years, the loss of food production was even more severe. For example, in 1988 the total cultivated area suffering from droughts and floods amounted to 50.7 mill. ha and accounted for a huge loss of production of 20 mill. t.

5.1. *Drought*

Drought is the most serious climatic disaster because the drought-stricken area is usually the largest and also drought events are most frequent in such areas. According to records, a total of 31 droughts occurred in China in the last 20 years, including the following six severe droughts: in North China an area of 30 mill. ha in 1972; in 1980 about 15 mill. ha; in 1981 and 1982, 7–14 mill. ha in each year; 1985/1986 (especially harmful to winter wheat grown in non-irrigated fields); and in 1988 more than 20 mill. ha or about one fifth of the total area of farmland in China.

The spatial distribution of drought indicates a very clear regionality with the most severe droughts concentrated in the region between 35–40 °N and 100–120 °E. This is the region of the North China Plain and the Loess Plateau where the interannual variability of precipitation for both, the annual total, and the spring and summer share are very large (see the values for Beijing, Jinan, Zhengshou, Yinchuan, Xi'an and Lanzhou in Table 5).

Irrigation is the most common counter-measure for relieving crops from drought. However, owing to the limitation of an irrigation system, sometimes other counter-measures should be considered and, where appropriate, carried out. The use of new technology to promote efficiency of water use and the selection of species and cultivars of crops with higher drought tolerance are potential alternatives.

Table 6. Some examples of rainstorms in China.

Date	Region	Area with precipitation > 200 mm (km^2)	Total amount of precipitation at the center of rainstorm (mm)
1–10 July 1935	Middle and Lower	very large	1,200
1–10 Aug. 1963	Hobei Province	100,000	1,329
24–30 May 1973	South China	local	1,268
5–7 Aug. 1975	Honan Province	43,675	1,631
9–14 July 1981	Sichuan Province	63,700	366
21 Aug. 1977	Shanghai	local	592

5.2. *Flood and rainstorms*

Rainstorms over a limited area, during a short period of a few days only, may cause floods and heavy crop losses. Floods are, next to droughts, a main climatic disaster in China. According to historical records, from 206 B.C. to 1949 A.D., 1029 heavy floods have occurred in China, about one in every two years, on average. In the Eastern Monsoon Region of China, all large river basins are likely to suffer frequently from floods. Table 6 shows some examples of heavy rainfall. Rainstorms may occur occasionally in both North and South China, but sometimes even in Northwest Arid China.

5.3. *Other climatic disasters*

In Northeast China (Helongjiang, Jilin and Liaoning Provinces) cold damage to crops in the growing season may cause great loss to grain production. For example, the cold damages, which occurred in 1969, 1972 and 1976, reduced the grain yield by around 5 mill. t. each in comparison with the previous year.

Severe frost usually causes great loss to crops, both to temperate-zone crops (as wheat) and subtropical crops (as citrus), and occasionally even to tropical crops (rubber trees). For example, an unusually late spring frost (10–12 April, 1953) heavily affected wheat cultivation in seven provinces in the North China Plain, and more than 2 mill. ha of farmland were destroyed, reducing yields by about 40–85% against normal. Even in subtropical and tropical China, severe frost may occur. In January 1955, for example, more than 3 mill. rubber trees were heavily damaged by frost.

Typhoons frequently hit China, mainly the coastal provinces. In 1989, three strong typhoons hit the Hainan Island and about 20 mill. rubber trees were heavily damaged.

Besides, hail storms and hot and dry winds also occur, sometimes causing heavy damage to agriculture.

6. Effects of topography on climate

Mountains and hills occupy about 70% of the total area of China. The effect of topography on China's climate may be studied in two aspects: the physical, thermal and dynamic aspect, and the geographical aspect.

6.1. *Impacts of thermal and dynamic effects of the Tibetan Plateau on climate and vegetation of China*

The Tibetan Plateau is characterized by two outstanding features: the high elevation of more than 4,500 m a.s.l and a large area, which covers over 2 mill. km^2 (2,000–2,500 km from east to west and 1,000–1,500 km from north to south). Appearing as a continent in an atmospheric ocean, the Tibetan Plateau imposes both, dynamic and thermal effects on the climate, influences the vegetation, and impacts the surrounding areas.

6.1.1. *Impacts of thermal and dynamic effects of the Tibetan Plateau on climate and vegetation in China*

One of the most conspicuous features of the climate of China is the presence of a humid subtropical zone in a latitudinal belt (20–33° N) which is normally arid or semi-arid. Recently, Ruddiman and Kutzbach (1991) gave the following explanation: "In the summer the sun heats the high (Tibetan) plateau which in turn rapidly warms the thin overlying atmosphere. The warm air is less dense, so it rises. As the air rises and cools, its ability to hold water vapor diminishes. This process ultimately leads to the formation of rain clouds and produces the seasonal monsoon rains that fall on the southeast margin of the (Tibetan) plateau, especially in Asia". The subtropical zone of China is just located in this region, and represents a very productive agricultural region due to favourable wet and warm conditions. It comprises a vast stretch of arable land, about a fourth of China's total, even though it is occupied by many mountains and hills. About half of China's entire grain crop is produced here.

6.1.2. *Northward displacement of the arid region*

China's arid zone is displaced to the north of the latitudinal belts and normally occupied by deserts. Unlike similar zones in the subtropics, China's arid lands are located in the temperate zone. "On the opposite, on the northern side of the (Tibetan) Plateau are the Kunlun mountains. These are the driest mountains in the world. Desert vegetation covers the mountain slopes from the foothills up to the snowline; at the base of Kunlun Mountains is a vast expanse of sand desert, the Taklamakan, which is the arid core of Asia, where annual precipitation drops below 10 mm" (Chang 1983).

The productivity of natural vegetation is much lower in the Northwest Arid Region compared to the humid subtropical China. The natural landscapes are desert, desert grassland, grassland and a small part of forest grassland. Many

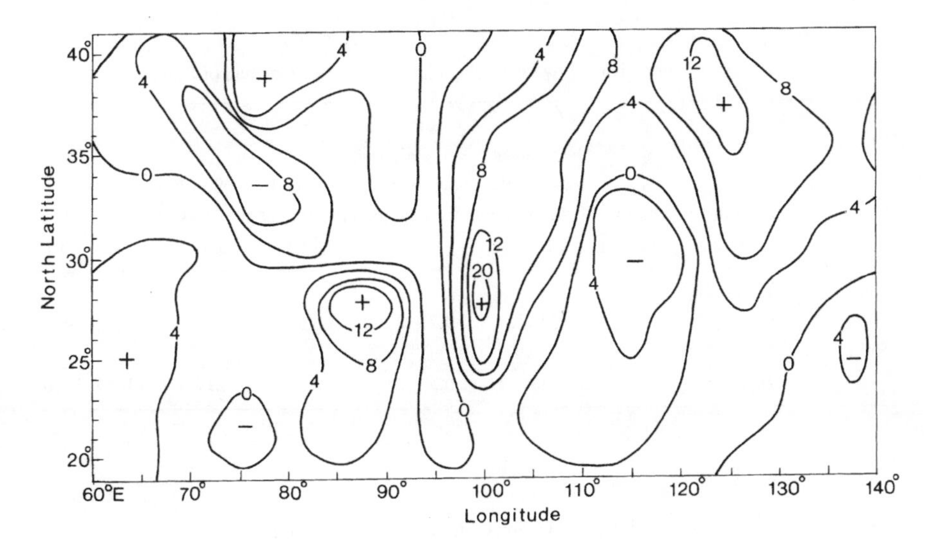

Figure 10. Distribution of monthly sunshine duration (hrs) in South China in January (1954–1960).

high mountains surround this Arid Region. The mountains get comparatively larger precipitation (snow) on their higher slopes. Melt water from snow and ice can be used in the warm summer season for irrigation and many aspects of human activities.

6.1.3. *Differentiation of winter climate in the western and eastern part of South China and its agricultural implication*

Based on the floristic characteristics, subtropical and tropical South China were divided into a western and eastern part (Commission for Natural Regionalization 1959). The boundary is located between 103° and 105° E. From the distribution of sunshine duration in South China in January (Figure 10), a sharp contrast between the eastern and western part can be seen. This may be caused by the dynamic effect of the Tibetan Plateau. The remarkable differences in sunshine duration between these two parts would cause significant differences in vegetation and agriculture. Rubber cultivation can be taken as a good example to illustrate the agricultural implication of this differentiation.

Owing to the urgent demand for rubber in the early fifties, China made great efforts and achieved success in cultivating rubber trees (*Hevea brasiliensis*) in South China. The total production of natural rubber increased from about 200 t (1950) to 237,600 t (1978). Despite this great success, China has also suffered great losses, due to climatic disasters, viz. cold injury and wind damage by typhoon. Only the cold injury is referred in this section. Since the rubber tree is indigenous to the tropical lowland rainforest near the equator,

Table 7. Comparison of overwintering climate of Guangzhou and of Ruili.

	Latitude Longitude	Altitude (m.a.s.l.)	Mon. mean temp. of the coldest month (°C)	Mon. mean of daily min. temp. of the coldest month (°C)	Cloud amount at 14 h	Sunshine hours of Jan.	Years of data
Guangzhou	23°08′ N						
	113°19′ E	7	13.3 (Jan.)	9.7	6.2	154	1951–1980
Ruili	24°01′ N						
	97°51′ E	776	12.6 (Jan.)	6.5	2.7	217	1957–1980

Table 8. Production of dry rubber (t/yr) in Ruili.

Year	1970	1973	1974	1975	1976	1978	1979	1980	1982	1983	1984	1985	1986	1987
Prod.	6	299	263	309	279	556	659	766	956	1046	1190	1338	1443	1660

such as the Amazon Basin, where the annual variation of temperature is very stable and the monthly temperatures vary between 26–28°C, the rubber tree is very sensitive to low temperature.

After introduction to South China, rubber trees encountered quite different geographical and climatic conditions from those of the Amazon Basin. As cold air advection is very strong and frequent in winter in South China, the daily average and minimum temperatures are much lower there than in the Amazon Basin. The mean temperature of the coldest month in some Rubber State Farms in South China is between 7 to 13°C lower than for Manaus, Amazon Basin. Consequently, overwintering is the main limiting factor for rubber tree culture in South China, especially in Guangdong and Yunnan Provinces. A severe cold wave invasion into South China causes severe injury to rubber trees.

But cold injury to rubber trees is quite a complex problem because it is difficult to define the specific low temperature threshold for such a damage. Based on field surveys and microclimatic observations in Rubber State Farms in Guangdong and Yunnan Provinces, sunshine appears to represent an important factor to the well being of rubber trees. In a climate environment with ample sunshine, rubber trees can resist lower temperatures.

For example, Guangzhou (23° N, 113° E, 7–30 m) and Ruili (24° N, 98° E, 780–850 m) were chosen as experimental sites for rubber tree cultivation. Thousands of trees were planted in Guangzhou in the early fifties and

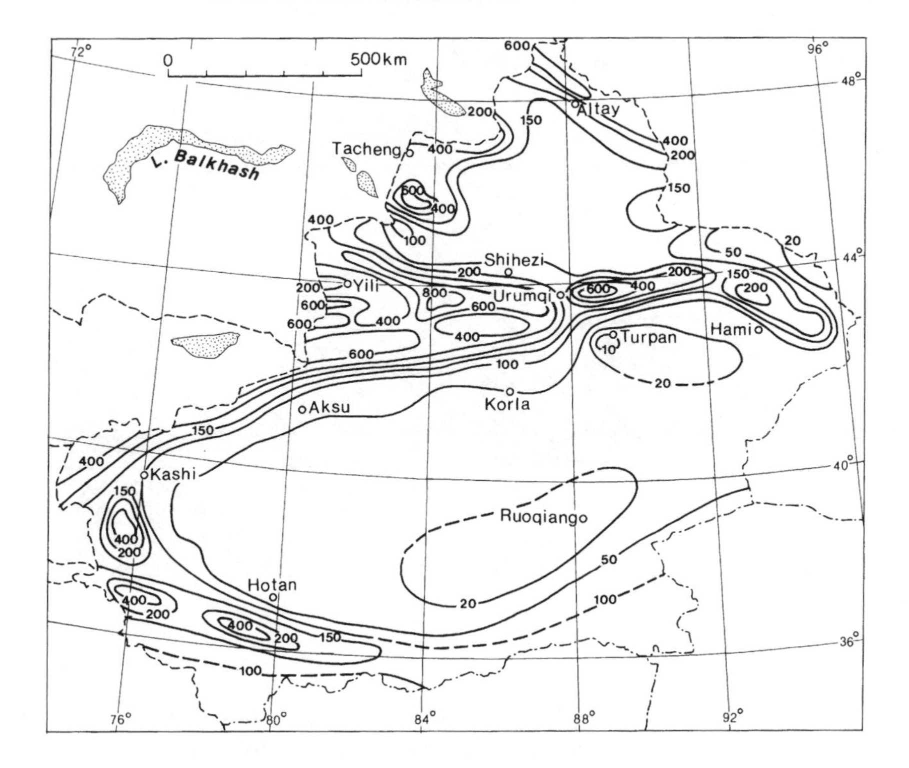

Figure 11. The distribution of annual rainfall (mm) in Xinjiang Uygur Autonomous Region.

in Ruili in the early sixties. Comparing the overwintering climate of these two sites (Table 7), more favourable growing conditions seemed to occur in Guangzhou than in Ruili. However, the present status of overwintering of rubber tree in both sites proved the contrary. Rubber trees in Guangzhou suffered from severe cold damage during the winters 1954/1955, 1956/1957, 1962/1963, 1967/1968, and 1976/1977. After cold damages in the winters 1956/1957 and 1967/1968, many rubber trees were replanted, but not a single one finally survived. In Ruili, however, growing of rubber trees is stable, in general, during the last 30 years, even in cases of cold damages in the winters 1973/1974 and 1975/1976. The annual production of dry rubber in Ruili is clearly thriving (Table 8). Hence, the experiment of rubber tree growing in Guangzhou completely failed due to intolerably low temperature and insufficient sunshine in winter, while the experiments in Ruili were quite successful, due to favourable climatic condition in winter featuring a lower temperature, but ample sunshine.

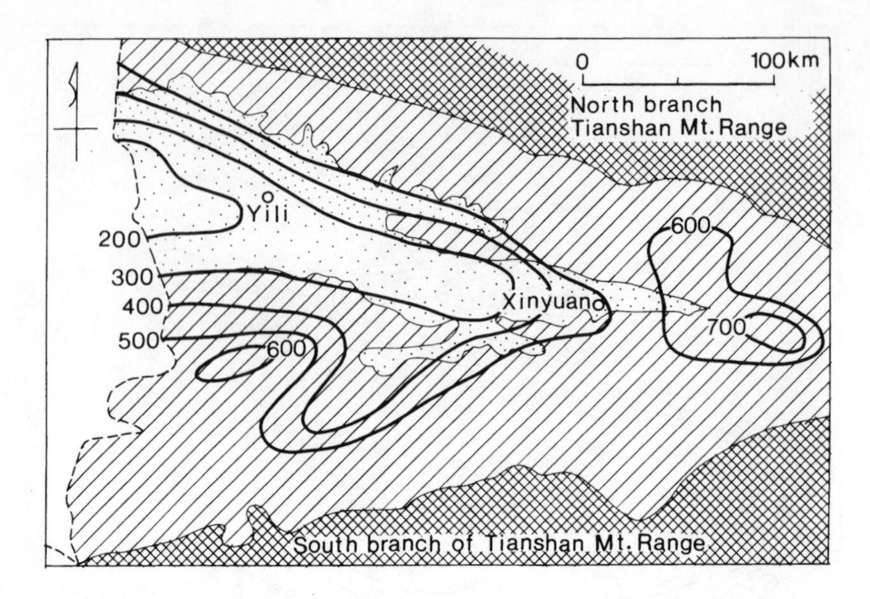

Figure 12. The annual distribution of precipitation in the area of Yili River Valley (14) (mm).

6.1.4. *Topography and climate at macro- and meso-scales*

Topography obviously influences precipitation distribution over space, as seen in arid Northwest China (Figure 11; Xu Deyuan 1990; Zhang and Dun 1987). In particular, high mountains greatly influence the regional precipitation pattern. For example, the annual precipitation totals for three stations located in river valleys, such as Yili (43°57′ N, 81°26′ E, 663 m), Urumqi (43°47′ N, 87°37′ E, 918 m) and Aletay (47°44′ N, 88 °05′ E, 735 m), account only for 258 mm (1951–1980), 278 mm (1951–1980) and 181 mm (1954–1980), respectively. But on the neighbouring high mountain slopes, precipitation increases rapidly, and totals between 600 and 800 mm are recorded. A good example for the meso-scale variation of precipitation is given for the Yili river valley (Figure 12). Also the distribution of annual accumulated temperature in Xinjiang caused mainly by the great differences in elevation mostly 1,000–4,000 m, shows the macro-scale relationship between topography and climate (Figure 13).

6.2. *Topography and climate at local- and micro-scales*

Local- and micro-scale topographies may cast very significant effects on some climatic factors, such as minimum temperature, sunshine, etc., accompanied by effects on vegetation and agriculture under some specific conditions. In this context, the local-scale topography refers to a horizontal scale of 1 to 100 km and a vertical scale of 100 to 1,000 m. Taking again the introduction of rubber cultivation in South China, as an example, the effect of the local

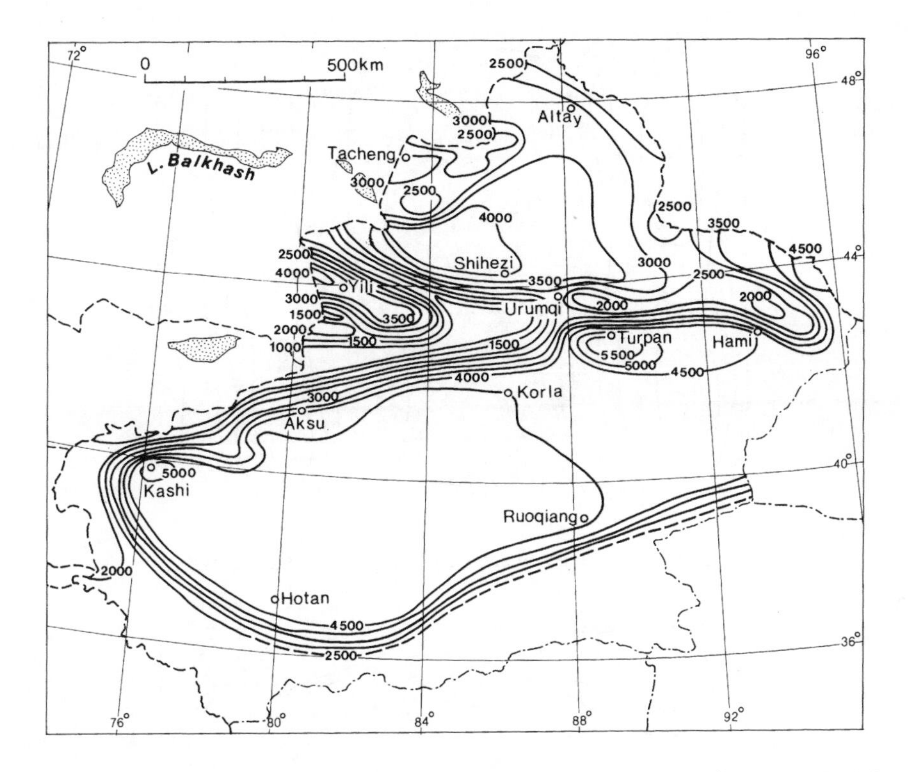

Figure 13. The distribution of annual accumulated temperature (GT > 0°C degree days) in Xinjiang Region.

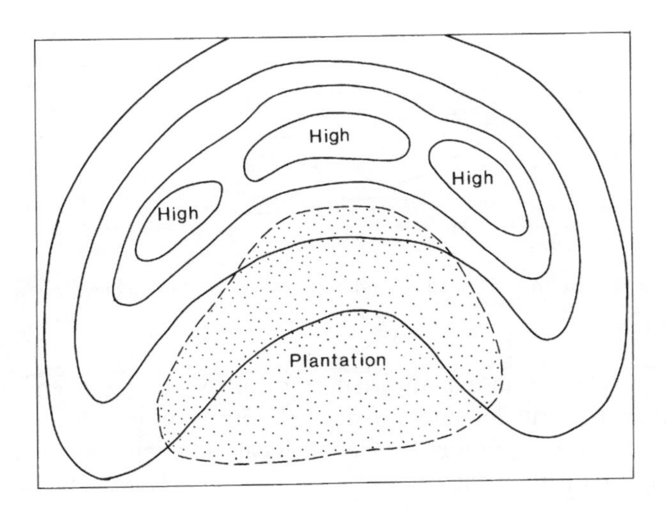

Figure 14. "Difficult-in and easy-out type" of topography.

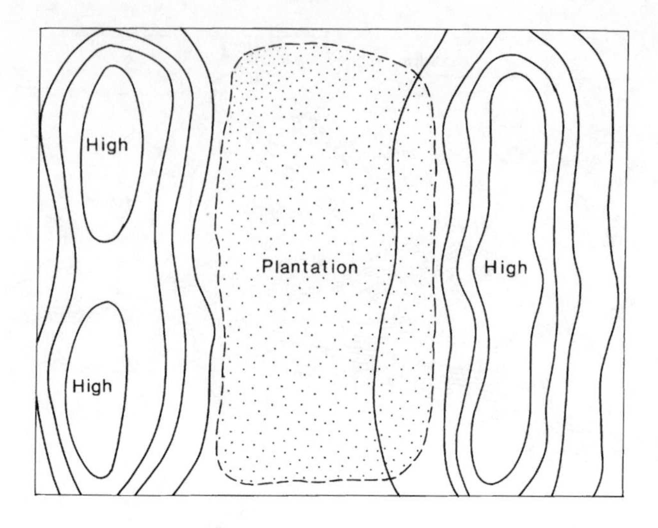

Figure 15. "Easy-in and easy-out type" of topography.

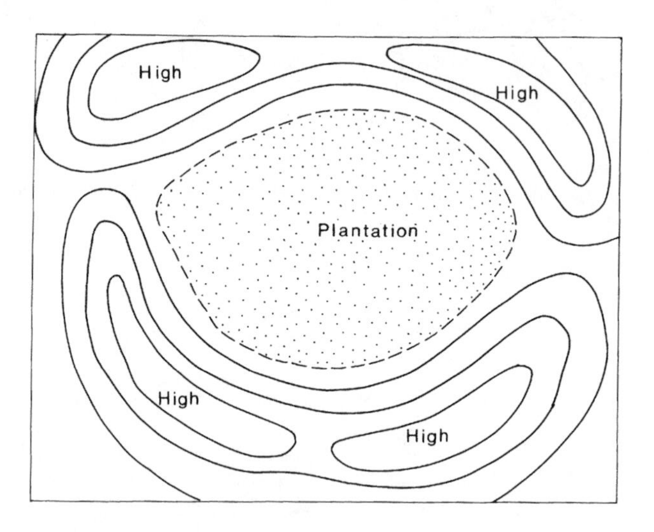

Figure 16. "Difficult-in and difficult-out type" of topography.

topography on temperature is shown during cold waves in winter. To function as a protection to rubber farms against the invasion of advective cold air from the north and of drainage of cold, local-radiational air, four types of topography are distinguished:

A. "Difficult to get in (for advectional cold air) and easy to get out (for radiation cooling air) type"; this type may be called "Difficult-in and easy-out type" (see sketch map, Figure 14);

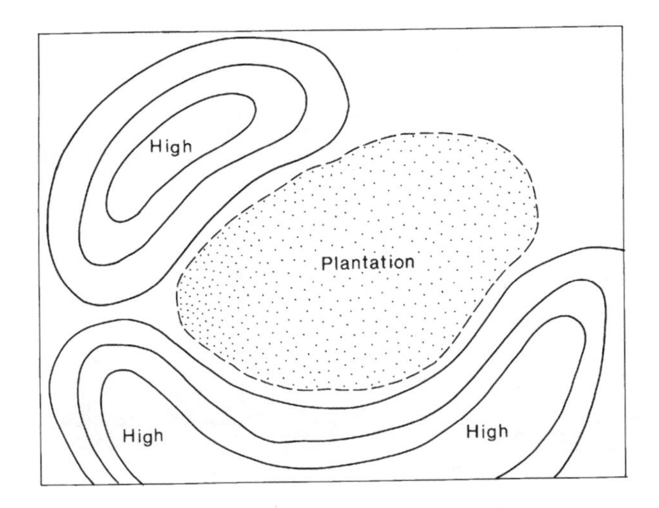

Figure 17. "Easy-in and difficult-out type" of topography.

Table 9. Comparison of climate of four counties in some severe cold winters.

Types of local climates		Monthly mean temp. (°C)					Absolute min temp. (°C)					Sunshine (hours)				
		Jan. '55	Feb. '57	Jan. '63	Feb. '68	Jan. '77	Jan. '55	Feb. '57	Jan. '63	Feb. '68	Jan. '77	Jan. '55	Feb. '57	Jan. '63	Feb. '68	Jan. '77
Gaozhou	A	12.7	11.7	12.8	10.8	12.3	1.6	0.5	1.0	5.5	3.8	176	31	233	25	66
Luchuan	B	–	11.6	11.0	9.6	9.7	–	0.3	0.1	3.1	1.2	–	26	211	10	53
Longzhou	C	11.3	11.8	11.9	9.8	9.9	–3.0	2.2	–1.3	2.4	1.4	79	31	198	10	28
Ningming	D	–	10.8	12.1	9.5	9.6	–	–	0.7	–1.0	1.3	–	28	200	13	33

B. "Easy-in and easy-out type" (Figure 15);
C. "Difficult-in and difficult-out type" (Figure 16);
D. "Easy-in and difficult-out type" (Figure 17).

Four counties (Gaozhou, Luchuan, Longzhou and Ningming) located in the same latitude (about 22° N) and at about the same altitude < 150 m were chosen as the areas for rubber cultivation trees in a large scale in the early fifties. Forty years of different growing conditions for rubber trees in the four counties have elapsed. In Gaozhou, about 80% of the rubber trees have survived, in Luchuan only 50%, in Longzhou less than 5% while no rubber tree survived in Ningming. An attempt is made to compare the severe cold winter climates in the four counties concerned (Table 9). As a result, the growing condition for rubber trees can be characterized as follows: Type A

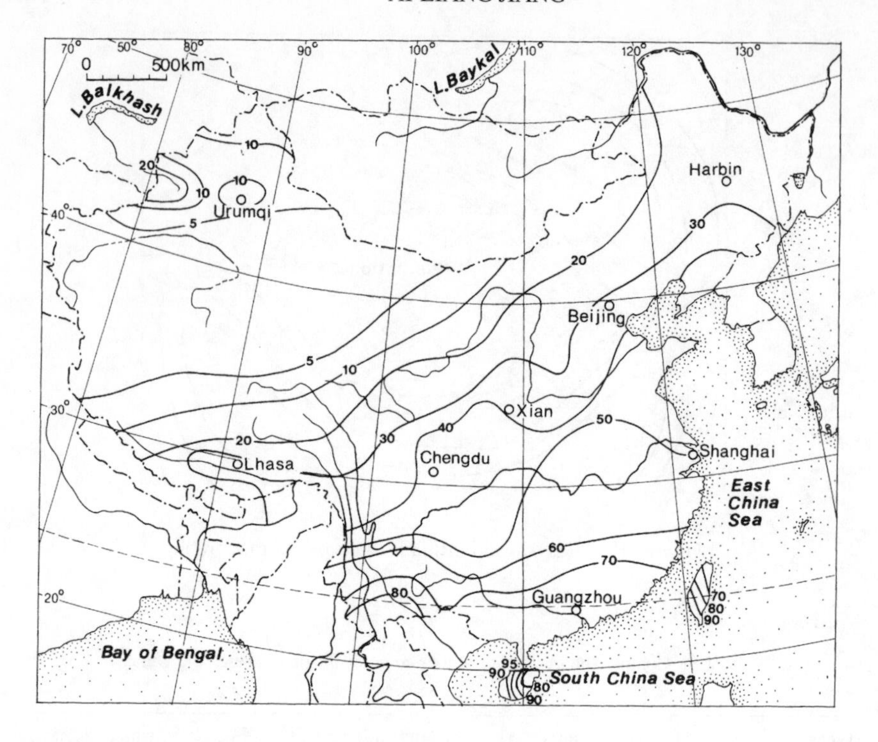

Figure 18. The annual distribution of the index of 'Climatic Productivity of Agriculture' (Y) in China (t/ha.yr).

represents the most favourable local climate for rubber tree overwintering while type D, together with type C, are the most unfavourable.

7. The climatic index of productivity of agriculture

Jiang (1984a) proposed an index to show the climatic productivity of crops in general. This index entitled "Climatic Productivity of Agriculture" was also expressed mathematically. Four years later, a modified expression was proposed as follows (Jiang and Zhang 1988):

$$Y = f_1\ (Q) * f_2\ (T) * f_3\ (W) = AQ * f_2 * f_3,$$

where Y stands for the climatic productivity expressed in t/ha.yr of dry biomass; Q is the global radiation (kilojoules/m^2); f_2 is the temperature reduction coefficient, f_2 = 0–1; and f_2 = 0 when the daily temperature <0 °C, f_2 = 1 when temperature > 30°C; f_3 is the moisture reduction coefficient, f = 0–1; f_3 = P/Eo, where P is the precipitation and Eo the potential evapotranspiration calculated by Penman's formula.

Figure 18 shows the distribution of Y values in China. Jiang and Wei (1985) checked the real conditions in Hainan Island. Similar checks were performed for the Inner Mongolia grasslands (Fan 1986), southern part of Jiangxi Province (Jiang 1990), and the North China Plain (Jiang 1990). It is found for some experimental sites that crop or grass yields correspond well with the Y values calculated under optimal agronomic conditions. The Y values may be regarded as an expression of the potential production of biomass under natural climatic and optimal agrotechnical conditions. The distribution pattern over space is somewhat similar between the Y-index (Figure 18) and the distribution of precipitation (cf. Figure 5). Therefore, precipitation can be considered the main factor influencing pattern of agricultural production in general.

Yoshino and Chiba (1984) have studied the "Regional division of China by precipitation", applying Principal Component Analysis. They found the square deviation contribution of the first principal component is 77%, of the second component 15%, and the cumulative square deviation contribution of the first and second components is 92%.

The author, applying the principal component analysis for the annual variation of the monthly Y values, found that the square deviation contributions of the first and second component are 92 and 4%, respectively. This result seems to show: (1) the pattern of annual variation of the monthly Y values can be considered as only one type; (2) Y may be regarded as a very good index to represent the agricultural productivity in China in general.

If the results given in Figure 18 are considered acceptable, the potential productivity for a given region can easily be followed. For example, the potential productivity, averaged for six provinces south of the Middle and Lower Yangtze Valley (Zhejiang, Jiangxi, Hunan, Guangdong, Guangxi and Fujian), can be estimated as 65 t/ha of dry biomass. But the actual averaged yield is equivalent to 13 t/ha dry biomass, by assuming that the economic coefficient of grain is 35%. There is a big gap between the potential and actual value. Still it seems possible to close the gap through a combination of hard work, advanced agrotechnics, a reasonably high input, and a suitable agricultural policy.

During all the development and achievements in agriculture, it is most important to pay attention to conservation of nature as well as to the natural eco-system and natural resources. Sustainable agricultural development must be the ultimate goal that "meets the needs of the present without compromising the ability of future generations to meet them".

8. Climate changes and agriculture

It was found that the 1980s are the most obvious warming decade in this century both for the globe and for China. However, in view of the impact on agriculture, the author holds that mean and minimum temperatures in some specific periods are often more important in agriculture than the annual mean (Jiang 1993). For example, mean air temperatures of October and November in Beijing and North China Plain during 1980s were remarkably higher than those in last three decades. Together with the improvement of mechanization and application of new cultivars, the higher temperature in this period will stabilize the double cropping system of winter wheat and maize in this region. This system was widely accepted by the farmers. For example, the area of double cropping of wheat and maize in Beijing was less than 1×10^3 ha in 1982, and it increased rapidly to 147×10^3ha in 1992. Another example is related to the minimum temperature. During last 40 years, it is found that the extreme minimum in winter in some localities in the subtropical zone of China shows a tendency to decrease from 1950s to 1980s, and frost and snow occurred more frequently. For example, the harmful extreme minimum temperatures ($< -4.5 or -0.5$°C) appearing in Kunming in 1950s to 1980s are as follows: In 1950s: −5.4°C (in the winter 1951/1952); in 1960s: −5.1°C (1960/1961), −4.6°C (1963/1964); in 1970s: −5.4°C (1973/1974), −4.9°C (1975/1976), −4.8°C (1978/1979), in 1980s: −6.8°C (1982/1983), −7.8°C (1983/1984), −5.2°C (1985/1986). During these severe cold winters mentioned above, crops and trees suffered heavy damages. Hence it is suggested that agricultural activities in a region should be arranged according to the regional climate change, especially the characteristics during the sensitive period of crop growing season, so as to avoid disasters and make use of the advantages of climate resources.

References

Chang, Devoid, H.S. (1983): The Tibetan Plateau in relation to the vegetation of China. *Ann. Missouri Bot. Gard.*, 70, 564–570.

Commission for Nature Regionalization of China (1959): *An integrated nature regionalization of China* (draft). Science Press, Beijing, 290 pp.

Domroes, M. and Peng, G. (1988): *The climate of China*. Springer-Verlag, Heidelberg, 356 pp.

Fan, J. et al. (1986): The calculation of the product of natural grazing grassland and the comment on regionalization. *Chinese J. Agri. Meteorol.*, 7, 53–56.

Ho, Do-Fen (1979): The natural rainfall condition in the growing season of winter wheat in five provinces of North China and its impact on the yield. *Agricultural Meteorology*, 2, 45–52 (in Chinese).

Jiang, A. and Li, S. (1981): Topo-micro-climate and cold-proof of rubber trees. *Agricultural Meteorology*, 2, 45–52 (in Chinese).

Jiang, A. (1984a): On the relation between the climate and the potential agricultural productivity of China. In: *Proceedings of Symposium on the Memory of Prof. Ko-Zhen Zhu*, 178–201.

Jiang, A. (1984b): Climate and agricultural land use in China. In: *Climate and agricultural land use in monsoon Asia*, ed. by M.M. Yoshino. University of Tokyo Press, Tokyo, 297–316.

Jiang, A. and Wei, L. (1985): On the agroclimatic productivity of Hainan Island. *Chinese Journal of Tropical Crops*, 6, 1–12.

Jiang, A. (1988): Climate and natural production of rubber (*Hevea brasiliensis*) in Xishuangbanna, southern part of Yunnan Province in China. *Int. J. Biometeorology*, 32, 280–282.

Jiang, A. and Zhang, F. (1988): A model of the climatic productivity of agriculture of China. *Chinese J. Agri. Meteorol.*, 9, 16–18.
Jiang, A. et al. (1990): Land supporting capacity and the transformation efficiency of natural resources. *Natural Resources*, 5, 36–40.
Jiang, A. (1993): On the climate change of China in recent 40 years and its impacts on agriculture. In: *Proceedings of International Symposium on Climate Change, Natural Disasters and Agricultural Strategies*, May 26–29. China Meteorological Press, Beijing, 102–107.
Jiang, A. and Chen, S. (1991): On the zonality and non-zonality of climatic elements of China. To be published.
Li, S. et al. (1988): *Agroclimatic resources and agroclimatic regionalization of China*. Science Press, Beijing, 341 p.
Liu, En-Lan (1936): Variation of rainfall in China. *Acta Geographica Sinica*, 3, 529–532.
Lu, Chi-Yao (1963): The influence of rainfall on the yield of winter wheat in North China. *Acta Meteorologica Sinica*, 33, 292–298.
Ruddiman, W.F. and Kutzbach, J.E. (1991): Plateau uplift and climatic change. *Scientific American*, March, 66–75.
Shen, Xiang-Yi (1984): A preliminary study on water condition of crops. *Agricultural Meteorology* Chinese), 5, 5–11 (in Chinese).
Xu, Deyuan (1990): *The agroclimatic resources and regionalization of Xinjiang*. Meteorological Press, Beijing, 326 pp.
Xu, Guo-Chang and Zhang Zhi-Ying (1983): The function of Qinghai-Xizang (Tibetan) Plateau to the formation the arid climate of northwest region of China. *Plateau Meteorology*, 2, 9–15.
Yeh, Du-Z., Yang, G.-J. and Wang, X.-D. (1979): The mean vertical circulation over East Asia and Pacific Ocean, (1) Summer. *Atmospheric Science*, 3, 1–11.
Yoshino, M.M. and Chiba, M. (1984): Regional division of China by precipitation. *Geogr. Rev. Japan, Ser. A*, 57, 583–590.
Yoshino, M.M. and Aoki, T. (1986): Interannual variations of summer precipitation in East Asia: Their regionality, recent trend, relation to sea surface temperature over the North Pacific. *Erdkunde*, 40, 94–104.
Zhang, Jia-Cheng and Lin, Zhi-Quang (1985): *Climate of China*. Shaanxi People's Press, Xi'an, 450 pp.
Zhang, Jia-Bao and Dun, Zi-Feng (1987): *General view of precipitation of Xinjiang*. Meteorological Press, Beijing, 400 p.
Zhao, S. (1983): A new scheme for comprehensive physical regionalization in China. *Acta Geographica Sinica*, 38, 1–10.

Ai-Liang Jiang
Commission for Integrated Survey of Natural Resources
Academia Sinica
Datun Road 917 Bldg.
Beijing 100101
P.R. China

3.7. THE CLIMATES OF THE "POLAR REGIONS"

TADEUSZ NIEDZWIEDZ

1. Climatic boundaries of the Arctic and Antarctic Polar Regions

Polar regions have some unique climatic peculiarities:

- perpetual polar night and day (at latitudes greater than Polar Circles $-66°33'\,03''$ N and $66°33'\,03''$ S);
- negative radiation balance over vast areas;
- very low air temperatures throughout the year;
- strong winds with blowing snow and blizzards;
- small absolute water vapor content in the air and considerably large relative humidity, especially in summer;
- small ice surfaces on seas and glaciers on land (areas with a great albedo of 70–90%);
- permafrost.

Limits of ranges of many flora and fauna species are found on the edge of these areas. Tundra with comparatively rich organic life (at least in summer) may be called a subpolar zone. But soils thaw only near the surface and permafrost remains at deeper levels for the whole year.

Areas of permanent frost or the so-called polar deserts (Korotkevitsch 1972), ice-covered for the whole year, prevail in this region where there is no possibility for plant growth; even mosses and lichens cannot exist because of the lack of heat. Extreme conditions in the northern hemisphere are found in the Arctic Basin, especially in the interior islands such as Spitsbergen, Zemlya Frantsa–losifa, Novaya Zemlya, Severnaya Zemlya (North Land), New Siberian Islands, Wrangel Island, the Canadian Archipelago, and the northern and central parts of Greenland. The zone of polar deserts in the southern hemisphere includes, in general terms, the whole Antarctic continent and some islands situated south of 63° S (Korotkevitsch 1972).

The precise climatic boundary of the polar regions is difficult to define. The occurrence of polar day and night begins with Polar Circles, but these

M. Yoshino et al. (eds.), Climates and Societies – A Climatological Perspective, 309–324.

mathematical lines cannot be treated as a climatic boundary because climate is influenced by many other geographical factors. In Köppen's climatic classification the $+10°C$ isotherm of the warmest month is considered the border of polar regions. It agrees with the tree-line in some places. However, it is not appropriate, for example, in the southern part of Chile. Polar regions within these borders cover 85–95 mill. km^2. The area of permanent frost, limited by the 0°C isotherm of the warmest month, covers an area of about 27.5 mill. km^2, of which 13.5 mill. km^2 is in the Antarctic. Polar climates reach 68° N in the northern hemisphere, and 48° S in the southern hemisphere, on average.

Boundaries of polar regions on land areas are best determined by the so-called Nordenskjöld line. It is calculated with the Vahl formula modified by Nordenskjöld (Baird, 1964):

$$W = 9 - 0.1\ C$$

where 'W' is the temperature of the warmest month (in °C), and 'C' the temperature of the coldest month (in °C). The area under the arctic climate, determined in this way, covers about 7.64 mill. km^2 of land (Baird, 1964). On the other hand, the land area of the Antarctic covers about 14 mill. km^2. The contact zone where Arctic waters mix with warmer waters from the Atlantic and Pacific Oceans is considered the boundary of Arctic climates in sea areas. The area of Arctic waters covers 14.65 mill. km^2 (Baird 1964).

The boundary of the polar zone in the southern hemisphere is situated in the oceans surrounding the Antarctic, represented by the convergence zone between relatively warm waters of the temperate zone and cool Antarctic water. Only a few subantarctic islands, such as Macquarie, Kergueles, Heard, Bouver, South Georgia, South Sandwich and South Orkneys, are situated in this zone. The limits of the polar climate are shown in Figures 1 and 2. The area of the Antarctic climate, accounting for about 60–70 mill. km^2, is 2.5 times larger than the area of Arctic climate. Moreover, the large variation in the distribution of lands and seas in the Arctic causes large differences in polar climate conditions. For instance, Arctic climates occur as far north as 54.5° N in an area near Labrador and the Hudson Bay. In the Norwegian Sea, they are found even farther north at 72° N. These areas experience such a mild climate that they cannot be included in the polar zone, even though polar day and polar night occur.

The polar regions preclude any human activities because of their climatic severity. However, numerous nomadic communities lived near the southern border of the tundra in Europe and Asia. The Inuits in Canada and on the coasts of Greenland have adapted very well to the hard life under the severe polar climate conditions. They travel in summer even to the areas near 82° N.

The Antarctic was for long entirely inaccessible to people and the first man arrived on its coast only in 1895 (Norwegian expedition of Christensen and Borchgrevink). The first station was built at Cape Adare in 1898.

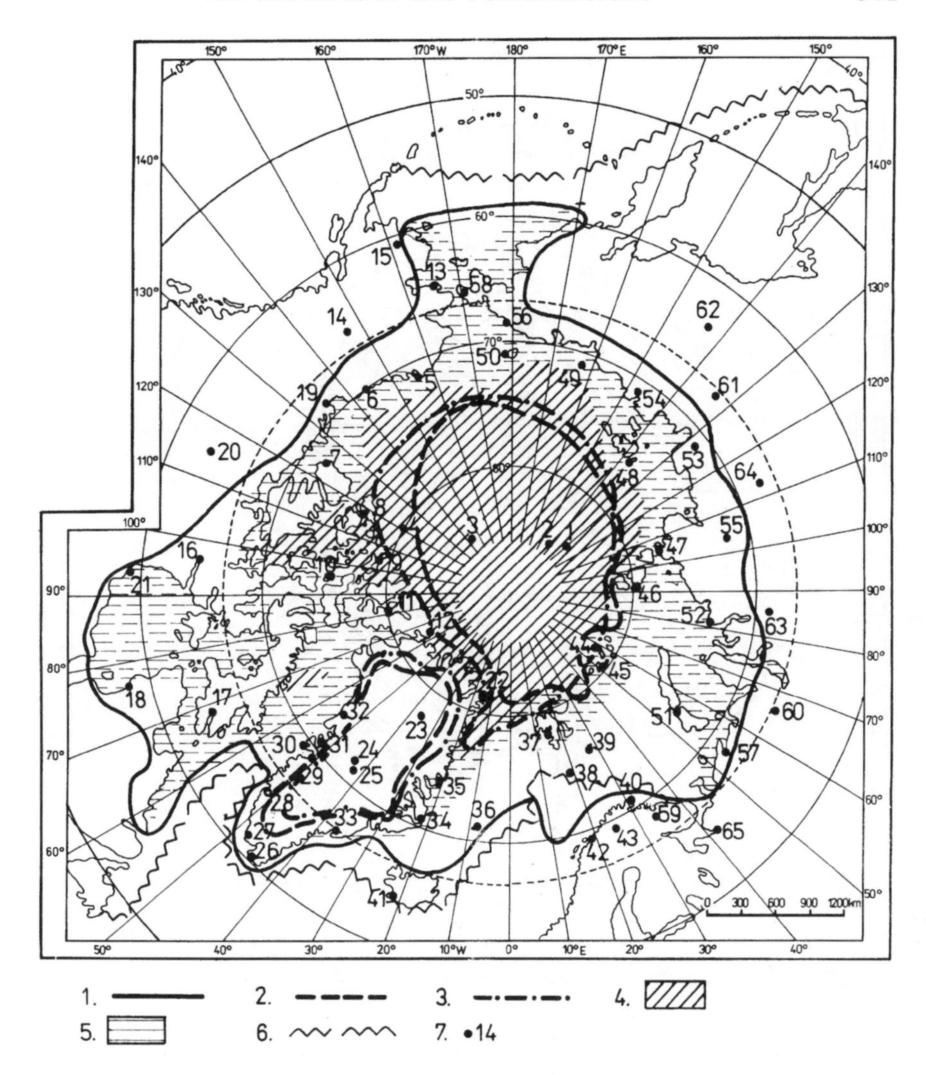

Figure 1. Arctic Polar Region: 1 – boundary of polar climate, 2 – 0° C isotherm of the warmest month, limit of the polar deserts, 3 – limit of the areas with annual negative net radiation budget, 4 – seas permanently covered with pack ice, 5 – sea areas covered by ice during winter, 6 – limit of icebergs, 7 – selected meteorological stations with data in Table 2 (numbers of stations in the table).

The central part of the Arctic and the Antarctic are currently investigated with a network of scientific stations. Drifting stations are also used in the Arctic. The largest part of the network was established during the International Geophysical Year 1957/1958 (Chapman 1959), including the first station on the ice sheet of the Antarctic. The inner part of Greenland ice cap was investigated from the climatologically by A. Wegener (expedition 1930/1931).

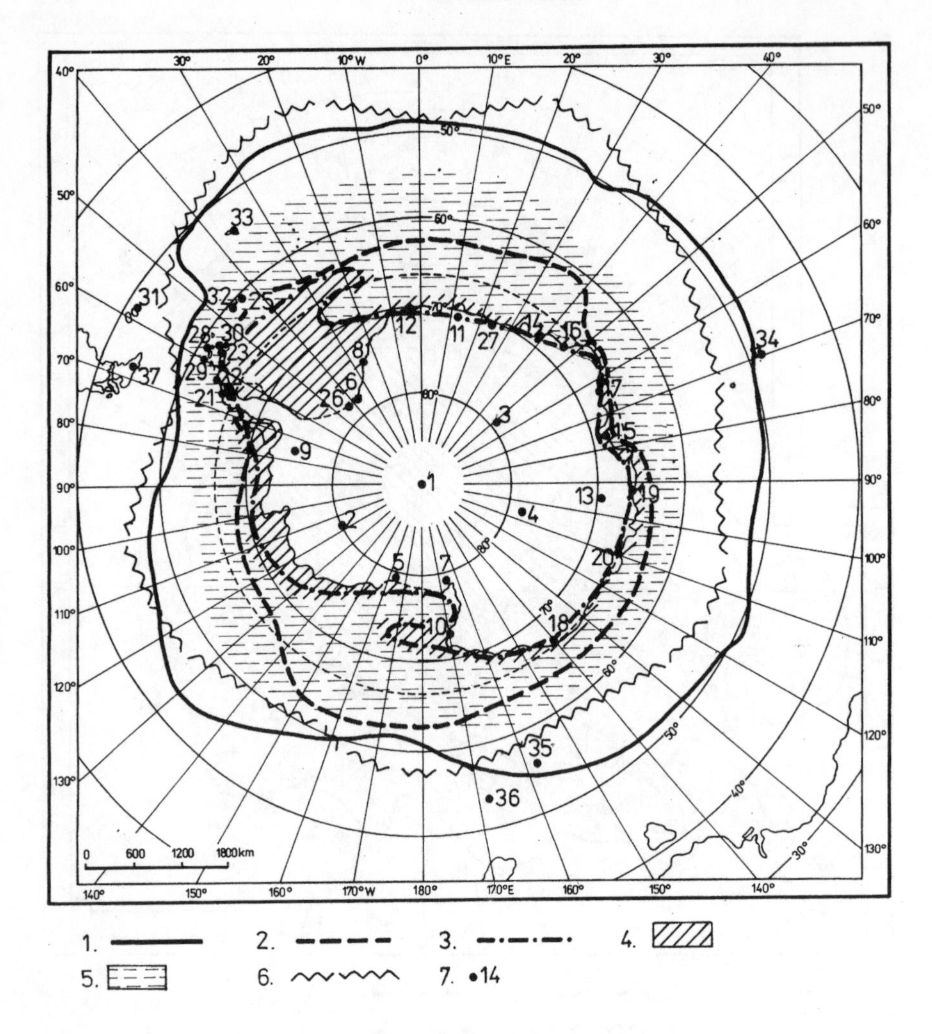

Figure 2. Antarctic Polar Region: 1 – boundary of polar climate, 2 – 0°C isotherm of the warmest month, limit of the polar deserts, 3 – limit of the areas with annual negative net radiation budget, 4 – seas permanently covered by pack ice, 5 – sea areas covered by ice during winter, 6 – limit of icebergs, 7 – selected meteorological stations with data in Table 3 (station numbers correspond with the number in the table).

Earlier, climatological data were collected during scientific expeditions, the most important of which was F. Nansen's drift on the ship "Fram" in 1893–1896. In 1898–1899, climatic records were obtained on board of the ship "Belgica" drifting on the ice near the Antarctic coasts. Many valuable data were collected also during the 1st 1882/1883 and 2nd International Polar Year, 1932/1933 respectively. In 1937, the Soviet Union established the first drifting scientific station "North Pole 1" (Treshnikov and Voskresensky 1976).

Table 1. Duration of polar day (PD) and polar night (PN) at different latitudes (f), taking refraction into account (after Martyn 1992).

Northern Hemisphere			Southern Hemisphere		
ϕ (N°)	PD	PN	ϕ (S°)	PD	PN
66	1 day	0	66	1 day	0
70	70 days	55 days	70	65 days	59 days
75	107	93	75	101	99
80	137	123	80	130	130
85	163	150	85	156	158
90	189	176	90	182	183

2. Astronomic and radiation factors of the polar climates

Polar regions are characterized by peculiar solar conditions resulting from purely astronomical reasons (situation behind the Polar Circles). The existence of the polar day and polar night for about half a year at the poles is the characteristic feature (Table 1). It is responsible for the absence of the daily cycle of many climatological elements and the large variations in solar energy. Humans under polar climate conditions are exposed to harsh natural biological rhythms due to the significant changes in photoecological conditions that are driven by daylight (Kwarecki 1987).

In summer, the central part of the Arctic Basin, covered with pack ice, is characterized by large cloudiness and the frequent occurrence of fog. These factors limit the inflow of direct solar radiation, resulting in an annual sunshine duration below 1000 h (only 800 h in Norwegian Sea and Barents Sea). On the islands of the Canadian Archipelago, duration of sunshine amounts to 1400–1700 h/yr, reaching 1800–2000 h/yr in the polar areas of Canada and Alaska (Dolgin 1971). In the tundra zone of Siberia, it is between 1200 and 1600 h/yr. The largest sunshine duration in the Arctic is recorded at the northwestern part of Greenland ice cap, amounting to 2300–2400 h/yr (Putnins 1970).

The large variation in sunshine duration depends also on circulation factors as well as cloudiness which occurs most often along frontal zones. Because of the large cloud cover which occasionally reach 90%, the annual totals of sunshine duration in the zone of the subantarctic front are estimated at about 300 h only. On South Orkney Islands, sunshine duration is only 517 h/yr, on Heard Island 495 h/yr. Low sunshine totals on the coasts of the Antarctic amount to 1600–2000 h/yr, while inside the continent it exceeds 3500 h/yr (3632 h/yr at Vostok) (Petrov 1976).

Unlike the Arctic, the center of the Antarctic experiences one of the largest totals of sunshine duration on earth while the annual sunshine total in the Arctic Basin does not exceed 1000 h/yr. The time of maximum sunshine also differs. While in the Arctic the maximum hours of sunshine occurs in April (250–350 h), in the Antarctic it is observed in December (up to 700 h in the center and 300–400 h on the coasts).

The annual totals of global radiation in the Arctic are lowest (2000–2300 MJ.m^{-2}) in Greenland Sea and Barents Sea between Norway and Spitsbergen (Marshunova and Chernigovskiy 1971). Near the North Pole it is estimated at 2900 MJ.m^{-2}/yr. The highest annual totals reach 3500–3800 MJ.m^{-2} in the continental part of Alaska and Canada, but they can rise to even higher values at the Greenland ice cap.

Annual totals of global radiation at the South Pole amount to 4446 MJ.m^{-2} (Marshunova 1980); Vostok has an annual value of 4650 MJ.m^{-2}. At some coastal stations, annual totals range between 3600 and 4375 MJ.m^{-2}; and are 1.3–1.4 times greater than in the Arctic, differing only slightly from values recorded in the equatorial zone (Marshunova 1976). They drop below 3000 MJ.m^{-2} at southern Shetland Islands.

Negative values of net radiation are experienced in the inner part of the Polar Basin covered with pack ice, in the inner part of Greenland ice cap and in areas of the Canadian Archipelago covered with glaciers. The annual net radiation budget near the North Pole is estimated at about -110 MJ.m^{-2}. The lowest values are recorded in the southwestern part of Kara Sea with only -330 MJ.m^{-2}/yr (Chernigovskiy and Marshunova 1965). Net radiation on the Greenland ice cap is very low -500 MJ.m^{-2}/yr (Dolgin 1971). Positive values of radiation balance reach 210 to 420 MJ.m^{-2}/yr on the coasts of Canada, Alaska and Siberia, increasing to 840 MJ.m^{-2}/yr in the tundra farther from the sea. High values of net radiation, amounting to 850–1050 MJ.m^{-2}/yr are recorded in the perenially ice free parts of the Norwegian and Barents Sea.

Annual net radiation balance in the Antarctic is negative and amounts from -200 to -400 MJ.m^{-2} (Marshunova 1980). In the coastal zone, net radiation is near zero. Annual radiation balance in the seas of the subantarctic zone ranges from 0–800 MJ.m^{-2}. It drops to negative values only in Weddel Sea which is very often covered by ice, amounting to -40 MJ.m^{-2} near the Filchner Ice Shelf. The value in Southern Shetlands reaches 730 MJ.m^{-2}/yr. Mean annual values of net radiation in some ice-free areas, called Antarctic "oases", exceed 1500 MJ. m^{-2} (Schwerdtfeger 1970).

3. Circulation factors

The arctic with its vast ice-covered Polar Basin is a source area for the Arctic air masses which, through cyclones, can reach areas situated far to the south thus influencing the climate of the temperate zone. With respect to the climate dynamics of the Arctic, Barry and Hare (1974) pointed out the great role played by the arctic jet stream which occurs along the well-marked arctic front. The Arctic front separates the arctic airstreams from the westerly current in the temperate zone. Its great variations from season to season and also from day to day, which cause great changes in the Arctic weather are very important features of the Arctic environment.

The Antarctic, on the other hand, is the source area of the Antarctic air masses. Cyclonic patterns are formed in the contact zone where airmass activity is pronounced. The Antarctic High is usually located over the eastern Antarctic in winter with a mean pressure above 1035 hPa. There are six centers of active cyclogenesis around the Antarctic continent; the largest ones are in the Weddel Sea, with a mean pressure < 970 hPa, and the eastern part of Ross Sea. These Lows often reach the inner part of the Antarctic, particularly its western part, carrying moisture and causing snowfall.

Uniform ocean area surrounding the Antarctic favours the development of strong zonal circulation in the atmosphere around the Antarctic continent (Treshnikov and Salnikov 1985). Numerous deep cyclones, moving from West to East along the Antarctic front line, are formed. Favourable conditions for the development of meridional circulation exist in the Arctic, especially in the Atlantic sector. Considerably warm, polar-maritime air masses from the Atlantic can travel far into the inner part of the Arctic even to Novaya Zemlya because of Icelandic cyclogenesis. Advection of the Arctic air far into southern Europe, North America and Siberia often occurs, especially in the spring. It leads to a greater variability of climate in the subpolar part of the Arctic (zone of the tundra) than in the subantarctic zone situated on the oceans. The largest contrasts of pressure occur in winter, when, on the one hand, pressure can exceed 1080 hPa in the center of the Siberian High (e.g. 1084 hPa was recorded at Agata, Tunguska Basin, in December 1969), while on the other, a decrease in pressure to 915 hPa was recorded at the center of one of the Lows near Iceland (December 15, 1986). Similar contrasts of pressure occur between the Aleutian Low and the North American High. Atmospheric pressure of 1078.4 hPa was recorded in Norway during an exceptionally intense anticyclone over Alaska on January 31, 1989. It is the highest value of pressure ever recorded in North America (Limbert 1990).

Contrasts in atmospheric pressure decrease greatly in the Arctic in summer. A local Low is formed in the center of the Arctic Basin whereas an area of higher pressure is formed near Alaska with a ridge extending over the Pole towards Spitsbergen. The inner part of the Arctic ocean is penetrated by moving Lows (Dzerdzeevskii 1975). There is no quasi-permanent Arctic anticyclone even above the Greenland ice sheet (Jones 1987). The cross-Arctic ridge is strongest in April and May and forms pressure patterns similar to those associated with the Arctic anticyclones (Barry and Hare 1974; Korwowa 1976). The anticyclone in the vicinity of the Beaufort Sea is apparent in April.

Annual sea level pressure maps were prepared from data obtained from the Arctic Ocean Buoy Program (Walsh and Chapman 1990).

Strong pressure gradients in the Arctic are the reason for the strong winds, especially in the fall, winter and spring. Katabatic (downslope) winds similar to bora and foehn, occur frequently on the coasts of Greenland and other islands. Foehns can cause a temperature rise of up to 20°C and a relative humidity decrease down to 13% during a short period of time (Martyn 1992). Severe winter conditions, due to strong winds, exist on the northern coasts of Siberia. Along the northwest coast of the Taimyr Peninsula, 25–30% of all winds exceed 11 $m.s^{-1}$. Wind speed maximum at Dickson Island is 39 $m.s^{-1}$, but on the shores of Novaya Zemlya extremes of 50 $m.s^{-1}$ are possible during boras (after Anapolskaia and Zavialova from Barry and Hare 1974). The mean wind velocity on the shores of eastern Siberia is much lower (4–6 $m.s^{-1}$) throughout the year. Mean annual wind velocities in the central Arctic vary from 3.1 $m.s^{-1}$ at Alert to 7.2 $m.s^{-1}$ at Rudolph Island in Franz Joseph Land archipelago (Korotkevitsch 1972).

Specific orography and circulation conditions cause the very strong, katabatic slope or gravity winds, which blow from the ice plateau down towards the sea, as observed on the shores of the Antarctic. They influence strongly the snow accumulation and are very unfavourable to humans because of their strong cooling effect. The largest mean annual wind velocities are 22 $m.s^{-1}$ at Cap Denison, 10.7 $m.s^{-1}$ at Dumont d'Urville in Adélie Land, and 11.9 $m.s^{-1}$ at Mawson. At Vostok, typical of the inner part of the Antarctic, wind velocity averages 5 $m.s^{-1}$ (Korotkevitsch 1972).

4. Temperature conditions as a bioclimatic factor

Thermic conditions are a very important part of the climatic characteristics of the polar regions. It is known that the possibility of plant development depends almost wholly on summer temperature. Temperatures of the warmest month below 0°C occur only in the central part of the Arctic Basin, in the inner part of Greenland (to −10°C) and on glaciers of Canadian Archipelago islands. Small daily variations in temperature of only 4–8°C are a peculiar feature.

Positive temperatures, occurring during 2–3 months as well as constant inflow of solar radiation, cause a relatively high development of vegetation in the tundra zone on Arctic islands and shores of continents. Microclimatic influence, connected with greater heating of soil and rock surface on slopes depending on exposure, is an important factor. Relief has a great impact creating favourable microclimatic conditions. In summer, tundra thaws only to a depth of 150–200 cm. Changes in soil temperature, its thawing and freezing, leads to significant frost movements of soil and the formation of stony circular structures (polygonal soils), characteristic of the permafrost zone (Czeppe 1961, 1966). Presence of permafrost must be taken into account by human activity in the tundra zone. Construction of houses, for example, requires special protection against deformation caused by thawing of the ground. Similar problems occur with the construction of roads. All changes in the natural environment, even as little as driving a car or tractor, leave a permanent trace in the tundra. Thus, human activity in these regions leads to irreversible disturbances in the natural environment.

Winter temperatures depend mainly on radiation factors, especially in the areas where high pressure centers are formed (Siberia, northern Canada and Alaska), declining to below −30°C, even to −50°C in the Greenland ice cap. Mean January temperatures drop to values below −34°C in the central part of the Arctic Basin between the North Pole, Greenland and the Canadian Archipelago. In the areas with great cyclonic activity, e.g. the Atlantic sector of the Arctic, temperatures are influenced by advection. Such advection of relatively warm air from the Atlantic causes large temperature anomalies on the western coast of Spitsbergen in winter (mean January temperatures from 0 to −10°C).

Extreme temperatures may be the best indicator of the severe climate conditions of the polar zones. The lowest temperature ever recorded in the Arctic (Table 2), −70°C, occurred on the Greenland ice cap, caused mainly by a combination of high altitude (about 3000 m.a.s.l.) and strong radiational cooling. Temperature inversion is a common feature of the polar atmosphere. There, the depth of the inversion layer averages 400 m. There are also significant differences in the strength of the inversion, 10°C or more is not uncommon (Putnins 1970).

In subpolar areas, low thermic minima occur in Siberia; the lowest temperatures (about −71°C) are recorded at Oimyakon, in a mountain basin. But on the surrounding slopes, mean temperatures rise about 20°C higher; the warmest level of inversion layer is occasionally found at 1,500 m.a.s.l. In January 1989, unprecedented cold temperatures were observed in Alaska. On January 26, the lowest ever recorded temperature of −60°C occurred at Tanana (Limbert 1990). Temperatures about −50°C were then recorded in the Yukon (Canada), but in February 1947 temperature at Snag Airport dropped to −62.8°C (Chapman 1959). However, there are no such low temperatures in the central part of the polar basin; the minimum observed at stations drifting on sea-ice is −53°C.

Temperature inversion is typical of this region in winter and occurs with a frequency of 80–90%. Ground inversion layer is commonly 300–500 m in depth (Barry and Hare 1974; Vovinckel and Orvig 1970). Because of the strength of the Icelandic Low and the influence of warm sea currents (Gulfstream) a large drop of temperature is usually not observed in winter. Absolute minimum on the western shores of Spitsbergen is −33.5°C, −31.6°C in Bear Island and −28.4°C in Jan Mayen Island.

Temperatures below −40°C are difficult for humans in polar regions. In January, in the Arctic region, according to Hastings (from Barry and Hare

Table 2. Extreme and mean temperatures for selected stations in Arctic and adjacent areas (MIN – absolute minimum, MAX – absolute maximum, Jan. – mean January, Jul. – mean July, Ann. – mean annual).

No	Station	ϕ (N)	λ	Hs (m)	MIN (°C)	MAX (°C)	Jan. (°C)	Jul. (°C)	Ann. (°C)
1	'North Pole' 6	81–86°	148–38°E	0	–50.0	1.4	–33.8	0.0	–20.0
2	'North Pole' 7	85–86°	150–35°E	0	–49.3	1.2	–36.3	0.1	–19.8
3	'Alpha'	82–86°	176–112°E	0	–49.4	2.8	–32.2	0.0	–18.3
4	Derfting station 'B'	75–82°	126–98°E	6	–53.3	5.6	–30.0	0.8	–18.3
5	Barrow	71°18′	156°47′W	7	–48.9	25.6	–26.2	4.3	–12.2
6	Barter Island	70°07′	143°40′W	15	–50.6	22.2	–26.4	4.9	–11.9
7	Sachs Harbour	71°57′	124°44′W	84	–47.8	17.8	–29.9	5.6	–14.4
8	Mould Bay	76°14′	119°20′W	15	–52.8	15.6	–33.1	4.0	–17.4
9	Isachen	78°40′	103°32′W	25	–53.9	18.9	–34.6	3.7	–18.7
10	Resolute	74°43′	94°59′W	64	–51.7	16.1	–31.8	4.6	–16.2
11	Eureka	80°00′	85°56′W	2	–52.8	19.4	–35.9	5.7	–19.1
12	Alert	82°30′	62°20′W	62	–47.8	20.0	–31.9	3.9	–17.8
13	Nome	64°30′	165°20′W	4	–41.1	27.2	–15.3	9.7	–3.3
14	Fairbanks	64°49′	147°52′W	133	–54.4	33.9	–23.9	15.4	–3.4
15	Bethel	60°47′	161°43′W	3	–46.7	30.0	–15.8	12.6	–1.8
16	Baker Lake	64°18′	96°00′W	4	–50.0	27.8	–32.9	10.7	–11.9
17	Frobisher Bay	63°45′	68°33′W	7	–45.6	24.4	–26.5	7.9	–8.9
18	Port Harrison	58°27′	78°08′W	6	–49.4	26.7	–25.0	8.9	–6.9
19	Aklavik	68°14′	134°50′W	10	–52.2	30.6	–28.6	13.8	–8.9
20	Yellowknife	62°28′	114°20′W	215	–51.7	30.0	–27.8	15.9	–5.5
21	Churchil	58°45′	94°04′W	11	–49.4	35.6	–27.5	12.0	–7.2
22	Nord	81°36′	16°40′W	35	–51.1	16.3	–29.6	4.2	–16.4
23	Northice	78°04′	38°29′W	2,343	–70.0	0.0	–41.0	–9.0	–29.6
24	Centrale	70°55′	40°38′W	2,993	–65.0	1.0	–36.5	13.0	–28.0
25	Eismitte	70°42′	40°42′W	3,000	–64.8	–3.0	–41.7	–11.2	–30.2
26	Nanortalik	60°08′	45°11′W	7	–17.2	20.1	–3.3	6.5	1.8
27	Ivigtut	61°12′	48°10′W	30	–28.9	23.1	–5.4	9.8	1.8
28	Godthaab	64°10′	51°45′W	20	–26.0	23.0	–7.7	7.6	–0.7
29	Jakobshavn	69°13′	51°03′W	31	–43.2	21.9	–13.5	8.2	–3.9
30	Godhavn	69°14′	53°31′W	11	–33.8	18.3	–11.8	8.0	–3.2
31	Umanak	70°41′	52°07′W	8	–35.2	18.0	–12.9	7.8	–4.0
32	Upernavik	72°47′	56°10′W	35	–40.0	19.0	–17.0	6.0	–6.4

Table 2. (Continued)

No	Station	ϕ (N)	λ	Hs (m)	MIN (°C)	MAX (°C)	Jan. (°C)	Jul. (°C)	Ann. (°C)
33	Angmagssalik	65°37′	37°39′W	29	–32.0	25.3	–6.8	7.7	–0.4
34	Scoresbysund	70°25′	21°58′W	17	–43.7	16.8	–15.3	4.7	–6.7
35	Myggbykta	73°29′	21°34′W	2	50.9	22.1	–20.2	3.7.	-10.0
36	Jan Mayen	70°59′	8°20′W	23	–27.9	18.1	–3.9	5.4	–0.2
37	Isfjord Radio (Spitsbergen)	78°04′	13°38′E	9	–33.5	17.0	–11.5	4.7	–4.7
38	Bear Island	74°31′	19°01′E	14	–31.6	23.6	–7.9	4.4	–2.1
39	Hopen	76°31′	28°01′E	10	–35.6	17.4	–13.7	2.0	–6.1
40	Vardö	69°36′	31°06′E	15	–23.7	25.8	–4.9	8.9	1.0
41	Reykjavik	64°08′	21°56′W	18	–17.1	23.4	–0.3	11.4	5.1
42	Tromsö	69°36′	18°57′E	115	–15.8	28.5	–3.5	12.4	2.9
43	Karasjok	69°28′	25°31′E	135	–47.2	31.1	–15.5	13.0	–2.3
44	Ostrov Rudolfa	81°48′	57°58′E	48	–42.8	11.7	–20.0	0.6	-11.9
45	Bukhta Tikhaya	80°19′	52°48′E	6	–39.4	11.7	–17.2	1.1	–9.3
46	Ostrov Domashniy	79°30′	91°08′E	3	–47.2	6.1	–25.6	1.1	–14.0
47	Cap Cheluskin	77°43′	104°17′E	6	–49.0	24.0	–31.1	0.8	–15.6
48	Kotelnyj	76°00′	137°54′E	11	–49.0	22.0	–29.5	2.5	–14.3
49	Ostrov Chety-rechstolbovoy	70°38′	162°24′E	30	–51.0	23.0	–29.5	2.5	–14.3
50	Wrangel Island	70°58′	178°32′E	3	–45.6	18.3	–23.9	2.8	–11.7
51	Malye Karmakuly	72°23′	52°44′E	16	–43.9	24.4	–15.0	6.7	–4.2
52	Dikson	73°30′	80°14′E	22	–52.0	27.0	–27.5	3.6	–12.3
53	Bulun	70°45′	127°47′E	37	–49.4	30.6	–41.4	10.8	–14.5
54	Chokurdach	70°37′	147°53′E	20	–52.0	32.0	–36.2	10.2	–14.2
55	Khatanga	71°50′	102°28′E	24	–61.0	34.0	–34.9	11.8	–13.8
56	Cap Schmidt	68°55′	179°29′W	7	–47.0	26.0	–26.4	3.6	–12.1
57	Narjan Mar	67°39′	53°01′E	7	–51.0	33.0	–17.3	12.0	–3.9
58	Uelen	66°10′	169°50′W	7	–45.0	21.0	–21.7	5.4	–8.2
59	Murmansk	68°53′	33°03′E	46	–38.0	33.0	–9.9	12.8	0.1
60	Selechard	66°32′	66°32′E	35	–54.0	31.0	–24.4	13.8	–6.7
61	Werkchojansk	67°33′	133°23′E	137	–68.0	35.0	–48.9	15.0	–15.6
62	Oymjakon	63°16′	143°09′E	740	–71.0	33.0	–50.1	14.5	–16.5
63	Dudinka	69°24′	86°10′E	20	–57.0	30.0	–29.5	12.0	–10.7
64	Olenek	68°30′	112°36′E	130	–65.0	36.0	–40.9	14.1	–13.3
65	Arkhangelsk	64°30′	40°30′E	4	–45.0	34.0	–12.6	15.6	–0.6

1974), such temperatures are very rare in regions above the Barents, Greenland and Chukchi Seas. Also in the central Arctic Basin, their frequency is about 5% only. Maximum frequency exists in eastern Siberia, up to 70% of the days; on the Greenland ice cap- above 40%, above 30% in northern parts of Canadian Archipelago, and 10–20% in western Canada and Alaska.

Much more severe thermic conditions exist in the Antarctic. Absolute minimum of air temperature reached an unimaginable value of −89.2°C at Vostok station (3,488 m) on July 21, 1983. It is the world's lowest recorded temperature. A strong inversion with a mean depth of 1000 m exists above the ice sheet 9 to 10 months a year. On the coastal zone, the lowest observed temperatures are from −40 to −60°C, and on Antarctic Peninsula and South Shetland Islands from −26 to −32°C (Table 3). The inner part of the Antarctic continent has never recorded positive temperatures. For example, the maximum observed temperature on the South Pole is only −15°C. The absolute maximum temperature in the coastal zone is from 5 to 9°C, but increases to 12°C in the snow free "oases" (Martyn 1985).

The best bioclimatic indicator for designating the total effect of wind and low temperatures is the wind chill factor K_o ($kcal.m^{-2}.hr^{-1}$):

$$K_o = [(v * 100)^{1/2} - v + 10.5]\,[33 - t]$$

where t is the air temperature (°C) and v the wind velocity, in $m.s^{-1}$ (Oliver 1973). Exposed flesh can freeze when Ko values are above 1400 $kcal.m^{-2}hr^{-1}$. Extremely dangerous sensation for humans exists at > 2000 $kcal.m^{-2}.hr^{-1}$, when exposed flesh freezes in 1 min. Such conditions occur on the coast of the Antarctic at Mirny during 59% of the days (Dubrovin 1976). The wind chill factor is often expressed in terms of equivalent temperature. For example, if air temperature is equal −40°C, the equivalent temperature is −73°C at a wind speed of 10 $m.s^{-1}$ and as low as −84°C at a wind speed of 20 $m.s^{-1}$.

5. Conclusion

Considerable differences can be seen between the climates of the polar regions. Climate of the Antarctic is more severe, characterized by larger thermic differences than the Arctic, with the exception of the subpolar zone where conditions are reversed. Moreover, stronger winds, such as katabatic winds, occur on the Antarctic coast (e.g. Cape Adare). The high altitude of the continental glacier surface (3000–4000 m), connected with low atmospheric pressure (about 600 hPa) and low oxygen content is also very strenuous for human organs (the hypoxy factor). The bioclimatic conditions of the interior Antarctic with equivalent temperatures of 130–150°C below zero are the extremes on earth (Tichomirov 1968).

Human life under polar conditions is greatly constrained by the effects of low air temperatures and strong winds. In such situations, cooling of human

Table 3. Extreme and mean temperatures for selected stations in Antarctic and adjacent areas (MIN – absolute minimum, MAX – absolute maximum, Jan. – mean January, Jul.– mean July, Ann. – mean annual).

No	Station	ϕ (S)	λ	Hs (m)	MIN (°C)	MAX (°C)	Jan. (°C)	Jul. (°C)	Ann. (°C)
1	Amundsen-Scott (South Pole)	90°00′	–	2,800	–80.6	–15.0	–28.7	–59.5	–49.3
2	Byrd	80°01′	119°31′W	1,533	–62.8	–0.6	–14.8	–35.7	–27.8
3	Plateau	79°15′	40°30′E	3,625	–86.2	–18.5	–33.9	–68.0	–56.4
4	Vostok	78°28′	106°48′E	3,488	–89.2	–13.6	–33.0	–67.2	–55.5
5	Little America	78°18′	163°00′W	40	–60.6	5.9	–6.6	–29.6	–20.9
6	General Belgrano	77°58′	38°48′W	50	–57.2	6.8	–6.0	–32.7	–22.3
7	McMurdo	77°53′	166°44′E	24	–56.9	5.6	–3.3	–26.0	–17.5
8	Halley Bay	75°30′	26°39′W	30	–52.5	4.0	–5.2	–28.1	–18.8
9	Eights	75°14′	77°10′W	421	–60.0	2.2	–10.0	–33.5	–26.0
10	Hallet	72°18′	170°19′E	5	–47.8	5.6	–1.1	–26.4	–15.3
11	Novolasarevskaya	70°46′	11°49′E	87	–41.0	9.9	–1.2	–18.1	–10.8
12	S.A.N.A.E.	70°19′	2°21′W	52	–51.0	7.3	–4.3	–26.7	–17.3
13	Pionierskaya	69°44′	95°30′E	1,740	–62.4	–15.0	–23.4	–47.3	–38.0
14	Syowa	69°00′	39°35′E	15	–42.7	9.5	–1.0	–18.3	–10.6
15	Davis	68°35′	77°58′E	12	–38.3	9.5	–0.2	–17.2	–10.3
16	Molodozhnaya	67°40′	45°50′E	42	–42.0	8.5	–0.8	–18.7	–10.8
17	Mawson	67°36′	62°53′E	8	–35.4	8.8	–0.2	–17.6	–11.2
18	Dumont d'Urville	66°42′	140°00′E	41	–36.5	6.1	–1.3	–17.1	–11.0
19	Mirny	66°36′	93°01′E	30	–40.3	8.0	–1.8	–17.0	–11.5
20	Wilkes	66°15′	110°35′E	12	–37.8	7.8	–0.2	–15.6	–9.4
21	Argentine Island	65°15′	64°15′W	11	–43.3	11.7	0.2	–11.2	–5.1
22	Melchior	64°20	62°59′W	8	–29.6	9.2	1.1	–8.9	–3.5
23	Bahia Esperanza (Hope Bay)	63°24′	56°59′W	11	–32.1	14.6	0.9	–10.1	–5.0
24	Decepción	62°59′	60°43′W	8	–30.0	10.0	1.4	–8.0	–2.8
25	Orcadas (Laurie Island)	60°44′	44°44′W	4	–40.1	12.2	0.5	–10.6	–4.2
26	Ellsworth	77°43′	41°07′W	42	–56.7	2.2	–7.9	–33.8	–22.5
27	King Boduen	70°26′	24°19′E	38	–48.0	8.0	–4.7	–23.6	–15.2
28	Admirality Bay	62°03′	58°24′W	9	–32.2	10.6	1.5	–6.6	–2.2
29	Bellingshausen	62°12′	58°56′W	16	–26.3	8.1	0.9	–6.7	–2.7
30	Berbardo O'Higgins	63°19′	57°54′W	10	–30.2	12.0	1.1	–8.4	–3.8
31	Port Stanley (Falklands)	51°42′	57°52′W	51	–11.1	24.4	9.0	2.1	5.5
32	Signy Island	60°43′	45°36′W	7	–34.8	13.9	1.1	–10.0	–3.6
33	Grytviken (South Gerogia)	54°17′	36°30′W	3	–19.2	28.8	4.8	–1.5	1.8
34	Port-aux-Français	49°21′	70°12′E	14	–8.9	22.4	6.9	1.6	4.3
35	Macqurie Island	54°30′	158°57′E	7	–8.9	12.4	6.7	3.2	4.7
36	Campbell Island	52°53′	169°09′E	15	–6.7	20.8	9.4	4.9	7.1
37	Ushuaia	54°48′	68°19′W	6	–19.6	29.0	9.2	1.6	7.7

organs occurs. This can be controlled with special clothes and diet. For this reason, research stations under the most hazardous climatic conditions are built under ice and snow. The famous Eskimo (Inuit) house "igloo" offers very good protection against strong wind and frost in the Arctic climate.

One could ask why investigating the climate of the polar regions is so important. Kellogg and Schware (1981) provide some answers: first, "most climate models suggest that climate changes will be greatest in the polar regions"; second, "feedback mechanisms between the cryosphere, atmosphere, and oceans would affect the climate outside the polar regions"; and third, "substantial change in ice sheet volume would affect sea levels with potentially drastic consequences for shoreline areas" (sea level can rise even by a few meters). The whole ice volume of Greenland and the Antarctic contains about 80% of the existing fresh water supply on Earth.

Increase in carbon dioxide content in the atmosphere may have a considerable impact on polar regions. Experiments show that average global increase of surface temperature for a doubling of carbon dioxide will probably be between 1.5 and 4.5°C (Kellog and Schware 1981). It may cause a considerable change of the global circulation of the atmosphere, the course of sea currents, and lead to changes in the spatial distribution of temperature and precipitation. However, the greatest changes would be caused by thawing of great ice masses in the Arctic and Antarctic. Opinions about projected scenarios of further climate changes in polar regions are contradictory. During the years 1966–1987, the greatest marginal ice zone was north central Asia while the greatest cooling occurred over Scandinavia (Walsh and Chapman 1990).

Considerable changes may also be expected in the subpolar tundra zone, where permafrost may retreat. As a result, ecosystems and river discharges will be affected. Such effects are, however, difficult to predict.

An increasing "ozone hole", especially above the Antarctic, leads to the penetration of harmful radiation to the surface of the Earth. Spring ozone depletion over the Antarctic is caused by emissions of chlorine-fluorine-mathanes. Abnormally low ozone contents were observed in 1987, 1989 and 1995.

One of the great environmental problems in Fairbanks, Alaska and in Whitehouse (Yukon Territory, Canada), is the large load of air pollution caused by fuel burning in automobiles and aircrafts. The pollutants are trapped in the temperature inversion resulting in persistent ice fog (Barry and Hare 1974). A similar situation is found at Norilsk, Siberia. But most aerosols in the polar atmosphere are produced by violent volcanic eruptions (Voskresensky 1988).

References

Atlas Antarktiki (1966): *Atlas of Antarctic*. Part 1, GUGK, Moscow–Leningrad (maps), 1969, Part 2, Gidrometeoizdat, Leningrad (text in Russian).

Atlas Arktiki (1985): *Atlas of Arctic*. GUGK, Moscow.

Baird, P.D. (1964): *The polar world*. Longmans, Green and Co., London.

Barry, R.G. and Hare, F.K. (1974): Arctic climate. In: *Arctic and alpine environments*, ed. by J.D. Ives and R.G. Barry. Methuen, London, 17–54.

Barry, R.G. (1989): The present climate of the Arctic Ocean and possible past and future states. In: *The Arctic Seas. Climatology, oceanography, geology and biology*, ed. by Y. Herman. Van Nostrand Reinhold Company, New York, 1–45.

Chernigovskiy, N.T. and Marshunova, M.S. (1965): *Klimat Sovetskoyi Arktiki – radiacionnyj rezim* (The climate of the Soviet Arctic-radiation regime). Gidrometeoizdat, Leningrad, 198 pp. (in Russian).

Chapman, S. (1959): *IGY: Year of discovery*. The University of Michigan Press.

Czeppe, Z. (1961): *Roczny przebieg mrozowych ruchów gruntu w Hornsundzie (Spitsbergen), 1957–1958* (Annual course of frost ground movements at the Hornsund, Spitsbergen). Zeszyty Naukowe UJ, Prace Geograficzne, Z. 3, Kraków, 75 pp.

Czeppe, Z. (1966): *Przebieg lównych procesów morfogenetycznych w pol undiowo-zachodnim Spitsbergenie* (The course of main morphogenetic processes in South-West Spitsbergen). Zeszyty Naukowe UJ, Prace Geograficzne, Z. 13, Kraków, 129 pp.

Dolgin, I.M. (ed.) (1971): *Meteorologiczeskij rezim zarubieznoj Arktiki* (Meteorological regime of the foreign Arctic). Leningrad, 227 pp. (in Russian).

Dubrovin, L. (1976): *Chelovek na ledjanom kontinente* (The man on the ice continent). Gidrometeoizdat, Leningrad, 158 pp. (in Russian).

Dzerdzeevskii, B.L. (1975): *Cirkulacionnyje schemy v troposferie centralnoj Arktiki* (The circulation patterns in the troposphere of the central Arctic). w: Izbrannyje Trudy, Izd. Nauka, Moscow.

Jones, P.D. (1987): The early twentieth century Arctic High – fact or fiction? *Climate Dynamics*, 1, 63–75.

Kellog, W.W. and Schware, R. (1981): *Climate change and society. Consequences of increasing carbon dioxide*. Westview Press, Boulder, CO, 178 pp.

Korotkevitsch, E.S. (1972): *Polarnye pustyni* (Polar deserts). Gidrometeizdat, Leningrad, 420 pp. (in Russian).

Korwowa, A.M. (1976): Nowyje karty sriedniemiesiacznogo dawleni ja wozducha w Arktikie (New maps of the mean monthly sea level pressure in the Arctic). In: *Voprosy Poliarnoj Klimatologii, Trudy AANII*, T. 328. Gidrometeoizdat, Leningrad, 22–43 (in Russian).

Kwarecki, K. (1987): Adaptacja i zdrowie cz owieka w strefach polarnych (Adaptation and human health in the polar zones). In: *Antarktyka-przyroda i cz owiek*. Ossolineum, Wrocław, 115–152 (in Polish).

Limbert, D.W.S. (1990): Weather events in 1989 and their consequences. *WMO Bulletin*, 39(4), 254–273.

Markov, K.K., Bardin, V.J.L., Lebedev, V.A., Orlov, I. and Svetova, I.A. (1968): *Geografija Antarktidy* (Geography of Antarctic). "Mysl" Publishing House, Moscow, 439 pp. (in Russian).

Marshunova, M.S. and Chernigovskiy, N.T. (1971): *Radiacionnyi rezim zarubeznoyi Arktiki* (Radiation regime of the foreign Arctic). Gidrometeorologiceskoje Izdatelstvo, Leningrad, 182 pp. (in Russian).

Marshunova, M.S. (1976): Ob izmencivosti solnecnoy radiacii v Antarktide (About the variability of solar radiation in Antarctic). In: *Voprosy Poliarnoj Klimatologii, Trudy AANII*, T. 328. Gidrometeoizdat, Leningrad, 93–105 (in Russian).

Marshunova, M.S. (1980): *Uslovija formirovani ja i charakteristiki radiaciyonnogo klimata Antarktidy* (Conditions of forming and characteristic of the radiation climate of Antarctic). Gidrometeoizdat, Leningrad, 214 pp. (in Russian).

Martyn, D. (1992): *Climates of the world*. PWN, Warszawa/Elsevier, Amsterdam, 435 pp.

Oliver, J.E. (1973): *Climate and man's environment. An introduction to applied climatology.* John Wiley and Sons, New York.

Prtrov, L.S. (1976): Prodolzitelnost' solnecnogo sijaniya v juznom poljarnom rayonie (Duration of sunshine in the southern polar region). In: *Voprosy Poliarnoj Klimatologii, Trudy AANII*, T. 328. Gidrometeoizdat, Leningrad, 116–130 (in Russian).

Putnins, P. (1970): The climate of Greenland. In: *World Survey of Climatology*, Vol. 14 (Climates of the polar regions). Elsevier, Amsterdam, 3–128.

Rakusa-Suszczewski, S. (1987): Srodowisko przyrodnicze i dzia alnosc cz owieka (Natural environment and human activity). In: *Antarktyka – przyroda i cz owiek.* Ossolineum, Wrocław, 11–113 (in Polish).

Schwerdtfeger, W. (1970): The climate of the Antarctic. In: World Survey of Climatology, Vol. 14 (Climates of the polar regions). Elsevier, Amsterdam, 253–355.

Spravocnik po klimatu Antarktidy (1977): *Antarctic climate record.* Gidrometeoizdat, Leningrad (in Russian).

Steffensen, E.L. (1982): The climate at Norwegian Arctic stations. Det Norske Meteorologiske Institutt, Klima No. 5, Oslo, 44 p.

Tichomirov, I.I. (1968): *Bioklimatologi ja centralnoj Antarktidy i aklimatizaci ja cheloveka* (Bioclimatology of central Antarctic and human acclimatization). Nauka, Moscow, 198 pp. (in Russian).

Treshnikov, A.F. and Salnikov, S.S. (1985): *Severnyj Ledovitj i Jushnyj Okeany* (Arctic and Antarctic Oceans). Nauka, Leningrad, 501 pp. (in Russian).

Treshnikov, A.F. and Voskresensky, A.I. (1976): Klimat zony dreyfujuscich ldov (Climate of the drifting ice zone). In: *Arkticheskie Drejfujuscije Stancii* (Arctic drifting stations), Problems of Geography, Vol. 101. "Mysl" Publishing House, Moscow, 87–97 (in Russian).

Voskresensky, A.I. (ed.) (1988): *Monitoring klimata Arktiki* (Arctic climate monitoring). Gidrometeoizdat, Leningrad, 205 pp. (in Russian).

Vovinckel, E. and Orvig, S. (1970): The climate of the North Polar Basin. In: *World Survey of Climatology*, Vol. 14 (Climates of the polar regions), Elsevier, Amsterdam, 129–252.

Walsh, J.E. and Chapman, W.L. (1990): Short-term climatic variability of the Arctic. *Journal of Climate*, 3, 237–250.

Tadeusz Niedzwiedz
Instytut Meteorologii i Gospodarki Wodnej
Oddziat W. Krakowie
ul. P. Borowege 14
30-125 Kraków
Poland

PART 4

LOCAL SCALE CLIMATES

4.1. INTRODUCTION TO THE PROBLEMS ON LOCAL CLIMATE AND MAN

JANUSZ PASZYNSKI

As an approach to the problems of the interactions of local climate and man, the papers contained in the fourth part deal with the way in which the local factors must be taken into account when attempting to assess the climatic potential of a small area. The general assumption is made that main features of climate depend on numerous factors of various kind.

It is generally recognized that different factors play leading roles in the formation of climate, depending on the spatial scale (Geiger 1969; Munn 1966; Oke 1987; Yoshino 1975). In the case of the global scale, astronomical factors are of primordial importance, because they control the amount of incoming solar radiation. They thus determine the geographical repartition of main climatic zones on Earth, such as polar, temperate, subtropical or equatorial zones. Regional differences within those zones then depend essentially on the main features of the circulation of the atmosphere. Finally, the type of the so called "active surface", forming the lower boundary of the atmosphere, is of crucial importance in view of small-scale climate differentiation.

Therefore, when large areas are being surveyed on small-scale maps, climate is likely to be a very important factor in determining behavior of man as well as his various activities: social, cultural, economic. When small areas are being surveyed on large-scale maps (1:200000 or greater) the interaction between climate and man becomes inversely related, and human influence upon local climatic differences becomes much more evident. The reason is that it is quite possible to alter the properties of the Earth's surface, over small areas at least, while it is impossible to change astronomical factors controlling global climate, or the atmospheric circulation which generates climatic differences at the regional scale.

Local climates owe their particularities to various physical and geometrical properties of the active surface. In reality, we commonly deal with a layer than a surface in the strict geometric sense. Such a layer comprises not only the lowest part of the atmosphere, including the vegetation canopy layer or

M. Yoshino et al. (eds.), Climates and Societies – A Climatological Perspective, 327–331.

urban canopy layer, but also the uppermost part of the materials forming the substratum of the atmosphere usually, extending to a depth where diurnal exchange of energy becomes negligible. Geometrical and physical properties of the layer so defined, may be divided into several groups, in regard to their importance to individual processes taking place there.

We can thus distinguish:

1. Radiative properties controlling absorption of solar radiation and emission of long-wave radiation by the Earth's surface: the reflectivity (albedo), the emissivity, the turbidity of the atmosphere due to air pollution, the slope (inclination) and aspect (exposure) of the surface.
2. Thermal properties controlling the conductive transfer of heat between the interface and the underlying material: thermal conductivity and heat capacity of soil, water, snow, etc.
3. Aerodynamic properties controlling the turbulent transfer of both sensible and latent heat between the interface and the atmosphere: roughness, zero-plane displacement; sometimes also the slope inclination and relative altitude above the bottom of the valley which control the gravitational drainage of cold air.
4. Hydrologic properties controlling the process of evaporation: the soil-humidity, and the so called "water covering factor" which expresses the share of the surface covered by water which could freely evaporate.

In this way various land forms, soils, vegetation, artificial constructions, and many other local factors control the process of energy exchange which takes place on the Earth-Atmosphere interface (Boyen et al. 1976; Deacon 1969; Miller 1981). This process can be represented quantitatively by the energy balance equation:

$$Q + G + H + E + A = 0,$$

where: Q is the net radiation, G is the conductive flux of heat to/from the substratum, H is the turbulent flux of sensible heat to /from atmosphere, E is the turbulent flux of latent heat of evaporation, melting, condensation or freezing, to/from the atmosphere, and A is the flux of heat of anthropogenic origin.

Some other forms of exchange of energy on the interface, as e.g. chemical heat of carbon assimilation by vegetation, or heat contained in precipitation, may be omitted here as negligible, because their quantitative share in the energy balance is very small, although they are sometimes very important as from the biological point of view.

Q in turn is the difference between all incoming and outgoing radiative fluxes of both solar (short-wave) and terrestrial (long-wave) origin, according to the equation of radiation balance:

$$Q = K + L,$$

where K is absorbed solar radiation and L is net long-wave radiation.

Each term of these equations, excepting K, may become either positive or negative. Conventionally, we assume that a given flux has a positive value when it is directed to the interface either from above or from below. Inversely, it is negative when directed from the interface to the atmosphere or to its substratum.

It is possible to modify the structure of the balance of energy on the interface Earth-Atmosphere by altering one or several physical properties enumerated. This, in turn, leads to changes of local climatic conditions. Such changes, firstly, impact the air temperature on the ground. This temperature must be thus considered as a final result of processes of transfer, exchange and conversion of energy, taking place in the interface layer between the atmosphere and its substratum.

Modifications in the structure of the energy balance, which are due to human activities, may be inadvertent or intentional in character or both. Inadvertent climatic changes occur often in heavy urbanized and industrialized areas, and are commonly a consequence of air pollution. Their consequences are very unfavorable, because they cause not only the deterioration of air quality, but they provoke sometimes a general worsening of climatic conditions. These problems are examined in details in the papers contained in this part of the book.

Purposeful changes of the structure of the energy balance through the intermediary of intentional modifications of properties of the Earth-Atmosphere interface, tend towards amelioration of local climate. Such interventions are especially important from the agricultural point of view. One of them are various measures designed to protect agricultural or horticultural fields against local frosts. Local "frost pockets" which appear often at night, may be caused either by especially strong outgoing long-wave radiation, by insufficient inflow of heat from deeper soil layers accumulated there during daytime, by heat losses due to strong evaporation, by total lack of vertical mixing of air in the lowest air layers, or by gravitational inflow of cold air from surrounding slopes. Sometimes, two or even more of these phenomena occur simultaneously, thus combining the final effect in form of very low surface and air temperatures.

The choice of adequate intervention depends on which term or component of the energy balance is responsible for the local frost. Tree-belts diverting the flow of cold air from its drainage area, or artificial water reservoirs heating the air above them as well as the surrounding areas as a consequence of the high heat capacity of water, belong to such climatic ameliorations. They are passive measures modifying local climatic conditions. Sometimes it is necessary to use active means to alter micrometeorological conditions during relatively short periods, such as mixing of the air with propeller-like devices, artificial screening with smoke, or direct heating of the air by small ovens. In these or similar ways humans are able to change local climate, according to their practical requirements.

In climatology the expression "topoclimate" is often used in the sense of "local climate" (Yoshino 1975; Geiger et al. 1995). The meaning of both terms is the same, because Greek "Topos" corresponds to Latin "Locus" – a very small area or a very small region. Therefore, both terms are used here interchangeably. In this way topoclimatology must be considered as a part of physical climatology which is concerned with the interaction between local factors such as relief, soils, and vegetation, on the one hand, and processes of transfer, exchange and conversion of energy and water, on the other.

A very important task of modern topoclimatology seems to be topoclimatological mapping. Detailed climatic maps, similar to large scale maps presenting the distribution of various components of geographical environment (e.g. geomorphological or soil maps), are indispensable to qualify the existing climatic conditions. Such topoclimatological mapping involves the use of modern measurement methods, and first of all remote sensing and satellite imagery. The satellite techniques have become recently a powerful tool in the climatologist's hand, enabling him to plot detailed climatic maps of relatively large areas.

It is also possible to establish some derived maps, drawn up for different practical purposes, and containing the evaluation of existing climatic conditions, carried out from various points of view. The aim of such derived maps is to determine the suitability of sites for various practical purposes and to delimit sites which are to be avoided for climatic reasons, as well as sites, where special measures for improvement of existing topoclimate could be introduced.

References

The scientific literature concerning problems of local climates and their modification is very ample. Therefore, our list publications contains just several fundamental works of rather general character.

Boyen, H., Dogniaux, R. and Paszynski, J. (1976): *Méthodes de deetermination du bilan thermique de la surface active de la Terre.* Institute Royal Météorologique de Belgique, Publications, Serie A, No. 99, Bruxelles.

Deacon, E.L. (1969): Physical processes near the surface of the earth. In: *World Survey of Climatology*, Vol. 2, *General climatology*, ed. by H. Flohn. Elsevier, Amsterdam, 39–104.

Geiger, R. (1969): Topoclimates. In: *World Survey of Climatology*, Vol. 2, *General climatology*, ed. by H. Flohn. Elsevier, Amsterdam, 105–138.

Geiger, R., Aron, R.H. and Todhunter, P. (1995): *The climate near the ground.* Vieweg, Braunshweig, 528 pp.

Landsberg, H.E. (1981): *The urban climate.* International Geophysics Series, Vol. 28. Academic Press, New York.

Miller, D.H. (1981): *Energy at the surface of the earth. An introduction to the energetics of ecosystems.* International Geophysics Series, Vol. 27. Academic Press, New York.

Munn, R.E. (1966): *Descriptive micrometeorology.* Academic Press, New York.

Oke, T.R. (1987): *Boundary layer climates.* Methuen, London.

Paszynski, J. (1991): Mapping urban topoclimates. *Energy and buildings*, 15–16, 1059–1062.
Yoshino, M.M. (1975): *Climate in a small area. An introduction to local meteorology*. University of Tokyo Press, Tokyo, 549 pp.

Janusz Paszynski
Zaklad Klimatologii
Instytut Geografii i Przestrzennego Zagospodarowania
Polish Academy of Science
Krakowskie Prz 30
00927 Warsaw
Poland

4.2. INTERACTIONS OF MAN AND CLIMATE IN THE URBAN WORLD – THE INFLUENCE OF URBANIZATION ON THE LOCAL CLIMATE AND THE INFLUENCE OF URBAN CLIMATE ON MAN

FRITZ WILMERS

Summary

Urbanization changes the local climate. It is useful in this regard to differentiate the development of cities from the construction of single houses or settlements in undisturbed areas. In this chapter, criteria for towns are given and the most important changes to the local climate due to their presence are presented. The major impacts of urbanization on local climates are expressed in urban heat island, net radiation and energy balance, windfield and water balance which are discussed with examples drawn mainly from the mid latitudes.

An important consequence of urbanization can also be found in the deterioration of air quality. Air pollution affects urban climates via heat island, energy balance and windfield, and should therefore be investigated as part of larger urban climate programmes.

The dynamics of an urban climate are governed in part by city structure. The urban mosaic classified into types of built up structures and open areas serves as poleotopes with corresponding climatotopes.

Changes in local climate increase with urban growth; similarly, conurbations can alter regional climates.

Urbanization and urban climate impact people who must adapt to the new living conditions created. These conditions can have positive effects like the moderation of the winter but more often they are a health hazard due to episodes of hot and inhospitable weather with which they are associated.

M. Yoshino et al. (eds.), Climates and Societies – A Climatological Perspective, 333–359.

Urban climate impacts upon humans are best demonstrated by considering the energy balance of the human body in outdoor conditions.

A town can be improved by climate sensitive planning and building designs. Urban parks and vegetation judiciously used can enhance the quality of urban climates.

1. Introduction

Every intervention on the earth's surface changes the climate. Depending on the character and dimension of the intervention, the alteration caused may produce small scale effects or larger ones.

Through the development of buildings and settlements, humans historically have influenced the evolution of local climates. In all continents, as civilization advanced, towns and cities developed into special types of human habitats. So the "urban landscape" was defined as "the kind of settlement which stands out in the natural landscape, is relatively compact, and purposely designed by humans to improve working and living conditions". ("Die aus einer natürlichen Landschaft sich hervorhebende, mehr oder minder verdichtete Siedlungsform, hervorgerufen durch eine sich bewußt vollziehende Zusammenballung von Menschen, um so besser ihren gewerblichen wie geistigen Beschäftigungen nachzugehen." cit. Eckert; Passarge (1930) after Kratzer (1956).)

Geographically it is not possible to separate a town from a rural community by the number of inhabitants (Eichler 1984). But the following four criteria used to justify the name "town" depend on the character and the structure of the population of the country. The criteria are (Tamms and Wortmann 1973):

1. size (proportion of population);
2. closed settlements;
3. urban density increasing towards downtown;
4. "centrality" – function to surrounding areas.

The significance of these criteria increases with increasing population, number of inhabitants and town size. Climatically, the properties of a town compared with the rural landscape may be summarized into the following three typical groups (Wanner 1986):

1. Surface type (sealed surface) and structures (increasing roughness);
2. Changes in type and number of biospheric life forms (humans, animals and plants);
3. Increase in number and size of technical establishments.

All these have some effects on urban climate as Landsberg (1976) pointed out. Urban climate can be defined as a special kind of mesoclimate or of topoclimate depending on the dimension of the settlement and the criteria used.

2. Changing of natural climate by urbanization

2.1. *The development of urbanization*

In the history of mankind the construction of buildings and settlements were important steps in the development process. They allowed people to live under extreme local climate conditions by modifying key variables such as temperature, wind and precipitation (Landsberg, 1976), while closed buildings enabled them to regulate indoor climates (Olgyay 1963; Reidat 1970; Evans 1980).

Urbanization affects all climatic elements although the magnitude of its impact and the complexity of the processes involved tend to be uneven across variables (Kratzer 1956; Landsberg 1981). Wanner (1986) identifies four major conditions which should be evaluated in urban climate studies, namely, the thermal structure, water balance, wind flow fields, and impact of aerosols.

1. Influence of thermal structure: amount of energy consumption for indoor climatic conditioning (heating and cooling); impact on human well being (warm and cold street); variation in snow melt and frequency of snow clearing; frequency of frost events and differences in phenological cycles.
2. Influence on the water balance: variations of liquid and solid precipitation; variations of evaporation and transpiration, and variations of runoff.
3. Influence on wind flow and structure of the wind field: transport of fresh air and renewal of air in the urban boundary layer (flow convergence and divergence); generation of funneling effects (for example, in street canyons).
4. Impact of aerosol: well being of people and animals (pulmonary and ophthalmonary diseases); plant diseases (wood decay); impact on inanimate matter (corrosion and decay of stones, bricks, metals, etc.).

Dense or closed settlements and building structures may have replaced the original natural cave of the primitive man. Improved thermal insulation of walls, ceilings and roofs help maintain indoor temperatures at a reasonably constant level (Givoni 1976). The invention of the single shelter wall by bushmen, for example, represented a step towards the present modern housing unit. The shelters acted as windbreaks and in so doing helped stabilize temperatures in the protected areas and prevented excessive nocturnal cooling.

In the course of human evolution, buildings and settlements of different types were developed as shelters against adverse weather conditions for people, animals and stores (Egli 1959).

The microclimatic effect can be measured at a single wall or building while its wider impacts reaches into the immediate neighborhood. Thus the accumulation of many individual effects results in a special urban climate. Clearly, the deviation of urban climate from that of the free or rural landscape is dependent on the size of a settlement or town (Flohn 1970).

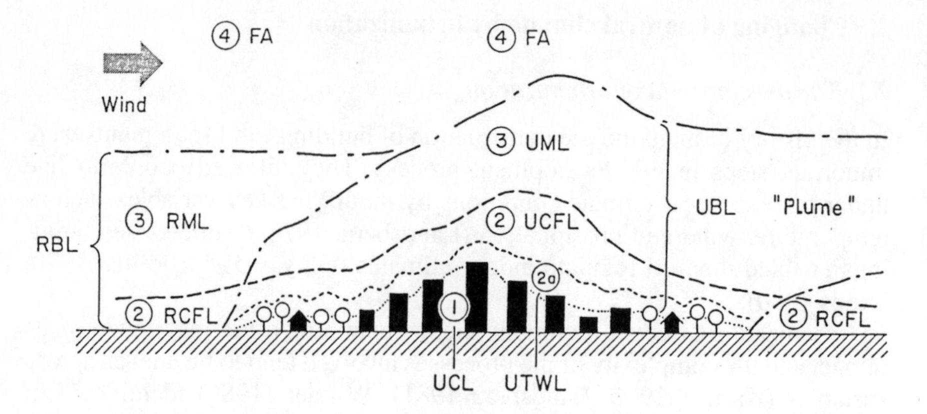

Figure 1. Planetary boundary layer changed by a town (Wanner 1976; Oke 1987). Urban U, Rural R. 1: Urban Canopy Layer UCL; 2: Surface Layer – Constant Flux Layer UCFL – RCFL; 2a: Turbulent Wake Layer UTWL; 3: Mixed Layer – Ekman – Layer UML – RML Urban "Heat-Plume"; 4: Free Atmosphere FA.

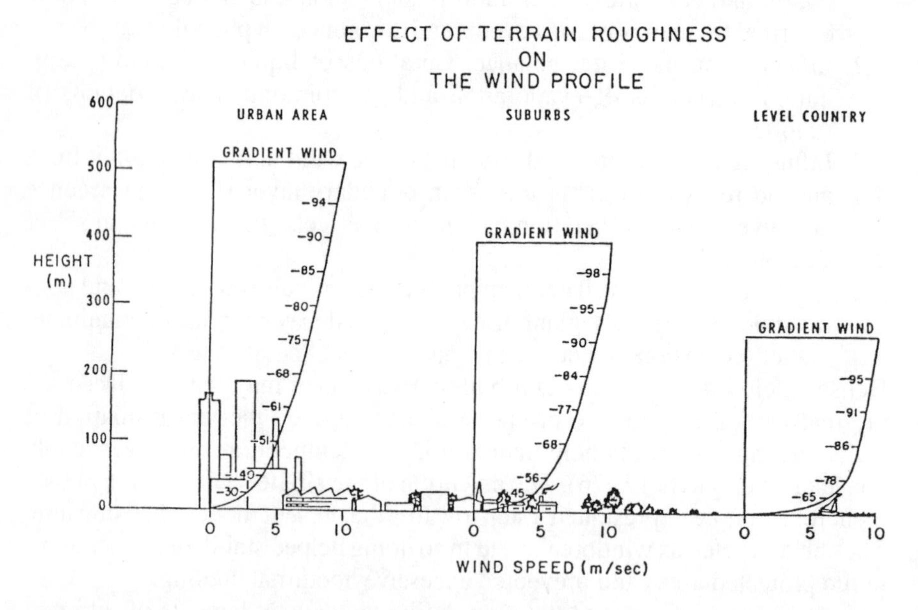

Figure 2. Boundary layer adiabatic wind-profiles (Davenport 1965, acc. to Equation (6). Covers show relative windspeed u(z) in the PBL Gradient Wind = Geostrophic wind U_G= 100%.

2.2. *The complex effect of urbanization on local climate*

The climatic effect of urbanization is very complex. Distinct climatic elements are influenced and altered in various ways as mentioned before. These alterations can be evaluated with planetary boundary layer (PBL) models

(Figure 1). In a city, the PBL is often referred to as the Urban Boundary Layer (UBL) (Oke 1978; Wanner 1986). The PBL is the layer of the atmosphere influenced by the ground surface, such influence is well demonstrated by the wind profile (Figure 2; after Davenport 1965). The "geostrophic wind" (U_G) above the PBL in the "free atmosphere" is controlled by pressure gradient (dp/dx) and the Coriolis parameter ($f = 2\,\omega \sin\phi$) as shown in Equation (1).

$$\text{(1)} \qquad U_G = (1/\langle r\, 2\, \omega \sin \phi\rangle)\,(dp/dx).$$

In the free atmosphere there is no friction force. In the PBL the wind profile at neutral thermal conditions has a typical logarithmic curve shape. In the so called "Ekman-" or "mixed-"layer, the mean velocity decreases from the geostrophic wind to the surface wind (see Section 2.2.3). Simultaneously, wind direction changes with increasing height in a counter-clockwise sense. The resulting wind field is shaped like a spiral, hence, the "Ekman-spiral" (Ekman 1905).

The UBL modifies the rural BL (RBL) hence the distinctiveness of the UBL in Figure 1. By this simple model the different properties of urban surfaces and structures and their behaviour within the UBL are analyzed. The following five phenomena can be distinguished too: the urban heat island, the radiation and energy balance, the urban wind field, the water balance involving humidity and precipitation, and air quality (air hygiene) (compare Section 2.1).

2.2.1. *The urban heat island*

The best known climatic effect of urbanization is the development of an "urban heat island" (Duckworth and Sandberg 1954; Mitchell 1962; Parry 1967). The increase of air temperature in towns found all over the world is the product of urban structure and surface change combined with the energy balance including anthropogenic heating of the air. The reflection of incoming solar radiation is less due to multiple reflection by vertical walls. Evaporation rates are also lower (Geiger 1961, 1965).

Urban heat islands are different in high and low latitudes (mostly stronger in high-sun season), day and night. They are dependent too on the humidity or dryness of the climate, and the size and structure of the city (density of built up areas; fraction of area occupied by vegetation).

The threshold wind speed (U_{crit}) required for heat islands was derived by Oke and Hannell (1970) from Equation (2).

$$\text{(2)} \qquad U_{crit} = 3.4 \log p - 11.6$$

where p is the population of the city. With $U_{crit} = 6\ ms^{-1}$, $p = 300{,}000$; at $12\ ms^{-1}$ $p =$ about 8 million.

They suggest 2000 as a minimum for the number of inhabitants required for the formation of a heat island. Equation (2) shows the increase in threshold windspeed with increasing size of a town. Oke (1973) demonstrated a positive

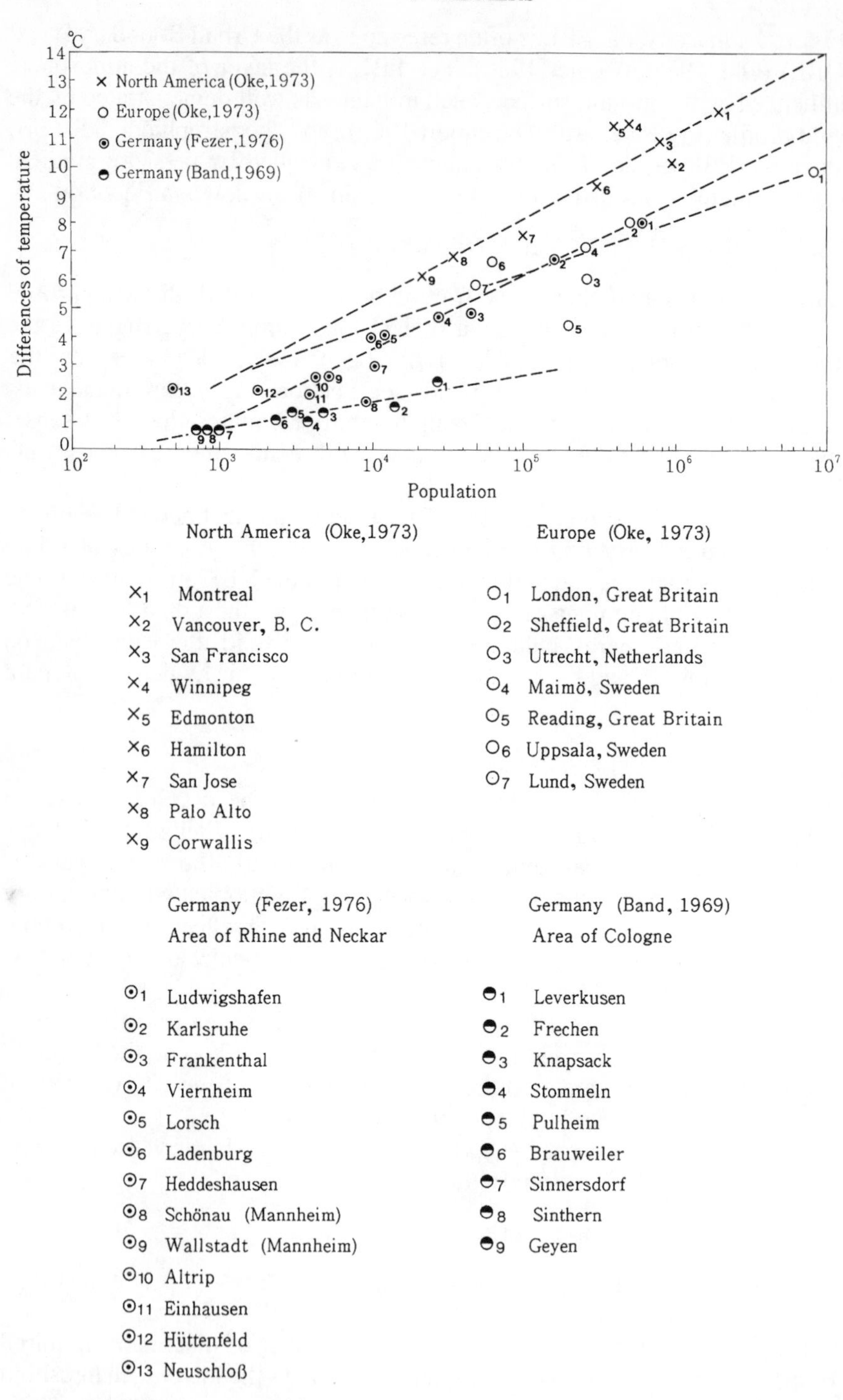

Figure 3. Increase of urban temperature difference according to number of inhabitants (Band 1969; Oke 1973; Fezer 1976).

relationship between heat island intensity and number of inhabitants. Fezer (1976) compared the dependence of heat islands on the number of inhabitants as well as the density of settlements (Figure 3). Heat islands vary considerably across settlement types including smaller ones (Band 1969).

Formulations by Sundborg (1951) showed the dependence of urban temperature anomaly on a number of meteorological factors including cloudiness and wind velocity. He obtained separate equations for the difference between urban temperature and the surrounding rural values for daytime and night time (Equations (3)).

$$D_d = 1.4° - 0.01N - 0.09V - 0.01T - 0.03e \quad (°C) \tag{3a}$$

$$D_n = 2.8° - 0.01N - 0.38V - 0.02T - 0.022e \quad (°C) \tag{3b}$$

where D_d, D_n = difference of temperature daytime (d) and night time (n); N = mean cloudiness (10/10); V = mean wind velocity (ms^{-1}); T = temperature (°C); e = vapour pressure (hPa).

The differences also vary across climate types. Dronia (1967) estimated that globally, increased urbanization caused a temperature rise of about 0.24°C between 1901 and 1940. In different climates, maximum temperature differences of more than 10°C have been measured between downtown and the surroundings at night. Downtown annual averages are also higher in towns than their surroundings by 1 to 2°C (Schmidt 1917; Chandler 1964).

The increase of temperature in towns is mainly dependent on the character and density of the built up areas. The orographic properties are also effective if there is any relief (Quitt 1960). This can be shown by several cross sections of temperature through towns using automobile transects. The first ones were done by Schmidt (1930) in Vienna, Austria. Others including Chandler (1965) for London, England and Kealoha (1968) at several towns of different sizes in the USA, also show these correlations.

More specific thermal analyses of a number of towns reveal heat islands differentiated by the town's structure. We may call them "heat archipelagos" (Farstang et al. 1975; Eriksen 1976). The heat islands conform to the general patterns discussed above. Taesler (1986) suggested that the effect of canyon geometry of the city streets on longwave radiation exchange was the main controlling factor in the development of urban heat islands. Oke (1981) proposed the use of canyon height to width ratio (H/W) or the sky view factor (F_s) seen from the midpoint of a street canyon for determining the maximum urban heat island intensity. The daily march of temperature in a street canyon is dependent on the whole geometry, also the direction to the sun. Times of shading or incident radiation are responsible for the amount of heat present in the system (Wilmers 1982).

2.3. *Radiation and energy balance*

The temperatures measured in cities have a strong correlation to the net radiation and the energy balance of the various surfaces (see Chapter 3). Equation (4) gives the heat balance for a surface as expressed by Albrecht (1940), Geiger (1965), and Oke (1978, 1987).

$$\text{(4)} \qquad R + E + H + G = 0,$$

where R = net radiation R = (I + D) (1 − a) + A + E′; I = direct solar radiation (sw); D = diffuse sky radiation (sw); a = albedo (reflectivity of the surface); A = atmospheric counter radiation (lw); E′ = outgoing surface radiation (lw); sw = shortwave (= 0.3 to 3 μm); lw = longwave (= 3 to 30 μm); E = evapotranspiration (flux of latent heat); H = turbulent heat transport (sensible heat flux); G = soil (ground) heat flux.

There are several measurements of the heat balance of cities exemplified by Bach (1970) and Oke (1982). Terjung et al. (1970), and Terjung and O'Rourke (1981) measured energy balance under a variety of conditions in towns and used it to classify urban types.

With the energy balance, a number of striking differences between various surfaces becomes apparent, especially surfaces with vegetation cover. Vegetation modifies the active surface and actual transpiration, increases surface roughness, and insulates the soil surface. Vegetation affects the wind field and the local energy consumption via the energy balance equation (Equation (4)). For urban climates, the diurnal march of evapotranspiration (E) and the soil heat flux (G) is very important. Table 1 shows a variety of urban surfaces and structures, and energy, temperature and roughness characteristics associated with them (Etling and Wilmers 1991). For each surface type the Bowen ratio of $\beta = H/E$ measures the fraction of energy which is consumed by E (Geiger 1965; Oke 1987). Since the built up areas mostly have "sealed" surfaces, water is not abundant and therefore little energy is used for E, and a greater fraction of the available energy is diverted to the thermal heat flux H.

Because of the lower insulation and the better heat flux into the soil beneath sealed surfaces, the soil heat flux (G) fraction is increased. Consequently, the sensible heat flux during the daytime period of warming is diminished, and the air temperature between buildings increases slower than that above vegetated surfaces such as lawn. At the same time, sensible heat flux densities differ between rural and urban surfaces. Thermal conductivity and heat capacity of surface types impact air temperature and consequently the heat island (Kraus 1987). During the cooling period the stronger soil heat flux to the surface reduces the cooling rate and so the night time temperature within the built up areas remains higher than outside them.

The thermal heat dome above built up areas reduces both the incoming shortwave radiation and the longwave reradiation. This results in generally stronger urban heat islands at night, but temperature differences between urban and rural areas can disappear during daytime (Karrasch 1986). The

Table 1. Surface types for urban-rural environments.

	Surface type	Energy balance		Temperature deviation (dT in °C) from grassed lawn of climate station		Roughness parameter Z_o (cm)
		G = %	β = (H/E)	day	night	
1.	*Open or vegetated surfaces*					
11	woods, closed	2	0.2	–3	2	300
12	woods, parks, groves, cemeteries	4	0.4	–2	1	150
13	public green areas, trees and shrubs rare	5	0.4	–1	0	5
14	trees and shrubs, alley trees	4	0.4	–1	1	50
15	home gardens, garden allotments	4	0.4	–1	1	50
16	greeneries, meadows, lawns (standard)	5	0.3	0	0	5
17	farmlands, agricultural lands	8	0.4	1	0	10
18	heathland	5	1.2	1	–1	5
19	marsh and moorlands	2	0.5	0	0	5
10	open water surfaces, rivers, lakes	10	0.2	–2	1	0.1
2.	*Sealed surfaces professional use*					
21	industrial areas and halls, sealed surfaces	60	20	3	4	300
22	industrial areas, small percentage vegetation	50	15	2	3	300
23	power plant	60	18	3	4	300
24	barracks, hospitals (schools)	50	15	2	3	300
25	railway stations, railways	50	15	2	3	100
26	trade centres, super markets with large parking lots	60	20	2	2	100
27	streets, freeways, motorways	60	18	2	2	10
28	airport buildings	60	20	2	2	300
29	aircraft runways, sealed	60	20	2	2	1

Table 1. (Continued)

	Surface type	Energy balance		Temperature deviation (dT in °C) from grassed lawn of climate station		Roughness parameter Z_o (cm)
		G = %	β = (H/E)	day	night	
3.	*Built up areas*					
31	office and commercial buildings, few vegetation	50	18	4	4	300
32	multi storage buildings, few vegetation	50	18	3	4	300
33	multi storage buildings, high vegetation perc.	10	0.8	2	2	300
34	closed built up areas high dwelling density, few vegetation	40	15	3	3	300
35	closed built up areas, high vegetation perc.	40	1	1	2	100
36	closed built up areas mixed with other buildings	40	10	2	2	100
37	dwelling 4–6 storeys, few vegetation	30	10	3	3	200
38	dwelling 4–6 storeys, more vegetation	20	5	2	2	100
39	dwelling single and two storeys buildings, high vegetation	10	0.8	1	1	50
4.	*Special surfaces, not allotted to others*					
41	asphaltous	60	20	4	3	1
42	concrete	60	20	2	2	1
43	natural stones and brick pavements	60	20	2	2	1
44	pavements with sand and gravel	50	15	1	1	1
45	glass areas, nurseries, glass roofs	40	15	2	2	1
46	quarries, sand areas	50	12	2	1	1
47	gravel, fallow land	50	12	2	1	100
48	refuse deposition	40	8	1	1	100
49	sporting areas without vegetation	40	8	1	1	100

Energy balance (Equation (4)) $R - E - H - G = 0$

mean temperature difference is maximized at temperature minima during the winter when anthropogenic heat production increases the heating effect of towns (Hanna 1969).

2.3.1. *Urban wind field*

"Moenibus circumdatis sequuntur intra murum arearum divisiones platearumque et angiportuum ad caeli regionem directions. Diregentur haec autem recte, si exclusi erunt ex angiportis venti prudenter". Vitruvius, liber primus, chapter vi (Fensterbusch 1964). Viturvius overcame strong winds in cities by planning them against the prevailing winds. Further: "When the winds are cold they cause harm; when they are warm they cause illness; when they are wet they impair health".

Built-up areas alter the physical properties of the surface and influence the wind field. The roughness creates eddies around buildings and streets, and free spaces channel the wind (Houghton and Carruthers 1976). Strong currents sometimes flow through some parts of a city. However, in smaller streets and in streets aligned at right angles to the mean flow, winds are relatively calm (Munn 1970; Werner 1979). Two-dimensional models of urban heat islands typically show a stream of rural air being transported from a rural area into a town and the changes that occur in the properties of the air as it travels from the former to the latter (Bornstein 1975, 1986) (Figure 1).

In cities, variations of winds are loosely correlated with the weather. Distinct weather situations can be arranged into a number of weather types following Schneider-Carius (1953) and Wilmers (1968, 1976) based on Fedorov (1931). Roughly four weather types (WT) can be distinguished. They are radiation WT (R), gust WT (G), cyclonic WT (C), and neutral WT (N). With gust and cyclonic WTs urban wind speed is high and differences at various parts of a city are small. Maximum differences in urban microclimate occur during anticyclonic weather types with bright sunshine (Geiger 1961, 1965). We shall call this event radiation WTs (R). In these cases, exposure to the sun or shade results in thermal differences within the urban microclimate (Jendritzky and Sönning 1979).

With differences of air density due to variations in energy fluxes and temperature, small scale wind systems are generated. These can be found in partly shaded streets (Albrecht and Grunow 1935; Georgii 1970) and are of great importance for the renewal of urban air (Georgii 1970). Kuttler (1988) showed by measurements and by wind tunnel experiments that air quality can be improved in traffic congested street canyons by an urban design that promotes air renewal.

Within built up areas urban structure guides and changes the wind field (Motschka 1964; Munn 1970). In the urban fringe the existence of an anticyclonic wind system as a rural-urban boundary wind is much more difficult to evaluate.

The wind speed inside the built-up area is reduced somewhat. The effect of urban structure on the PBL is underscored by wind profiles measured within the boundary layer. These profiles show that the height at which geostrophic flow is encountered increases with increasing roughness (Figure 2). Under neutral conditions the wind profile in the surface layer is given by Equation (5).

$$(5) \qquad U(z) = (U^*/K) \ln (z/z_0)$$

where $U(z)$ = windspeed in x-direction at height z; $U^* = (\tau/\rho)^{1/2}$ friction velocity; K = von Karman constant = 0.4; z_0 = roughness length; τ = shearing stress; ρ = air density.

In the Ekman-layer power law provides a better fit for the wind profile (Equation (6) and Figure 2 after Davenport 1965)

$$(6) \qquad [u(z)]/U_G = (z/z_G)^a$$

where U_G is the geostrophic wind speed (see Equation (1)) and z_G is the height of geostrophic wind.

Davenport (1965) gave mean values of the exponent a and z_G at neutral (adiabatic) conditions as

	Exp. a	z_G (m)
flat, open area	0.16	270
suburban	0.28	390
downtown (high buildings)	0.40	420

The renewal of air in a town is dependent on the effect of the building structure on the air flow. Air flow is regulated by both the structures and the open spaces in the built up areas. The effect can be estimated by the roughness parameter z_0 which ranges from about 1/10 to 1/30 of the mean height of the obstacles (Baumgartner et al. 1977). For vegetated surfaces z_0 may be estimated from Equation (7). For problems encountered in cities see Wilmers (1988), mean values of z_0 for key urban surfaces and structures are listed in Table 1 (Etling and Wilmers 1991).

$$(7) \qquad \log z_0 = 1.03 \log h_0 - 0.86$$

where h_0 = average height of plants (m); and z_0 = roughness length (m).

For solid obstacles with known height, Lettau (1969, 1970) gives the following estimate for z_0.

$$(8) \qquad z_0 = 1/2\ (Ha)/A$$

where H = height of obstacles (m); a = silhouette of area (m^2); and A = area covered by obstacles (m^2).

Changes of roughness involve changes of the wind field too; in wind flow an intermediate layer is generated which vanishes after some distance, destroyed by the turbulence (Figure 1).

Besides more gustiness the mean wind speed is reduced by built-up areas and settlements. This is demonstrated by both the lower value of the mean wind speed and the higher frequency of weaker winds less than 3.0 ms^{-1}, which can reach 30 to 40% of all winds. For this reason, good ventilation is essential for combatting air pollutants in cities.

2.3.2. *Urban water regime*

This discussion of the climatology of urban water regime will be restricted to atmospheric humidity and precipitation. However, mention should be made of two other variables and their tendencies. Urban areas tend to be generally more cloudy than their rural surroundings. The number of days with fog may increase or decrease in cities depending on atmospheric humidity and heat content.

2.3.2.1. *Humidity in towns*. The surface energy balance shows that air humidity is lower above sealed surfaces than over vegetated ones (Kratzer 1956). Many research studies show a humidity stress within built up areas (Brahe 1970). On the other hand, following some precipitation events in the summer, evaporation can increase significantly for a short time because the large amount of energy present in the urban fabric of stone, asphalt and concrete provide energy for the quick evaporation of the water standing on them. But since these high rates of evaporation are short lived, urban evaporation during summer on the whole, is less than that of the rural surroundings (Höschele and Schmidt 1974).

During the heating period, the amount of water vapour from fossil fuel pumped into the urban atmosphere is large (Hanna 1969). Therefore some analysis show smaller relative humidity in winter in a town but nevertheless the vapour pressure can be higher (Kratzer 1956). The relative humidity in urban areas is normally lower than outside the city because its relative humidity is inversely related to temperature (Chandler 1967). But the water vapour pressure can be an important factor because it plays a role in the energy balance of the human body (see 5.2).

2.3.2.2. *Precipitation in cities*. The amount and distribution of precipitation is profoundly influenced by the urban structure. The sealed surfaces and the urban structures increase convection in urban areas and enhance the generation of clouds. This development is enhanced by the anthropogenic heating of towns. The increased dust and gaseous and fluid aerosol loads are effective condensation and sublimation nuclei which enhance the generation of precipitation (Changnon 1981).

The development of small cloud droplets into larger particles requires time. According to Diem (1969) a single thunderstorm cell has a mean development time of about 20 minutes and a life time of about 45 minutes. With a mean geostrophic wind of 10 ms^{-1}, an air parcel would travel 12 km from the beginning of droplet formation to the first precipitation. That means that the precipitation field is shifted downwind of settlements as demonstrated by Eriksen (1972) and Changnon (1982). Therefore, the impact of cities on precipitation is confined to cities covering sufficiently large areas. Landsberg (1976) estimates a population threshold of 100,000 inhabitants for significant impacts on precipitation to occur.

Thunderstorms may develop out of heat islands (Band 1969; Havlik 1981). Changnon (1979) suggested a relationship between increased thunderstorms and urban population size. However, METROMEX showed the increase of precipitation occurred downwind of St. Louis (Braham 1981).

2.3.3. *Urban air quality*

One of the best known aspects of urbanization is the enormous production of air pollution. There are many sources for combustion derived air pollutants as well as other gaseous, fluid and dust forms of aerosols. Pollutants reduce visibility and bright sunshine in cities while creating urban dust and heat plumes (see fumifugium by Evelyn, 1661 and the first books on urban climates by Howard 1818, 1833). Because of the potentially severe harmful effects of the different pollutants, air quality also called air hygiene, is one of the most important aspects of urban climate (Wanner 1986, 1991) (see Section 5.3). Landsberg (1976) suggested the following relationship between air pollution and several variables in an urban environment

$$X = (Nq)/(du) \quad (9)$$

where X = is pollutant concentration; N = population size; q = mean per capita production of specific pollutant; d = diameter of the urban area; and u = wind speed.

In an urban area there are three typical vertical zones of maximum pollutant enrichment namely the streets, the roofs, and the height of industrial chimneys.

With weak winds concentration of aerosols is high. For this reason in many conurbations or urban agglomerations during synoptic situations with small exchange capacity such as neutral weather types, smog occurs either in a classical form like the London smog darkened by effluents from combustion, or the Los Angeles type which is a chemical smog caused by the combination of pollutants and bright sunshine. Most of these conurbations need to develop strategies against smog events. Urban dust plume aerosols also enhance heat island effects (Viskanta et al. 1977).

3. The urban mosaic

Urban climates result from multiple interactions of the different microclimates created by the mosaic of surfaces and structures found in a city. Parts of these interactions have already been shown in the preceding sections.

3.1. *Components of urban mosaic*

Kraus (1987) suggested the use of the term "special surface climates" (SSC) when dealing with specific microclimates due to each of the surface types which together form the urban mosaic. Each specific surface has its own properties of roughness, albedo, heat capacity, thermal conductivity, etc.

For practical use, especially for classification and calculation purposes-typical surfaces and structures were defined as listed in Table 1 (Etling and Wilmers 1991). The thermal effect of the different surfaces and structures may be evaluated from the temperature patterns revealed by different measurements and authors preferably those from middle latitude climates (Geiger 1961, 1965; Kessler 1971; Oke 1978, 1984; Wilmers 1978, 1988, 1991). Measured deviations of surface temperature for day and night and the corresponding air temperatures are listed too. Each deviation is the difference between the specific surface and a standard surface, namely, a grassed lawn of a climate station.

Examination of Table 1 shows the thermal importance of vegetation in urban climates. They control the temperature and humidity of their surroundings through the energy balance. Extensive green areas direct and clean air currents (Fezer 1976; Finke, 1876; Horbert et al. 1980, 1983, von Stülpnagel 1987; Wilmers 1978, 1982).

The effect of a single "green" space with vegetation cover can be measured locally, however, only large green areas have the capacity to influence their surroundings (Fezer 1976). Consequently, better results can be achieved in urban agglomerations by connecting the green spaces to produce a "network of biotopes" (Richter 1981). He proposed special systems of green spaces which could act as aisles for air currents through a city. Moreover, the biological value of green spaces can be enhanced by such a network because the areas occupied by the various biotopes may expand and in so doing ensure the survival of some botanical and zoological species (Sukopp et al. 1980).

Also connecting a network of green areas with foot and bicycle paths may broaden the opportunities for recreation for urban dwellers (Sukopp et al. 1980; Richter 1981).

3.2. *Urban poleotopes and climatotopes*

Urban structures, the growing size of settlement areas, and the special surfaces in an urban environment are changing the nature of the atmosphere near the earth's surface (Geiger 1961, 1965). Macroclimate specific climatotopes may be defined according to the respective ecotopes, the poleotopes and chorotopes of an urban agglomeration (Wilmers 1991). The size of these special environments vary from city to city hence their sealing (Bründl et al. 1986). In some models, a scale of 2 to 4 m is adequate (Jendritzky and Sievers 1989), in others, depending on the circumstances, a horizontal scale of 100 m up to 1 km may be satisfactory (Stock and Beckröge 1983; Wilmers et al. 1987).

Weischet (1979) proposed special urban structure complexes which Stock et al. (1983, after Brahe unpubl.) called climatotopes. In Tables 2 and 3, the most important urban and rural ecotopes are listed along with their respective climatotopes (Wilmers 1991). Each type is identified with a density unit for storeys and other structures, as well as percentages of open spaces between built up areas (Table 4).

With various urban inquiries it might be possible to identify more types of urban structures (Auer 1981).

4. The complex of urban climate and its impact on local climate

Urbanization changes the climate in a complex way. It began with the first wall built by people, and has progressed gradually since. All elements of climate are involved to some degree. Relationships with urbanization vary from element to element, the most striking effects being detected through research. At first, air temperature and changes in it were relatively easy to seize upon. Consequently, both variables have been valuable as instruments for measuring change in urban environments.

The essential but complex effects of urbanization on the local climate are expressed in the wind field and water regime.

An important consequence of urbanization can be found also in the deterioration of air quality. Air quality affects urban climates via heat islands, the energy balance and the wind field. These effects have to be investigated within a wider urban climate programme (Wanner 1991).

The complexity of urban climate is intensified by urban structures. Urbanization changes the local climate, increasing with urban growth. Temperature is rising, the energy balance is changing, and the wind field is modified by the structures. Similarly, conurbations can alter the regional climate.

5. Influence of urban climate and urbanization on mankind

In many countries, urban agglomerations develop from increased population. These agglomerations are also called conurbations, urban regions or metropolitan areas (Boustedt 1952). When large cities grow into conurbations they may be recognized as one unit for the purpose of urban climate (Eriksen 1971, 1980).

The development of cities has influenced and altered their inhabitants. The citizens are adapted to their special climate. In some cases, they are subjected to extremes of adverse weather conditions while in others, the urban climate can protect the inhabitants from too severe frost events.

Table 2. Classification of urban ecotopes in middle latitudes.

Structures and use POLEOTOPES	Vegetation	Corresponding Climatotopes POLEOCLIMATOTOPES
Industrial		*Industrial Climate*
Car, steel and fabrication industries, industrial parks, minor fabrications, railways and railway stations, freeways, train and truck depots, warehouses, generally 3 to 5 storey buildings, flat roofs.	Vegetation $< 5\%$, and parks very limited, trees rare.	Very strong industrial climate with dense structures and pavements; Radiation: less shortwave; more longwave by urban dust layer. Energy balance: low evapotranspiration, strong soil and sensible heat flux. Active layer for energy balance: urban canopy layer, flat roofs, sealed pavement. Anthropogenic heat production. Temperature high, strong diurnal amplitude. Bioclimatic condition: oppressive. Melioration by vegetation; reduction of pavements; reduction of emissions.
City		*City Climate*
Commercial: office and apartment buildings, warehouses, hotels, residential block buildings; up to 6 storeys, mostly flat roofs.	Grass and trees limited, some small parks; vegetation $< 10\%$.	Very strong climate with dense buildings and pavements; similar to industrial. Radiation: shortwave less, longwave more, urban dust layer. Energy balance: low latent, high sensible heat flux and soil heat flux. Active layer for energy balance: canopy layer roofs and roads and other pavements, different heat capacity. Strong anthropogenic heat production. Temperature high, strong interdiurnal amplitude. Bioclimatic condition: oppressive. Melioration by vegetation, green walls and roofs, reduction of pavements; reduction of emissions.
Compact Residential		*Urban Climate*
Single family dwellings with close spacings and multiple family dwellings, generally 2 storeys, pitched roof structures, narrow streets, few garages, some small house gardens and courtyards.	Grass and trees limited, some small parks; vegetation $< 15\%$.	Extreme climate by dense structures pavements; Radiation: shortwave reduced, longwave enhanced according to strength of dust layer. Energy balance: small latent heat dependent on amount of vegetation, more soil and sensible heat flux. Small anthropogenic. Active surface: mainly canopy layer, roofs, some parts of roads and walls. Temperature high, according to density of build up areas and percentage of vegetation. Bioclimatic condition: stressful. Melioration by vegetation, reduction of pavements; reduction of emissions.
Common Residential		*Urban Border Climate*
Single family dwellings with close spacings and multiple family dwellings, generally 2 storeys with flat roofs, garages or car pits, house gardens and some common green.	Limited grass and trees, some small parks, vegetation $< 30\%$.	Transient climate partly urban climate; cases of low-density built up areas preserve positive aspects of situation. Radiation: incoming shortwave and longwave radiation reduced according to urban dust layer, the same with outgoing radiation. Energy balance: latent heat flux in relation to amount of vegetation, corresponding less sensible and soil heat flux, less anthropogenic heat production. Active layer for energy balance: canopy layer, partly roofs, roads and pavements, vegetation. Temperature: small diurnal change of temperature. Bioclimatic condition: preferably comfortable.

Table 3. Classification of urban ecotopes in cities of middle latitudes.

Structures and use CHOROTOPES	Vegetation	Corresponding Climatotopes CHOROCLIMATOTOPES
Woods		*Wood Climate*
Few developed municipal, state and private woods; recreation areas.	Vegetation > 90%; heavily wooded.	Good bioclimate, partly positive effect on surroundings.
Metropolitan Natural		*Climate of Parks*
Parks, cemeteries, horticultural zones, private gardens, sport campuses; few small houses and structures.	Abundant grass, flowers or shrubs; lightly wooded; vegetation > 65%.	Good bioclimate, partly positive effects on surroundings, transport of fresh air.
Agricultural Rural		*Rural Topoclimate*
Agricultural farms and green areas for cattle, some single farm houses and storage buildings.	Grass, some horticulture and local crops (e.g. wheat, barley, sugar beet, etc.); vegetables; vegetation > 85%.	Topoclimate of fields and meadows, generation and transport of fresh air.
Wet Green Valleys and Bogs		*Topoclimate of Green Valleys*
Agricultural grass, some undeveloped; wasteland.	Preferably grasslands, some wild grasses, weeds and heath; lightly wooded; vegetation > 85%.	Generation and transport of fresh air.
Water Surfaces		*Topoclimate of Water Surfaces*
Open water surfaces, some small artificial and natural lakes, rivers and channels, used for transport and sports.	No vegetation but some abundant vegetation at the water edges and banks	Balancing effect on surroundings.

The effect of climate on the human beings is profound. In biometeorology, this effect has been divided into the following four effective complexes: the thermal, the photoactinic, the air chemical and the meteorotropic (Linke 1940; Flach 1981).

The thermal complex is very important in urban climates but it varies with the level of thermal thresholds established in various climates. A good part of urban climate measurements deals with the thermal complex (Bornstein 1986). The human effects increase with the higher temperatures found in built-up areas. In Tokyo, the number of hot nights (temperature above 30°C) increased with the increase of population (Nakamura 1980).

The photoactinic complex is strongly dependent on the air quality or air hygiene, as is the air chemical complex (Flach 1981). Both effects are connected with the serenity of the air, especially with respect to UV-radiation, which is strongly diminished by the urban heat plume. For this reason, a sufficiency of direct incident solar radiation on dwelling quarters during winter

Table 4. Legal structure and density of urban ecotopes in Germany.

POLEOTOPES (as in Table 2)	Number of storeys	Surface area number	Storey space area number	Building mass number
Industrial	1–2	0.8		9.0
Light-moderate industrial (commercial)				
,,	1	0.8	1.0	
,,	2	0.8	1.6	
,,	3	0.8	2.0	
,,	4–5	0.8	2.2	
,,	6, > 6	0.8	2.4	
City	1	1.0	1.0	
,,	2	1.0	1.6	
,,	3	1.0	2.0	
,,	4–5	1.0	2.2	
,,	6, > 6	1.0	2.4	
Residential	1	0.4	0.5	
,,	2	0.4	0.8	
,,	3	0.4	1.0	
,,	4–5	0.4	1.1	
,,	6, > 6	0.4	1.2	
Estate residential	1	0.2	0.3	
,,	2	0.2	0.4	
Rural and farm settlements	1	0.4	0.5	
,,	2, > 2	0.4	0.8	

was an important recommendation of the international architects league (cf. CIAM charte of Athen's 1933, written by Le Corbusier, 1932). Next to it the thermal complex remains very important as a bioclimatic factor of urban climate. But its importance varies with climatic zones.

5.1. *The versatile parts of urban climate*

From a historical viewpoint, cities were the germ-cells of civilization over all continents (Egli 1959). Cities developed variously in different cultures and historical periods but their basic format is often similar: the original development was the sacerdotal city, a residence or a citizen's town (Egli

1959). Modern towns offer several services to both their near and distant neighbourhoods. They constitute centres for transport and commerce, their importance depending on their site and centrality relative to the areas served (Tamms and Wortmann 1973). Special functions such as harbours, specialized industry and in the modern world, tourism, can favour urban growth (Tamms and Wortmann 1973).

Ancient cities were designed to meet specific needs of the population. Their architecture and building materials reflected the culture and civilization of the time. Unlike modern cities which bear a greater expression of the technology of modern materials, architecture and town planning in ancient cities were primarily climate driven. However, increasing sensitivity to problems associated with energy usage is engineering a return to climate based urban plans. This will not only save energy but also improve environmental quality (Hardenberg 1980).

The most severe menace of our time, namely, the rapid growth of world population, has increased the problem of urbanization. In 1700 the number of large cities (cities with more than 100,000 inhabitants) was 41 worldwide, in 1951 they were 879, 69 of them had more than 1 million inhabitants (Fischer 1951, 1990). Meanwhile, the number of megalopolis with more than 10 million inhabitants has reached 25 (Haber 1989; UNESCO 1987). Most of them are located in the Third World which has the most rapid population growth.

The largest growth areas within the megalopolis are unplanned slums (bidonvilles, fr.; favelas, port.). Slums are hardly adapted to the climate except in some hotter and dryer climates where huts can be built from light material since there is little requirement to protect against cold weather or precipitation.

The climatic requirements of cities in the various climatic zones were given by Egli (1951):

(a) tropics: prevention against precipitation, humidity and heat, incident solar radiation; necessity for ventilation and cooling.
(b) subtropics: prevention against too strong insolation; necessity for cooling.
(c) temperate zone: prevention against rain and snow, alternating against too cold and too hot weather; necessity of exposure to solar radiation.
(d) dry zone: prevention against dryness, blowing sand and dust, mostly against too strong solar radiation or heat. Monsoon zone: prevention against rain and humidity.
(e) arctic zone: prevention against cold and wind.

Givoni (1989) has given a comprehensive list of recommendations for urban design in different climatic zones.

5.2. *The complex effect of local climate and microclimate on humans*

The changing influence of urban climates on the inhabitants can be evaluated from statistics. The basic requirement for such evaluation is long sequence climatic observations from which statistical frequencies and persistence can be complied and used for the important task of assessing the impact of daily weather on urban dwellers (Landsberg 1981). Impact assessment should include the timing of pertinent weather phenomena, their frequencies and threshold values at which certain effects are triggered.

Humans have the capacity to regulate the energy balance of their body. Through regulation, thermal comfort can be maintained within relativly narrow limits. Further controls can be imposed through physical activity and thermal insulation (by clothing). The physical connections of human thermal regulation can be demonstrated by the energy balance of a human body. This balance is an enlarged version of the surface energy balance accomplished by the inclusion of energy production and utilization terms of the body (Equation (10); Höppe 1984; Jendritzky et al. 1990).

$$(10) \qquad W + M + Q_{SHIV} + Q + H + E_L + E_{SW} + E_{Re} + N + S = 0,$$

where W = work (physical, mechanical activity); M = metabolic rate (basic metabolism); Q = net radiation; Q_{SHIV} = relative heat production by shivering; H = sensible heat; E = latent heat with: E_L water vapour diffusion, E_{SW} transpiration by sweat, E_{Re} energy release by respiration; N = energy input by nutrient (e.g. surplus created by the higher temperature of nutrients); S = net energy storage.

The energy balance has been developed into a base for numerical calculations for thermal comfort, the PMV (predicted mean vote) by Fanger (1972) (Equation (11)). It was transformed into SI-Units by Jendritzky et al. (1990).

$$(11) \qquad PMV = f(H/A_{DU}, I_{cl}, t_l, t_{MRT}, e, V_r),$$

where H/A_{DU} = internal heat production per square unit; H = m (1−h); m = H + W (Wm^{-2}); h = W/m = mechanical effectively; I_{cl} = insulation by clothes (clo-units); t_l = air temperature (°C); t_{MRT} = mean radiant temperature (°C); e = actual vapour pressure (hPa); V_r = mean wind velocity (ms^{-1}).

Equation (11) was initially used to calibrate numerical models for measuring human comfort in urban surroundings (Jendritzky et al. 1990; Höppe 1984). It was developed for indoor climates and for steady state conditions. But the parameterization of the MRT (Jendritzky and Sönning 1979) and its modelling by Sievers and Zdunkowski (1986) provided an opportunity to model outdoor climates for humans in urban environments (Jendritzky and Sievers 1989).

The complex incorporation of the whole urban surroundings into the human energy equation can be realized through numerical models such as MUKLIMO (Sievers and Jendritzky 1986).

6. A town suitable for man

There are great expectations for town planning worldwide. In all countries urbanization is increasing. In particular, the megalopolis of the Third World poses a challenge for more humane architecture. Town planners are challenged to build climate sensitive cities to make life in towns as human as possible. Givoni (1989) proposed some rules for the design of towns in different climatic zones; see also Olgyay (1963), Chandler (1976), Evans (1980), Mayer and Höppe (1987), and Page (1976).

The most important mean to meliorate urban climate is given by vegetation (Wilmers 1978, 1989, 1991). In nearly all climates parks, green alleys and gardens can meliorate urban climates (Landsberg 1981). A good help for planners in this regard is given by meso- and micrometeorological models (Gross 1991, 1993). Applied meteorology gives some opportunities to optimize urban planning (Schirmer 1988, 1993).

7. The human behaviour according to the climate

Human response to climate includes adaptation to the prevailing climate. Auliciems (1981) proposed the climatization of buildings based on mean conditions of the different climatic zone instead of worldwide use of uniform thermal limits. Experience shows that the inhabitants of various climates adapt to their surroundings; they do not tolerate widely deviating conditions very well. Air conditioning might be unnecessary in hot climates. If it is too strong it can impair health because of the dramatic change between temperatures inside and outside air conditioned rooms (Auliciems 1981).

On the other hand people can work and indulge in recreational activity commensurate with the actual climatic condition. The choice of clothing may be important when deciding on what physical activity is appropriate under a given weather condition. Such activities should be reduced if the weather is uncomfortably hot (Weihe 1986). Within street canyons it might be possible to choose the parts with direct solar radiation if one feels cold, or a shade if the incoming radiation is too strong. Parks and other green areas and shade trees can also be helpful in hot climate (Barradas 1991).

References

Albrecht, F. (1940): Untersuchungen über den Wärmehaushalt der Erdoberfläche in verschiedenen Klimagebieten. Reichsamt f. Wetterdienst, Wiss. Abh. 8, No. 2, Berlin.

Albrecht, F. and Grunow, J. (1935): Ein Beitrag zur Frage der vertikalen Luftzirkulation in der Grossstadt. *Meteor. Z.*, 35, 103–108.

Auer, A.H. Jr. (1981): Urban boundary layer. In: *METROMEX*, 3, ed. by S.A. Changnon Jr. AMS, Boston, 41–62.

Auliciems, A. (1981): Towards a psycho-physiological mode of thermal perception. *Int. J. Biometeor.*, 25, 109–122.

Bach, W. (1970): An urban circulation model. *Arch. Meteor. Geophy. Biokl. B.*, 18, 155–168.

Band, G. (1969): Der Einfluss der Siedlung auf das Freilandkima. Mitt. Inst. Geophys., Meteor., Univ. Köln, H. 9, Köln.

Barradas, V.L. (1991): Air temperature and humidity and human comfort index of some city parks of Mexico City. *Int. J. Biometeor.*, 35, 24–28.

Baumgartner, A., Mayer, H. and Metz, W. (1977): Weltweite Verteilung des Rauhigkeitsparameters z_0 mit Anwendung auf die Eneriedissipation an der Erdoberfläche. *Meteorol. Rdsch.*, 30, 43–48.

Bornstein, R.D. (1975): The two-dimensional URBMET urban boundary layer model. *J. Appl. Meteor.*, 14, 1459–1477.

Bornstein, R.D. (1986): Urban climate models, nature, limitations and applications. WMO No. 652, 237–276, Geneva.

Boustedt, O. (1953): Die Stadt und ihr Umland, Raum und Wirtschaft. 1952, Bad Godesberg.

Braham, R.R. Jr. (1981): Urban precipitation processes. In: *METROMEX*, 5, ed. by S.A. Changnon Jr., AMS Meteor. Monogr. 18, No. 40. AMS, Boston, 75–115,.

Brahe, P. (1970): Klimatische Auswirkungen von Gehölzen auf umbauten Stadtplätzen. *Das Gartenamt*, 2, 57–70.

Bründl, W., Mayer, H. and Baumgartner, A. (1986): Untersuchung des Einflusses von Bebauung und Bewuchs auf das Klima und die lufthygienischen Verhältnisse in bayerischen Grossstädten. Stadtklima Bayern, Absch. über Klimamessungen München, Lehrstuhl Bioklim., angewandte Meteor., München.

Chandler, T.J. (1964): City growth and urban climates. *Weather*, 19, 170–171.

Chandler, T.J. (1965): *The climate of London*. Hutchinson, London.

Chandler, T.J. (1967): Absolute and relative humidity in towns. *Bull., Amer. Soc.*, 48, 394–399.

Chandler, T.J. (1976): Urban climatology and its relevance to urban design, WMO No. 438. WMO, Geneva.

Changnon, S.A. Jr. (ed.) (1981): *METROMEX. A review and summary*. Meteorol. Monog., 18, No. 40. AMS, Boston.

CIAM (Congrès International d'Architecture Moderne) (1933): Charte d'Athènes. Athens, (Fundamentals of modern architecture, proposed by Le Corbusier, 1932).

Davenport, A.G. (1965): The relationship of wind structure to wind loading. In: *Proceed. Conf. Wind Effects on Structures, Nat. Phys. Lab. Sympos.*, 16, Vol. 1. HMSO, London, 53–102.

Diem, M. (1969): Physik der Niederschlagsbildung. *Meteor. Rdsch.*, 22, 134–138.

Dronia, H. (1967): Der Stadteinfluss auf den weltweiten Temperaturtrend. *Berliner Meteor. Abh.*, 74, 4.

Eckert, M. (1930): Die Entwicklung der kartographischen Darsellung von Stadtlandschaften. In: *Stadtlandschaften der Erde*, ed. by S. Passarge. Hamburg, 1–14,.

Egli, E. (1951): *Climate and town districts – Consequences and demand.* Architektur, Erlenbach–Zürich.

Egli, E. (1959): Geschichte des Städtebaues. In: *Die Alte Welt*, ed. by E. Rentsch. Erlenbach-Zürich.

Eichler, H. (1984): *Geographisches Hand- und Lesebuch*. Touristbuch, Hannover.

Ekman, V.W. (1905): On the influence of the earth's rotation on ocean currents. *Ark. Mat., Astron. Fys.*, 2(11).

Eriksen, W. (1971): Die stadtklimatischen Konsequenzen städtebaulicher Entwicklung. *Städtehygiene*, 22, 259–262.

Eriksen, W. (1972): Zur Niederschlagsmodifikation im Bereich von Großstädten. *Städtehygiene*, 23, 164–166.

Eriksen, W. (1976): Die städtische Wärmeinsel. Neuere Erkenntnisse zur Gliederung, Genese und Bedeutung des innerstädtischen Temperaturfeldes. *Geogr. Rdsch.*, 9, 368–373.

Eriksen, W. (1980): *Klimamodifikationen im Bereich von Städten, Grundlagen und städtebauliche Aspekte*. Veröff. Joachim-Jungius-Ges. Wiss. Hamburg, 44, 161–175.

Etling, D. and Wilmers, F. (1991): unpubl. Hannover.

Evans, M. (1980): *Housing, climate and comfort*. The Architectural Press, London.

Evelyn, J. (1661): Fumifugium, or the inconvenience of the air and smoke of London dissipated. Oxford (Repr. National Smoke Abatement Societey, Manchester, 1933), (cit. after Landsberg, 1981).

Fanger, P.O. (1972): *Thermal comfort, analysis and applications in environmental analysis*. McGraw-Hill, New York.

Fedorov, E. E. (1931): Beispiel eines Vergleichs der Klimata zweier Ortslagen mit Hilfe der komplexen Methode (Sloutzk und Djetskoje Sjelo). *Meteor. Z.*, 48, 306–314.

Fensterbusch, C. (1964): cf. Vitruvii de architecture.

Fezer, F. (1976): Wieweit verbessern Grünflächen das Siedlungsklima? *Ruperto Carola*, (Z. Univ. Heidelberg), 57, 77–79.

Fezer, F. and Karrasch, H. (1985): Stadtklima. *Spektrum der Wissensch.*, August.

Finke, L. (1976): Zuordnung und Mischung von bebauten und begrünten Flächen. Schr. R. Städtebaul. Forsch., 03.044, BMRBS, Bonn.

Flach, E. (1981): Human bioclimatology. In: Landsberg, H. (ed.): *World Survey of Climatology*, Vol. 3, I, ed. by H. Landsberg. Elsevier, Amsterdam, 1–187,

Fischer Weltlexikon (1951): Fischer, Frankfurt.

Fischer Weltalmanach (1990): Fischer, Frankfurt.

Flohn, H. (1970): Produzieren wir unser eigenes Klima? *Meteor. Rdsch.*, 23, 161–164.

Garstang, P.D., Tyson, G. and Emitt, E. (1975): The structure of heat islands. *Rev. of Geophysics and Space Physics*, 13, 139–165.

Geiger, R. (1961): *Das Klima der bodennahen Luftschicht.* 4th ed., Vieweg, Braunschweig.

Geiger, R. (1965): *The climate near the ground.* 4th ed. Harvard University Press, Cambridge, MA.

Georgii, H.W. (1970): The effects of air pollution on urban climate. WMO Tech. Note 198. WMO, Geneva, 114–137.

Givoni, B. (1976): *Man, climate and architecture.* Appl. Sci. Publishers, London.

Givoni, B. (1989): Urban design in different climates. WCAP-10, WMO/TD-NO, 346. WMO, Geneva.

Gross, G. (1991): Anwendungsmöglichkeiten mesoskaliger Simulationsmodelle dargestellt am Beispiel Darmstadt. *Meteorol. Rdsch.*, 43. 97–112.

Gross, G. (1993): *Numerical simulation of canopy flows*. Springer, Heidelberg.

Haber, H. (1989): *Eiskeller oder Treibhaus*. Herbig, München.

Hanna, S. R. (1969): *Urban meteorology*. ATDL Contr. 35, Air Res. Lab. Oak Ridge, Tenn.

Hardenberg, Joachim Graf von (1980): *Entwerfen natürlich klimatisierter Wohnhäuser für heisse Klimazonen am Beispiel des Iran*. Werner-Verlag, Düsseldorf.

Havlik, D. (1981): Großstädtische Wärmeinsel und Gewitterbildung – ein Beispiel anthropogener Klimamodifikation. *Aachener Geogr. Arb.*, 14, 91–109.

Höppe, P. (1984): Die Energiebilanz des Menschen. Wiss. Mitt. Meteor. Inst., München, No. 49.

Höschele, K. and Schmidt, H. (1974): Klimatische Wirkungen einer Dachbegrünung. *Garten und Landsch.*, 84, 334–337.

Horbert, M. and Kirchgeorg, A. (1980): Stadtklima und innerstädtische Freiräume. *Stadtbauwelt*, 71, 67, 270–276.

Horbert, M., Kirchgeorg, A. and Stülpnagel, A.V. (1983): Ergebnisse stadtklima-tologischer Untersuchungen als Beitrag zur Freiraumplanung. Umweltbundesamt (ed.) Texte 18, Berlin.

Houghton, D.D. and Carruthers, N.B. (1976): *Wind forces on buildings and structures*. Edward Arnold, London.

Howard, L. (1818, 1833): *Climate of London deduced from meteorological observations*. 1st, 3rd ed., Harvey & Darton, London (cit. after Kratzer, 1956).

Jendritzky, G. and Sievers, U. (1989): Human biometeorological approaches with respect to urban planning. In: *Proceed. of the 11th Int. Congr. Biometor.*, ed. by D.M. Driscoll. SPB, Academic Publishing, The Hague, 21–35.

Jendritzky, G. and Sönning, W. (1979): Der Einfluss der Strahlung auf die thermischen Bedingungen in der Klimatherapie. *Z. f. Phys. Med.*, 8(6), 283–291.

Jendritzky, G. et al. (1990): Methodik zur räumlichen Bewertung der thermischen Komponente im Bioklima des Menschen. Fortgeschriebenes Klima-Michel-Modell, Akad. Raumforsch. Landespl. Beiträge 114, Hannover.

Karrasch, H. (1986): Trends der urbanen Modifikation des Klimas. *Heidelberger Geowiss. Abh.*, 6, 205–220.

Kessler, A. (1971): Über den Tagesgang von Oberflächentemperaturen in der Bonner Innenstadt an einem sommerlichen Strahlungstag. *Erdkunde*, 15, 13–20.

Kratzer, P.A. (1956): *Das Stadtklima*. 2nd ed., Vieweg, Braunschweig.

Kraus, H. (1987): Specific surface climates. In: Landolt-Börnstein, New Series V/4cl. 14. Springer, Berlin, 29–92.

Kuttler, W. (1988): Spatial and temporal structures of the urban climate – A survey. In: *Environmental Meteorology*, ed. by K. Grefen and J. Löbel. Kluwer, Dordrecht, 305–333.

Landsberg, H.E. (1976): Weather, climate and human settlements. WMO, Special Environmental Rep. 7, WMO No. 448. WMO, Geneva.

Landsberg, H.E. (1981): *The urban climate*. Internat. Geophys. Ser., Vol. 28. Academic Press, New York.

Landsberg, H.E. (1985): The biometeorology of urbanization and housing in developing countries. In: *Biometeorology 9, Proceed. Intern. Biometeor. Conference*, New Delhi, 1983, ed. by H. Lieth and S.C. Pandeya. Swets & Zeitlinger, Lisse.

Lettau, H.H. (1969): Note on aerodynamic roughness parameter on the basis on roughness element description. *J. Appl. Meteor.*, 8, 828–832.

Lettau, H.H. (1970): Physical and meteorological basis for mathematical models of urban diffusion processes. In: *Sympos. Multiple-Source Urban Diffusion Models*, 2, ed. by A.C. Stern. Nat. Air Pollut. Control Admin., Washington, D.C., 1–24.

Linke, F. (1940): *Das Klima der Großstadt. Biologie der Großstadt.* Dresden, 75–90.

Ludwig, F.L. and Kealoha, H.S. (1968): Urban climatological studies. Final Report, Work Unit 1235, A. Stanford-Research Institute, Menlo Park, CA.

Mayer, H. and Höppe, P. (1987): Thermal comfort of man in different urban environments. *Theor. Appl. Climatol.*, 38, 43–49.

Mitchell, J.M. (1962): The thermal climate of cities. In: *Sympos. Air over Cities*. Washington, 131–153.

Motschka, O. (1964): Ergebnisse von Windregistrierungen in Wiener Strassen. *Wetter und Leben*, 16, 139–146.

Munn, R.E. (1970): Airflow in urban areas. WMO Tech. Note 108. WMO, Geneva, 15–39.

Nakamura, Y. (1989): Grasp of correlation between climate and number of human deaths in summer based on day-to-day observation data. In: *Proceed. IFHP/CIB/WMO/IGU Internat. Conf. on Urban Climate, Planning and Building*, Kyoto, Nov. 6–11.

Oke, T.R. (1973): City size and the urban heat island. *Atm. Environment*, 7, 769–779.

Oke, T.R. (1987): *Boundary layer climates*, 2nd ed. Methuen, London.

Oke, T.R. (1981): Canyon geometry and the nocturnal urban heat island: comparison of scale model and field observations. *J. Climatol.*, 1, 237–254.

Oke, T.R. (1982): The energetic basis of the urban heat island. *Q.J. Roy. Meteor. Soc.*, 109, 1–24.

Oke, T.R. and Hannell, F.G. (1970): The form of the urban heat island in Hamilton, Canada. WMO Tech. Note 108. WMO, Geneva, 113–126.

Olgyay, V. (1963): *Design with climate*. University of Princeton, New York.

Page, J.K. (1976): Application of building climatology to the problems of housing and building for human settlements. WMO Tech. Note 150, WMO No. 441. WMO, Geneva.

Parry, M. (1967): The urban heat island. *Biomet.*, 2, 616–624.

Quitt, E. (1960): Die Erforschung der Temperaturverhältnisse von Brno und Umgebung. *Wetter und Leben*, 12, 311–322.

Reidat, R. (1970): The present situation, prospects and problems of building climatology. WMO Tech. Note, 109. WMO, Geneva, 1–7.

Richter, G. (1981): *Handbuch Stadtgrün – Landschaftsarchitektur in städtischen Freiräumen*. BLV, München.

Schirmer, H. et al. (eds.) (1988): *Stadtklima und Luftreinhaltung*. Springer, Heidelberg.

Schirmer, H. et al. (eds.) (1993): *Lufthygiene und Klima, ein Handbuch zur Stadt- und Regionalplanung*. VDI-Verlag, Düsseldorf.

Schmidt, W. (1917): *Zum Einfluss grosser Städte auf das Klima*. Naturwissenschaft, Vol. 5, 494 pp.

Schmidt, W. (1930): Kleinklimatische Aufnahmen durch Temperaturfahrten. *Meteor. Z.*, 47, 92–106.

Schmidt, R.D. (1980): Das Klima im Städtebau. Referatebl. *Raumentwicklung*, 1980, Sonderheft, 26, 30–39.

Sievers, U. and Zdunkowski, W.G. (1986): A microscale urban climate model. *Beitr. Phys. Atmosph.*, 59, 13–40.

Stock, P. and Beckröge, W. (1983): Synthetische Klimafunktionskarte für das Ruhrgebiet. Kommunalverband Ruhrgebiet, Essen.

Stülpnagel, A. von (1987): Klimatische Veränderungen in Ballungsgebieten unter besonderer Berücksichtigung der Ausgleichswirkung von Grünflächen, dargestellt am Beispiel von Berlin (West). Diss., FW 14, TU Berlin.

Sukopp, H. et al. (1980): Beiträge zur Stadtökologie von Berlin (West). Schr.reihe des FB Landschaftsentwicklung der TU Berlin, Bd. 3, Berlin.

Sundborg, A. (1951): Climatological studies in Uppsala with special regard to the temperature conditions in the urban area. Geografiska Instit., Uppsala, Geographia No. 22.

Taesler, R. (1986): Urban climatological methods and data. WMO/WCP. In: *Proceed. Tech. Conf. Urban Climatology and Its Applications with Special Regard to Tropical Areas*, Mexico, 26–10. Nov. 1984, 199–236. WMO No. 652, Geneva.

Tamms, F. and Wortmann, W. (1973): *Städtebau. Umweltgestaltung, Erfahrungen und Gedanken*. C. Habel, Darmstadt.

Terjung, W.H. et al. (1970): The energy balance climatology of a city-man system. *Ann. Assoc. Amer. Geogr.*, 60, 466–492.

Terjung, W.H. and O'Rourke, P.A. (1981): Energy input and resultant surface temperature for individiual urban interfaces, selected latitudes and seasons. *Arch. Meteor. Geophys. Bioklimat. B.*, 29, 1–22.

UNESCO (1987): *Statistical Yearbook*. Paris.

Viskanta, R., Bergstrom, R.W. and Johnson, R.O. (1977): Effects of air pollution on thermal structure an dispersion in an urban planetary boundary layer. *Beitr. Phys. Atmos.*, 50, 419–440.

Virtruvii de architectura libri decem. (1964): Fensterbusch, C. ed., Wiss. Buchges., Darmstadt.

Wanner, H. (1986): Die Grundstrukturen der städtischen Klimamodifikation und deren Bedeutung für die Raumplanung. *J.B. Geogr. Ges. Bern.*, Bd. 55/1983–1985, 67–84.

Wanner, H. (1991): Immissionsökologische Untersuchungen in der Region Biel (Schweiz). *Freiburger Geogr. H.*, 32, 19–54, Freiburg.

Weischet, W. (1979): Problematisches über die städtische Wärmeinsel und die Notwendigkeit einer Baukörperklimatologie. Siedlungsgeographische Studien, 407–423.

Weihe, W.H. (1986): Life expectancy in tropical climates and urbanization. In: *World Climate Technical Conference*, Mexico, WMO No. 652. WMO, Geneva, 313–353.

Werner, G. (1979): Regionale Luftaustauschprozesse und ihre Bedeutung für die räumliche Planung. BMRBS Schr.-R. 06, H, 032, Bonn.

Wilmers, F. (1968): Wettertypen für mikroklimatische Untersuchungen. *Arch. Meteor. Geophys. Bioklimat. B.*, 16, 144–150.

Wilmers, F. (1976): Die Anwendung von Wettertypen bei ökoklimatischen Untersuchungen. *Wetter und Leben*, 28, 224–235.

Wilmers, F. (1978): Grünflächen-Meteorologie. AKUMET Koll. DMG. *Angewandte Stadtklimatologie*, 19, Jan. 1978, 7.1–7.6.

Wilmers, F. (1982): Small-scale thermal effects of "green" and "shade" in an urban environment and their bioclimatic significance. In: *Sympos. on Building Climatology*, September 20–24, 1982, Proceed. P., 2, IB, Moscow, 624–643.

Wilmers, F., Scholz, K.-D. and Katzchke, D. (1987): Stadtklima und räumliche Planung – Klimaökologische Funktion und Beurteilung der Freiräume der Kernrandzone des Grossraums Hannover. Beitr. z. Regionalentw., ZGH. Hannover.

Wilmers, F. (1988): Green for melioration of urban climate. *Energy and Buildings*, 11, 289–299.

Wilmers, F. (1991): Effects of vegetation on urban climate and buildings. *Energy and Buildings*, 15, 507–514.

WMO (ed.) (1986): *Urban climatology and its applications with special regard to tropical areas*. Proceed. Technical Conf. Mexico, 26–30 Nov. 1984, WMO No. 652, WMO, Geneva.

Fritz Wilmers
Institut für Meteorologie und Klimatologie
Universität Hannover
Herrenhäuserstr. 2
Hannover 21
Germany

4.3. CLIMATES OF TROPICAL AND SUBTROPICAL CITIES

ERNESTO JAUREGUI

1. Introduction

During the last decades the uncontrolled flow of migrants to cities led to the proliferation of large metropolitan areas in the tropics. By the year 2000, the number of cities with more than one million inhabitants will increase threefold from the 52 that existed in 1982. Even in Africa, the least urbanized region in the developing world, the rate of population increase is the highest. By the end of the 1990s more than 340 million (42% of the total population) will be living in cities (U.N. Centre for Human Settlements 1987).

The accelerated urban growth has been putting pressure on such high priority services as sewage, water supply, education, and health. The mainly chaotic urbanization is leading to general environmental degradation including climate-related problems such as air pollution, heat stress, and poor ventilation.

Despite the acclimatization to heat by inhabitants of the tropics, increased morbidity and loss of productivity may result from the extra stress contributed by heat island growth in large cities. Even though tropical regions are projected to experience relatively small changes in temperature resulting from greenhouse warming, this increase in temperature will exacerbate the above mentioned health related problems.

In this chapter, after examining the main characteristics of the tropical environment, we deal with modifications of the climate/bioclimate of tropical cities due to the urbanization process, and how these changes, together with air pollution, are likely to affect the comfort/health condition of the population.

M. Yoshino et al. (eds.), Climates and Societies – A Climatological Perspective, 361–373.

2. The general climate of tropical / subtropical cities

Except for those cities located in the arid subtropics (where seasonal thermal amplitude is large) small annual temperature range, permanent (or seasonal) high temperature and humidity, high elevation of the sun and marked seasonal (or all year) rainfall characterize the bioclimates of tropical towns. Landsberg (1986) listed for 1981, 52 tropical metropolis with more than one million inhabitants. About two-thirds of these cities are located on the coast (and therefore ventilated by sea breeze) but the rest lie on elevated valleys or on coastal plains where ventilation is sometimes restricted.

Even though the term "tropical" is usually applied to the climate where temperature, humidity and radiation combine to create a feeling of sultriness, four basic types may be differentiated.

(a) Low-latitude (mostly coastal or lowland) equatorial climate where sultry conditions dominate during the whole year.
(b) Inner-tropical (wet/dry) monsoon climate where heat prevails for several months mainly before the rainy season.
(c) Tropical highland where the bioclimate is tempered by altitude.
(d) Subtropical dry desert bioclimate characterized by stress conditions caused by heat in summer and by cold in winter. This type is similar to mid-latitude bioclimate although more stressful in summer.

3. Health implications of bioclimate modifications by urbanization in tropical cities

3.1. *Local scale*

3.1.1. *The heat stress*

As a consequence of accelerated urbanization in the tropical world urban effects such as the heat island are gaining more relevance in large tropical cities. While this phenomenon has positive economic aspects (i.e., saving of energy for space heating in inner cities in winter) in mid-latitude urban areas, the urban excess temperatures observed in an increasing number of tropical cities (see WMO Tech. Note 652) only serves to increase the already prevailing high heat loads.

The rapid increase in population in cities in the arid/semiarid subtropics (i.e., the sun belt in the U.S.) is producing a substantial increase of weather stress in summer nights associated with an intensification of the heat island (Balling and Brazel 1986). In some cases the increasing urban/rural thermal contrasts in arid environments are moderated by evapotranspiration from trees in parks and streets as documented by Spronken-Smith (1990) for the city of Tucson, Arizona. But green areas are usually scarce in these cities, especially those in developing countries where water is limited to more basic needs.

As pointed out in a paper by Oke et al. (1989) new buildings in the urban tropics show a tendency to move away from traditional passive means of cooling and as cities and their heat islands grow, the demand for power to run air conditioning equipment will increase. This will imply for many developing countries increasing expenditure on imported energy supplies for air conditioning and, as these authors point out, this heat-island related additional power consumption would also contribute to problems of increasing carbon dioxide emission and air pollution.

3.2. *Global scale*

Although tropical regions are projected to experience relatively small changes in temperature resulting from greenhouse gas warming, this increase in temperature will exacerbate the heat-stress/health effects on the population.

3.3. *Air pollution*

Notwithstanding their relevance in urban bioclimate, the heat island and global warming effects in tropical cities seem at present to pose less of a problem than the one related to the ever increasing deterioration of air quality in these urban centers.

An increasing trend observed in the incidence of respiratory illness and bronchial asthma seems to be related to the seasonally high levels of air pollution reported in some tropical metropolis (Galindo 1987; Jauregui 1987).

Even though tropical cities are in general less industrialized, many of them are plagued by high levels of air pollution sometimes due to inadequate location of industry with respect to prevailing winds and unfavorable geography. This is particularly true for those large cities located in inland valleys or coastal plains. Here the examples are many: Delhi, Kuala Lumpur, Mexico City, Sao Paulo, Bogota, Dacca, Santiago and the Mexican cities of Guadalajara, Monterey, Tijuana and Ciudad Juarez, all of them with more than one million inhabitants and moderate to critical air pollution conditions.

Only in the last decade and in a few number of these cities have programs to monitor air quality (the necessary first step to assessing the air pollution problem) been set up to subsequently enforce air pollution standards designed to reduce emissions.

As pollution problems have worsened in such large cities as Santiago (Chile), Sao Paulo (Brazil), and Mexico City (Mexico), environmental authorities are beginning to enforce more stringent measures such as a work-day without a car, the banning of traffic in the downtown area at peak hours, the shutting down of some industries for violation of pollution regulations, introduction of unleaded gasoline, and changeover of some industries to low sulphur fuels. But in some cases, even though all measures listed above have been adopted in cities where air pollution has become critical, as in Mexico

Table 1. Mean annual values of main pollutants observed in Mexico City in 1990 (at downtown merced market station) as compared with corresponding U.S. and Canada standards. (Source: Federal District Office, Annual Report 1991.)

Pollutant	Standard (Annual) (1)	Observed Annual avg. (2)	2/1
SO_2	0.02 ppm	0.069 ppm	3.5
NO_2	0.05 ppm	0.054 ppm	1.08
O_3	0.15 ppm	0.470 ppm	3.1
TSP[a]	70 microg/m^3	198 microg/m^3	2.8
PM 10[b]	50 microg/m^3	120 microg/m^3	2.4

[a] TSP – Total suspended particles.

[b] PM 10 – Respirable fraction.

City, daily concentration of pollutants like ozone and nitrogen oxides (its precursor) have continued to rise.

To illustrate the air pollution problem in a tropical city, in what follows we examine some recent available data from Mexico City, one of the major metropolitan areas in the world plagued with a critical air pollution condition.

The Mexican capital is located in an elevated valley (2,250 m) in central Mexico (lat.: 20° N) and surrounded by mountains (except to the northeast) that extend another 800 m higher. Since the central plains to the NE of the city are semiarid (precipitation: about 450 mm/year) with scant vegetation, the eastern suburbs are subject to dust storms at the end of the dry season (February to April) (Jauregui 1989).

At night and early morning (especially during the dry season) the downslope winds from surrounding mountains and the converging circulation induced by the heat island combine to restrict the lateral dispersion of pollutants (Jauregui 1989). The vertical dilution of urban gases is further restricted by the frequent (more than 70%) surface inversions observed during this period.

Once the stable layer is heated from below by abundant insolation, turbulent mixing dilutes the pollutants in the vertical, while the regional winds, that descend to the ground (usually with a northerly component) transport pollutants to the southern suburbs, where usually the highest levels of ozone are observed.

Table 1 shows the mean annual values of the main pollutants observed (in 1990) at the merced market Station in downtown Mexico City. It is clearly

Table 2. Three-months average of atmospheric lead (microg/m^3) in Mexico City for the period 1988–1990. (Source: Federal District Office, Annual Report 1991.)

	NE industrial sector			Average for city		
	1988	1989	1990	1988	1989	1990
Jan–Mar	4.7	2.9	2.4	1.8	1.7	1.4
Apr–Jun	2.2	1.8	1.7	1.2	1.1	1.1
Jul–Sep	2.4	1.4	1.6	1.5	1.5	1.3
Oct–Dec	3.2	2.2	2.1	2.8	1.3	1.4

Three-months Average. Standard: 1.5 microg/m^3

Table 3. Number of days/yr when standard (mainly O_3) was exceeded in Mexico City for the period 1986–1990. (Source: Federal District Office, Annual Report 1991.)

	1986	1987	1988	1989	1990
Days/yr	274	314	329	339	329
Percent	75	86	90	93	90

seen that most contaminants exceeded by several times the standards in the inner city in that year.

Among the measures to abate emissions the city authorities have introduced (since 1988) a new gasoline with low lead content. This course of action seems to have reversed the growth of two of the most dangerous pollutants, lead and sulphur dioxide.

Table 2 shows that atmospheric lead levels, although still high, display a decreasing trend for the three-year period. However, despite recent efforts to reduce emissions, these measures do not seem to have been sufficient since air pollution has reached record levels, mainly due to ozone (except for the eastern sector where dust is the main pollutant) particularly in the southwest sector where it is accumulated driven by the prevailing northeasterly winds during the afternoon (Table 3). Recently, an increasing trend of the levels of ozone has been pointed out (Jauregui 1992).

4. The urban climate of tropical cities

4.1. *General*

Modification of the various components of climate by urbanization has been studied extensively, especially in mid-latitude cities (see bibliographies by Chandler 1970; Oke 1974, 1979). During the last two decades a growing number of studies on tropical urban climatology has emerged. Although most of this work has been based upon data from standard stations and near-surface temperature/humidity traverses, the results available for the tropical urban environment indicate that there is a resemblance to those obtained in mid latitudes. This was to be expected since the urban/atmosphere system in the tropics is essentially similar to that of temperate cities. With some variations, both large cities in the tropics and mid latitudes display similar physical characteristics in the central business area with clusters of high-rise concrete/glass/iron structures, parking lots and few green areas.

The greatest tropical/mid latitude contrasts are likely to be found in: (a) Land-use characteristics of the suburbs and nearby rural area. The suburbs of large tropical cities are generally to be distinguished by large squatter settlements of low-cost one/two story dwellings of the working class flanking mostly unpaved roads with little vegetation. Contrasting with this condition, a large proportion of the suburbs in mid latitude cities is occupied by vegetation (Oke 1982); and (b) The regional or synoptic control. While in most cases the suburbs and the adjoining rural areas of mid latitude cities are usually covered by vegetation all year round, those corresponding to some tropical cities, (except in equatorial and desert climates) undergo a significant annual variation in their physical properties imposed by the alternation of the wet/dry seasons.

Given the diversity of climates within the tropical world, it may not be an easy task to identify the features that distinguish the effects of urbanization on the climates of tropical cities from those corresponding to the mid-latitudes. In what follows we will try to summarize the effects of urbanization from mainly descriptive work done in the tropical/subtropical cities of Lagos, Ibadan, Delhi, Shanghai, Mexico City, Guadalajara, Phoenix and Tucson.

4.2. *The heat island*

Modification of air temperature by urbanization (the heat island) has been a central subject in urban climate studies in the tropics. It has also been the easiest one to undertake by simply comparing temperature differences between the observatory (urban) and the airport or by making traverses with a thermometer.

When hourly temperature observations are available at both urban/rural sites, the daily and seasonal development of the heat island intensity may be assessed. As mentioned in the previous section, the results in estimation

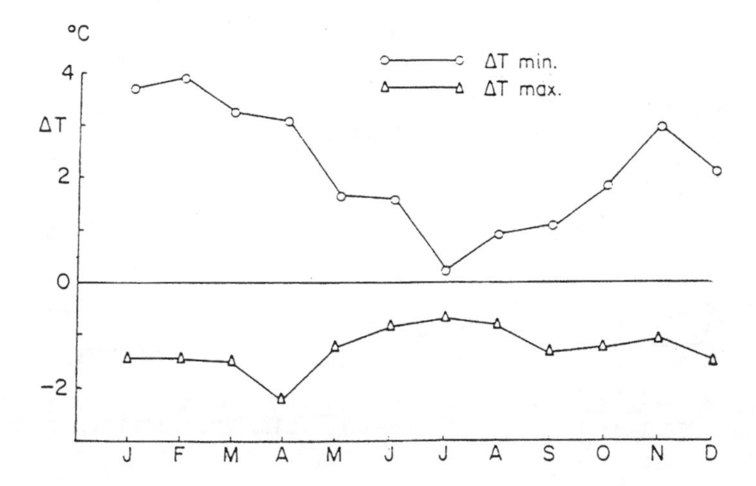

Figure 1. Mean monthly heat/cool island intensity between Rayon station (urban) and Expreriencia station (rural) in Guadalajara (Jauregui et al. 1991).

Table 4. Typical values of heat island intensity in some tropical/subtropical cities (°C).

City	Climate	Season dry	Season wet	Pop. millions	Lat. deg.
Ibadan	equatorial	6	2	1.8	7
Lagos	equatorial	4	2	5	6
Delhi trop.	w/d	6	4	8	28
Mexico City	trop. w/d	9	–0.5	16	20
Guadalajara	trop. w/d	4	1	2.7	20
Shanghai	trop. w/d	6	2	12	31
Phoenix	desert	2.5	(summer)	1.2	33
Tucson	desert	2.5	(summer)	0.7	32

References: Ogunyoyimbo 1983; Adebayo 1991; Padmanabhamurty 1986; Jauregui 1986; Chow 1986; Balling and Brazel 1987; Spronken-Smith 1990; Jauregui et al. 1991.

of the heat island intensity may differ in accordance with the wide range in physical characteristics of both urban and rural sites.

Notwithstanding the above-mentioned limitations, some generalizations of the heat island phenomenon as observed in the tropics are possible: (a) Although apparently at a less marked rate than in mid-latitudes the heat-island intensity seems to grow with city size in the tropics (see Oke 1973; Jauregui 1986). But more critical factors such as length of fetch and/or urban

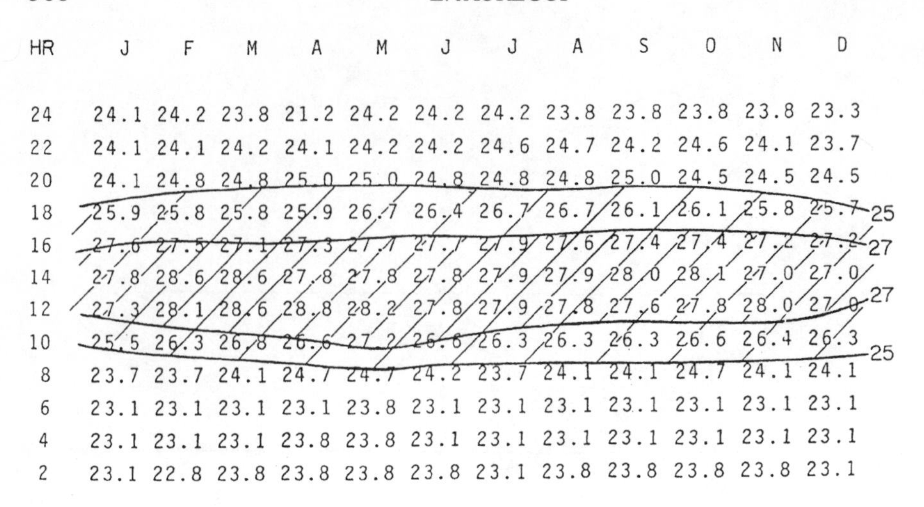

HR	J	F	M	A	M	J	J	A	S	O	N	D
24	24.1	24.2	23.8	21.2	24.2	24.2	24.2	23.8	23.8	23.8	23.8	23.3
22	24.1	24.1	24.2	24.1	24.2	24.2	24.6	24.7	24.2	24.6	24.1	23.7
20	24.1	24.8	24.8	25.0	25.0	24.8	24.8	24.8	25.0	24.5	24.5	24.5
18	25.9	25.8	25.8	25.9	26.7	26.4	26.7	26.7	26.1	26.1	25.8	25.7
16	27.6	27.5	27.1	27.3	27.1	27.1	27.9	27.6	27.4	27.4	27.2	27.2
14	27.8	28.6	28.6	27.8	27.8	27.8	27.9	27.9	28.0	28.1	27.0	27.0
12	27.3	28.1	28.6	28.8	28.2	27.8	27.9	27.8	27.6	27.8	28.0	27.0
10	25.5	26.3	26.8	26.6	27.2	26.6	26.3	26.3	26.3	26.6	26.4	26.3
8	23.7	23.7	24.1	24.7	24.7	24.2	23.7	24.1	24.1	24.7	24.1	24.1
6	23.1	23.1	23.1	23.1	23.8	23.1	23.1	23.1	23.1	23.1	23.1	23.1
4	23.1	23.1	23.1	23.8	23.8	23.1	23.1	23.1	23.1	23.1	23.1	23.1
2	23.1	22.8	23.8	23.8	23.8	23.8	23.1	23.8	23.8	23.8	23.8	23.1

Figure 2. Distribution of effective temperature for Kuala Lumpur, Malaysia, 2° N (adapted from Sani, 1980).

geometry and building materials would be a better choice (although more difficult to assess) than defining the city size by its population, as pointed out by Oke (1982); (b) The heat island in tropical cities usually reaches its peak at the end of the nocturnal cooling period. In contrast, this maximum value occurs 3 to 5 hours after sunset (Oke 1982) in mid-latitudes due in part to release of heat for space heating in winter; (c) A seasonal variation in the heat island development is often evident with the highest values being observed towards the end of the dry season, when clear, calm dry weather prevails. In contrast, the heat island is best defined in mid-latitudes in calm, clear summer nights (Oke 1982). Table 4 shows typical values of heat island intensity for several tropical cities. During the wet season the presence of prevailingly unstable moist air masses results in weak urban/rural thermal contrasts; and (d) In tropical wet/dry or arid/semiarid subtropical climates thermal contrasts reverse as soon as turbulent mixing is promoted by solar heating and a "cool" island is sometimes established (Figure 1) in the afternoon, especially during the dry season, when the city seems to have more availability of water (for evapotranspiration) than the rural surroundings (see Jauregui 1986). Indeed, energy balance measurements undertaken in Mexico City (Tacubaya Observatory) by Oke et al. (1992) show that the low Bowen ratio values observed in a densely built-up district (in March) confirm the large contribution of evaporation to the turbulent heat transport, suggesting that the city is, in these dry months at least, a relative source of moisture.

Since the heat island is likely to be more intense at night and early morning in the tropics, the excess urban heat is bound to affect only the night comfort, especially at the end of the dry season which for some cities is the warm/hot season. In contrast with the nocturnal heat island, the "cool" island is usually a weak (1 or 2°C) phenomenon (Figure 2); limited to the vicinity of urban parks nevertheless it helps reduce locally, the discomfort of the prevailingly high afternoon temperatures in tropical cities. The radiation from high sun and urban surfaces and the weak ventilation, however, tend to offset the beneficial effects of the relatively lower afternoon temperatures.

4.3. *Humidity*

From the few available data on urban/rural humidity differences that have been undertaken in tropical cities, it follows that at night relative humidity is, as would be expected, lower than the rural surroundings due to the warmer air of the heat island. In tropical wet/dry and sub/tropical dry climates, as noted in the previous section, the city has been shown to be moister than the adjoining rural environment (see Chow 1986; Jauregui 1986; Padmanabhamurty 1986; Jauregui 1994).

4.4. *Wind*

Again, the scanned data on urban/rural wind contrasts indicate that, as in the case for mid latitude cities, the heat island acts as the forcing mechanism to induce a convergent surface flow at night and early morning (Jauregui 1988). During that period the prevailing light winds are stronger in the built-up area than in the rural perimeter whereas the reverse relationship exists when winds are stronger in the afternoon (see Padmanabhamurty 1986).

5. The bioclimate of tropical cities

The last decades of this century have witnessed the accelerated urbanization of those regions with the most oppressive climates in regard to human comfort. The prevailing excessive heat and humidity that characterize the tropical environment place a strain on the physiology of humans. No unique scheme of bioclimatic classification has been devised with regard to human comfort/health. We will use here the simple effective temperature (ET) concept in order to illustrate similarities or contrasts in the bioclimate of the tropical world. Admittedly, the ET index gives only a crude approximation to prevailing physioclimatic conditions since it leaves out the effect on humans of radiation and wind.

Moreover, in applying the ET index one should be aware of variation in response due to particular cultural, ethnic, health conditions implicit in the average person. According to Landsberg (1972) the boundary of sultriness may be set at 24 ET, while 35 ET is considered the upper limit of tolerance by this author. However, this scheme is derived from mid-latitude experience, being related to a corresponding scale of sensations of comfort/discomfort experienced by persons acclimatized to temperate climates.

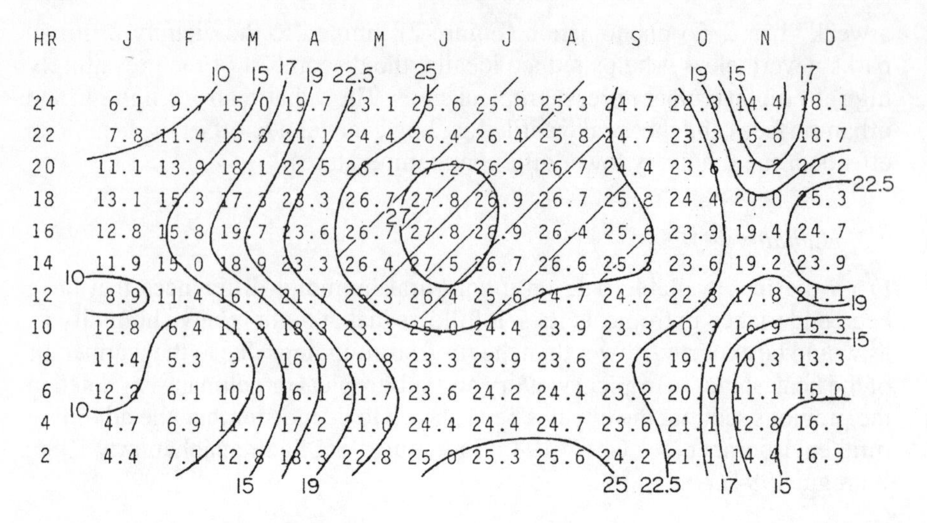

Figure 3. Distribution of effective temperature for New Delhi, India, 20° N (adapted from Chowdhuri and Ganesan, 1983).

In trying to determine the limits of comfort for India, Lahiri (1984) arrived at the conclusion that high humidity (and temperature) experienced in subtropical New Delhi (lat. 29° N) during monsoon months (ET = 27) is not considered unpleasant (under fan). She suggests that threshold thermal sensations should be higher than those for temperate lands. On the other hand, Chaudhuri and Ganesan (1983) postulate a value of 25°C ET as a limit for comfort more realistic, since hours with ET greater than 25 are those hours when air conditioning is actually used in Indian cities.

Evidently, more surveys are needed in order to define more realistic limits of bioclimatic categories for tropical cities. In concluding, we shall use the ET index for different hours of the day and for each month to illustrate the generalized patterns of the distribution of this index and identify periods of environmental comfort/discomfort in some representative cities in the tropics/subtropics (for a detailed account see Jauregui 1991b).

(a) *near the equator bioclimates*

Figure 2 shows values of ET greater than 25°C ET prevailing during most of the day and throughout the year for Kuala Lumpur. But nights are in these equatorial locations, and especially after midnight less stressful (ET less than 25°C ET) and more congenial. This characteristic is common to most tropical cities and becomes more marked as one moves inland or away from the equator.

(b) *inner tropics*

At higher latitudes within the tropics sultry conditions (ET = 25°C to ET = 30) are limited to the afternoon hours of the summer or warmer season while nights are lightly warm or cool, as is the case of Delhi (Figure 3).

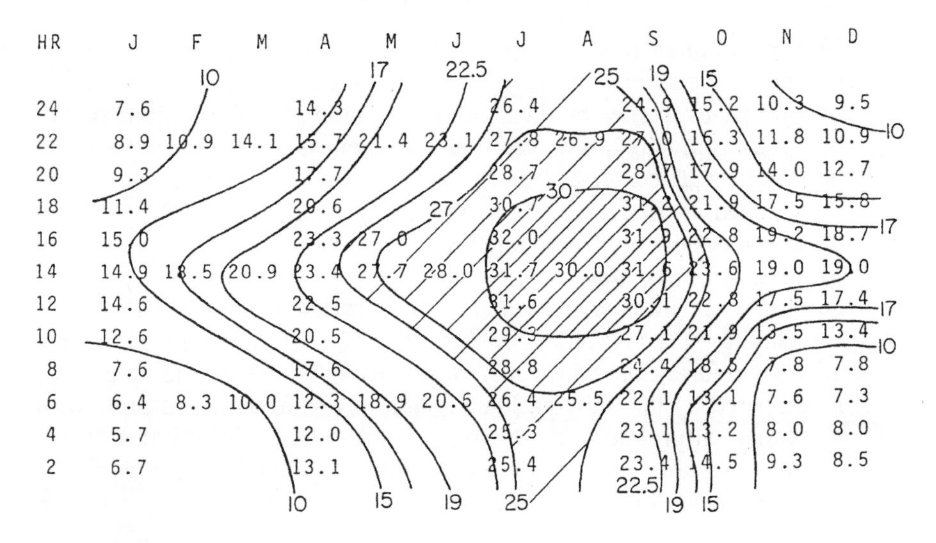

Figure 4. Distribution of effective temperature for Mexicali, Mexico, 32.5° N (precipitation 60 mm/yr., 40 m a.s.l.).

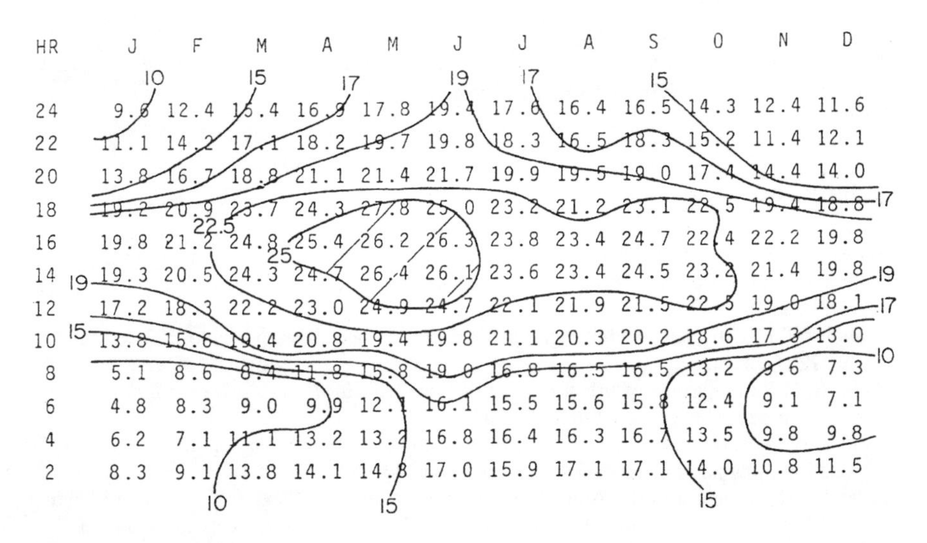

Figure 5. Distribution of effective temperature for Guadalajara, Mexico, 21° N (1,590 m a.s.l.).

During the cool part of the year days and nights are either warm or cool (ETs less than 25).

(c) *subtropical desert*

Extreme bioclimatic conditions are illustrated in Figure 4 for the city of Mexicali, on the U.S. Mexico border (pop. 1 million inh., precip. 60 mm/yr, alt. 4 m a.s.l.) where prevailing afternoon ET values are above

27°C in summer reaching up to 32 ET in July. Radiational nocturnal cooling to clear skies in the dry desert air brings back the environment to comfort levels (i.e., below 25 ET) after midnight. The combination of hot air and blowing dust in summer afternoons is not an unusual event in desert cities. The excessive heat load and dust content of the air during these storms (with ETs between 32–35) are likely to affect the health of the urban population (Jauregui 1989).

(d) *mountain tropical bioclimates*
Quite a large number of tropical cities are located in the mountains. Depending on the altitude, comfortable conditions are perceived in the afternoon during the warm period in these cities where the bioclimate is tempered by the altitude. Ideal conditions from the point of view of bioclimate are found in those tropical cities within the 1,300–1,800 m altitude range like the Mexican city of Guadalajara (see Figure 5) where stress from either heat or cold are relatively low during most of the year.

Acknowledgements

The author acknowledges Mr. A. Estrada's assistance with some of the drawings and Ms. M.L. Maya for collection of data.

References

Adebayo, Y. (1991): Heat island in a tropical city. *Theor. and Appl. Climatology*, 43, 137–147.
Balling, R. and Brazel, S. (1987): The Phoenix urban heat island. *J. Ariz.-Nev. Academy Sci.*, 21(1), 75–81.
Chandler, T. (1970): Selected bibliography on urban climate. Tech. Note No. 155, WMO 276, WMO, Geneva.
Chow, S. (1986): Aspects of urban climate of Shanghai. In: *Proceed. Tech. Conf. on Urban Climatology*, Mexico City, 1984, 87–109.
Choudhuri, A. and Ganesan, H. (1983): Meteorological requirements on air conditioning in relation to human habitat for comfort. *Mausam*, 34, 281–286.
Federal District Dept. (1991): Annual report on air quality for Mexico City. Urban Planning and Ecological Protection. Office of Ecological Planning, Mexico City (in Spanish).
Galindo, I. (1987): Evaluation of impact of some atmospheric elements on health. In: *Proceed. Symp. on Climate and Human Health*. WCAP No. 1, Leningrad, Vol. 11, Geneva, 17–46.
Jauregui, E. (1986): The urban climate of Mexico City. In: *Proceed. Tech. Conf. on Urban Climatology*. WMO/WHO.
Jauregui, E. (1987): Air pollution and incidence of respiratory illness in Mexico City. In: *Proceed. Symp. Climate and Human Health*. WCAP 1, Vol. 11. Leningrad, Geneva, 180–183.
Jauregui, E. (1988): Local wind and air pollution interaction in the Mexico Basin. *Atmosfera*, 1, 131–140.
Jauregui, E. (1989): The dust storms of Mexico City. *Int. J. of Climatology*, 9, 169–180.
Jauregui, E. (1989): Meteorological and environmental aspects of dust storms in Northern Mexico. *Erdkunde*, 43, 141–147.

Jauregui, E. (1991): The human climate of tropical cities: An overview. *Int. J. of Biometeorology*, 35, 151–160.
Jauregui, E. (1992): Mexico City – A critical zone in global environmental change. A paper presented at Meeting Com. Climatology, 27th IGC, University Park, U.S.A., 3–8 Aug. 1992, 14 pp.
Jauregui, E. (1994): Areal and temporal humidity variations in Mexico City. In: *Proceed. Com. of Climatology*. IGU, Brno, Czeck Republic, 287–292.
Jauregui, E., Godinez, L. and Cruz, F. (1991a): The heat island of Guadalajara, Mexico. *Atmos. Environment*, 26B(3), 391–396.
Landsberg, H. (1972): The assessment of human bioclimate. WMO Tech. Note 123, WMO, Geneva.
Lahiri, M. (1984): Indices for comfort analysis. *Mausam*, 35, 275–276.
Landsberg, H. (1986): Problems of design for cities in the tropics. In: *Proceed. Tech. Conf. on Urban Climatology*. WMO/WHO, Tech. Note 652.
Oke, T. (1984): City size and the urban heat island. *Atm. Environment*, 779–783.
Oke, T. (1974): Review of urban climatology 1968–73. Tech. Note 134, WMO, 383, 132 pp.
Oke, T. (1979): Review of urban climatology 1973–76. Tech. Note, 169, WMO, 539, 100 pp.
Oke, T. (1989): The energetic basis of the urban heat island. *Q. J. Roy. Meteor. Soc.*, 108, 1–24.
Oke, T., Taesler R. and Olsson, L. (1989): The tropical urban climate experiment. In: *Proceed. Conf. on Urban/Building Climatology*, WMO/CIB/IFHP, Kyoto, Japan.
Oke, T., Zeuner, G. and Jauregui, E. (1992): The surface energy balance in Mexico City. *Atmosph. Environment*, 26B(4), 433–444.
Oguntoyimbo, Y. (1986): Aspects of urban climates in tropical Africa. In: *Proceed. Tech. Conf. on Urban Climatology*, Mexico City, Tech. Note, 652, 110–135.
Spronken-Smith, R. (1990): Air temperature survey of urban parks in Tucson, Arizona. In: *Proceed. Int. Symp. on Urban Climate, Air Poll. and Planning in Tropical Cities*, WMO/OMMAC, Guadalajara, Mexico.
Sani, S. (1980): The urban climate of Kuala-Lumpur. Department of Geography, University of Malaysia.

Ernesto Jauregui
Centro de Ciencias de la Atmosfera
Dept. Meteorologica General
Circuito Exterior
Cd. Universitaria
Mexico D.F.
C.P. 04510 Mexico

4.4. AIR POLLUTION: A LOCAL PROBLEM BECOMES A GLOBAL PROBLEM

HEINZ WANNER

1. From high smoke-stacks to the "1950s syndrome"

In the first half of the 20th century air pollution policy was still based on the use of high smoke-stacks. This was a "smoke-stack policy" aimed to rarefy trace constituents or stir pollutants emitted into the atmosphere by using high chimneys, so that living organisms (humans, animals, and plants) and natural transport and storage media (atmosphere, hydrosphere, cryosphere, and lithosphere) received nominal damage.

As a result of both population explosion and industrialization, however, consumption of energy (obtained partly from fossil sources) reached levels that very rapidly led to severe local pollution (catastrophic smog in London, Los Angeles, etc.).

Because air is a very turbulent medium, pollutants emitted into the atmosphere are transported over relatively wide areas and are well mixed. In addition, the ratio of the mass of air pollutants (gases, liquids or solids) to the mass of the air is very high by comparison with the ratio in the case of water, ice or soil. Thus it was determined around 1950 that air as a storage medium could no longer sufficiently purify itself on a global scale. This led to the phenomenon known as the "1950s syndrome", a term used to describe the progressive increase in the concentrations of important trace constituents or pollutants (e.g. carbon dioxide, CO_2; and nitrous oxide, N_2O) in the atmosphere after 1950 (Figure 1).

M. Yoshino et al. (eds.), Climates and Societies – A Climatological Perspective, 375–380.

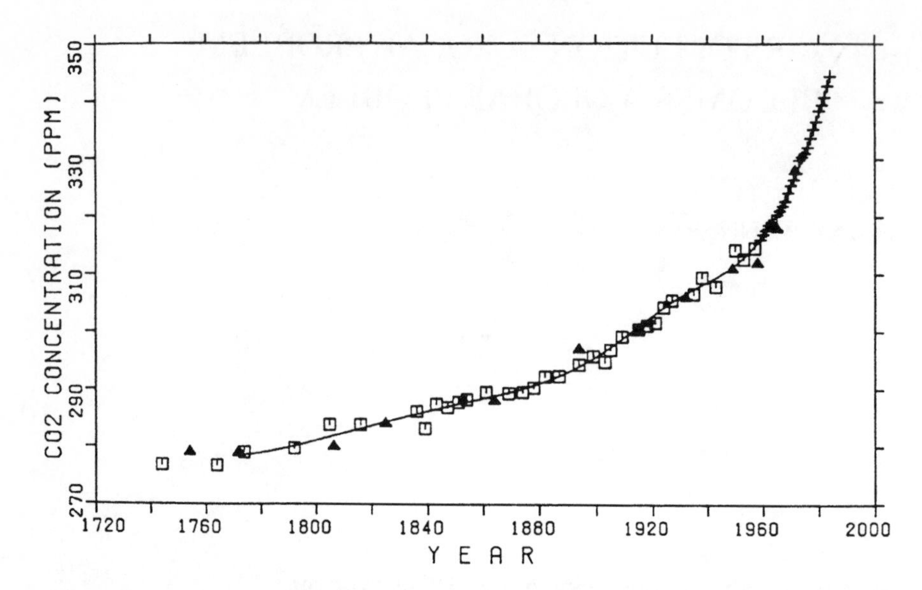

Figure 1. The 1950s syndrome, showing the increase in CO_2 in the atmosphere since 1740 and the marked change occurring in the middle of the twentieth century. Ice core sample measurements were done on air bubbles (squares and triangles) in Siple, Antarctica, and measurements were made at a station in Mauna Loa, Hawaii starting in 1958 (crosses). Source: Neftel et al. (1982); Friedli et al. (1986).

2. Fundamental processes

In order to understand how measurements and models of the origin, diffusion and spatio-temporal evolution of air pollution are formulated, four fundamental processes must be considered simultaneously:

(a) *Source term (emission).* This involves knowledge of the types of pollutants emitted and their proportions, in both spatial and temporal terms, and as precisely as possible.

(b) *Transport and turbulent diffusion of air pollutants.* The meteorological branch of air pollution control is primarily concerned with horizontal wind and temperature stratification. It attempts to describe and model the advance and retreat (advection) as well as the diffusion of pollutants.

(c) *Chemical transformation of air pollutants.* Since most air pollutants are unstable and subject to numerous chemical reaction mechanisms (with many intermediate products), these products, their reaction rates, and their concentration changes over time must be known precisely.

(d) *Removal and deposition: air pollution sinks.* Most pollutants which have been transported and chemically transformed following emission into the atmosphere are usually not destroyed but rather are stored in other media (soil, water, biosphere) as a result of deposition. We still have very little knowledge of deposition and uptake processes (e.g. in stomata of leaves or in the human lung).

Meteorologists and specialists in air chemistry occasionally argue over the significance of these four processes. They are quite certain, however, that only simultaneous consideration of all these processes on different spatial and temporal scales will lead to approaches useful in explaining and modelling air pollution.

It is also important to note that air pollutants or trace constituents have a specific life and residence times in the atmosphere, determined by the speed of destruction mechanisms (rate of destruction). This lifetime in turn has a decisive influence on the spatial diffusion of pollutants. Air pollutants with low rates of destruction, and hence long life-times, are transported over great distances and diffuse to some extent beyond the troposphere into the lower stratosphere, where, for example, they could deplete the ozone layer.

Figure 2 is an attempt to present diffusion processes as a time-space diagram. The ellipses show interactions as well as spatio-temporal correlation of meteorological processes. Typical spatial categories (listed above the ellipses) as well as characteristic air pollutants (beneath the ellipses) can be assigned to each of these processes. As previously mentioned, the spatio-temporal dimension in which an air pollutant has an influence grows in proportion to its lifetime. For example, sulphurous compounds (ammonium sulfate, for instance), which have long lifetimes, are condensation products in acid rain. Other trace gases with long lifetimes contribute to the greenhouse effect or are catalysts in the depletion of the ozone layer.

We increasingly expect science to provide us with diagnoses and to make forecasts. At the bottom of Figure 2 is a list of processes correlated with four typical time scales. These processes must be taken into account in the construction of a suitable model. But the problem is that these processes are governed by very different temporal and spatial scales (consider, for example, the rapid chemistry of radicals in the extensive transport of photochemical oxidants). Model-builders are thus forced to deal with very limiting simplifications.

3. The scientist's fear of nonlinearity and deterministic chaos

Scientists who are concerned with making predictions about dynamical systems have made many statements that reveal a considerable fear of the future. In part, this fear results from the characteristics of these systems as well as from processes that take place within them, which are very often linked with the concepts of nonlinearity and deterministic chaos. Why is this significant?

Nonlinearity in the present context refers to processes that do not proceed in regular (linear) fashion. Instead, their spatial and temporal evolution is governed by irregular (nonlinear) increases or decreases (as in the case of

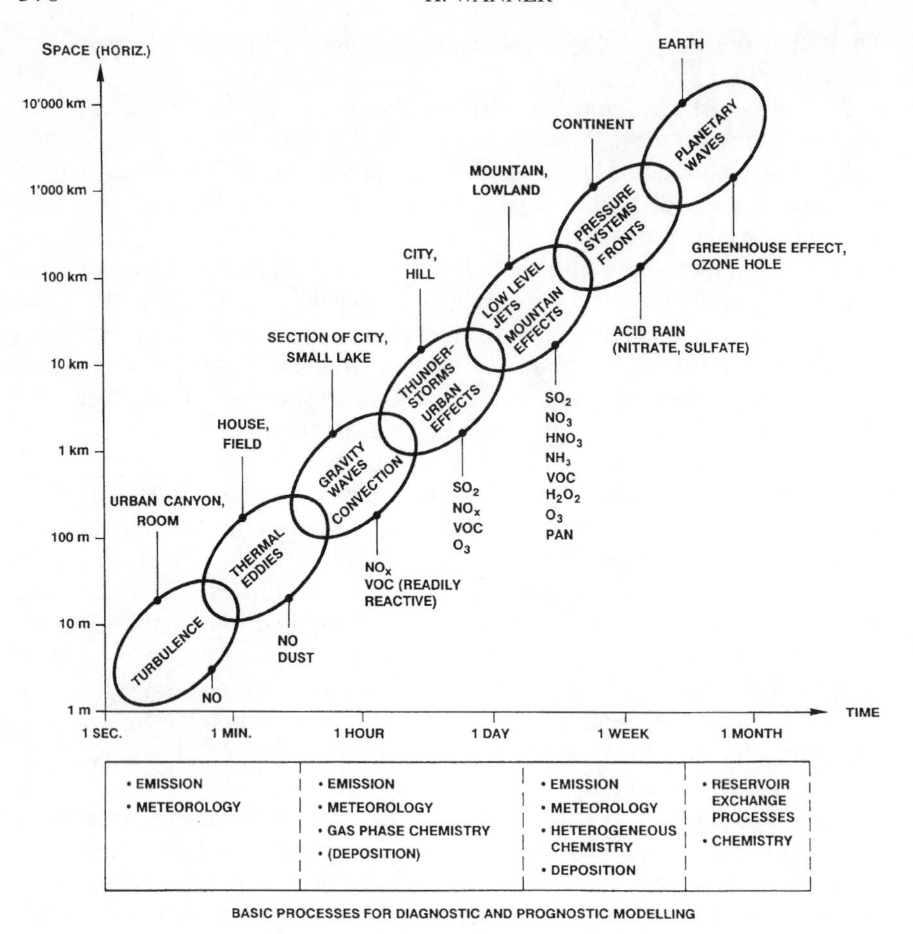

Figure 2. A time-space diagram of important diffusion processes.

nonlinear expressions in describing differential equations). This means that certain controlling factors in a system sometimes make a sudden evolutionary "jump" (see the 1950s syndrome, Figure 1). For example, nonlinearity is a necessary but not a sufficient condition for the evolution of chaotic movements that take the form of turbulent vortices of different magnitudes. Unfortunately, it is virtually impossible to forecast these chaotic movements (a problem in long-range weather forecasting). They are so greatly dependent on a system's initial state that there is no way in which we can either determine or measure this state with sufficient accuracy.

Some controlling elements of balance are comparable to a sand-heap whose sides suddenly cascade downwards like an avalanche when a few more grains of sand are added to the heap. In other words, there are many cases in nature in which we deal with processes where we often do not know when the

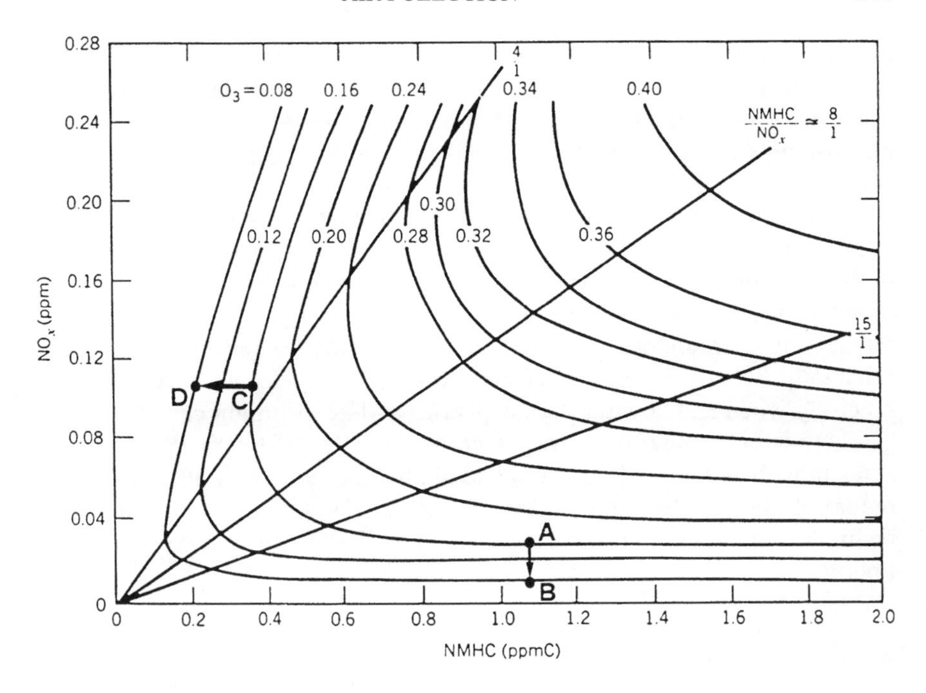

Figure 3. A diagram of ozone isopleths, showing maximum ozone concentrations using the empirical kinetic modelling approach (EKMA). Ozone concentrations (in ppm) developed at different concentration ratios of NMHC to NOx on the lee of a metropolitan area on a sunny day. Source: Dodge (1977).

slightest impacts on the environment will produce collapse, i.e. cause the "environmental sand-heap" to cave in.

Those involved in atmospheric research are in a corresponding position. For example, they are expected to be able to predict where and when certain ozone precursors (nitrogen oxides, hydrocarbons) will produce a certain type of ozone reduction. Yet, even here there are nonlinear relationships. Moreover, the models employed (in this case, the frequently used EKMA – Empirical Kinetic Modelling Approach) grossly oversimplify natural conditions. Thus application of such modelling techniques to the complex conditions of Switzerland (diverse topography, wide distribution of pollution sources) must, at the very least, be undertaken with great caution.

Figure 3 shows maximum hourly mean values for ozone, compiled on the basis of an "initial emission" of nitrogen oxides (NO_X) and non-methane hydrocarbons (NMHC) on a sunny day on the lee of a metropolitan area in the atmospheric boundary layer. High concentrations of ozone are extremely undesirable at this atmospheric level. Yet, even in this limited situation, it is clear that ozone behaves in nonlinear fashion (represented by the curved lines known as isopleths), and also exhibits different gradients (represented by the distances between isopleths) with different concentration ratios of NMHC to NO_x. Thus there are areas where minimal concentration changes in one initial substance produce great concentration changes in ozone, while a simultaneous increase in both initial substances always produces a proportional increase in ozone.

Forecasters are now confronted with the question of when a nonlinear or even chaotic evolution, collapse or extreme event might occur, and how many of these events are likely to occur during a given period of time as the result of anthropogenic, i.e. human-induced influences. There are certainly enough threatening phenomena, such as climatic breakdown, extreme weather, desertification, the ozone hole, etc. But who or what model will predict for us precisely when and how the "atmospheric sand-heap" might cave in?

There is only one proven way to combat this fear. Effective control measures must be taken early enough and applied in the right ways. In the case of air pollution control, this is a simple matter: The sources of pollution must be limited, and there must be a corresponding limit to our sense of individual liberty.

Acknowledgements

Our gratitude goes to Ted Wachs for translating and typing the manuscript and Res Brodbeck for the drawings.

References

Becker, K.H. and Lobel, J. (eds.) (1985): *Atmosphärische Spurenstoffe und ihr physikalisch-chemisches Verhalten* (Atmospheric trace constituents and their physical-chemical behaviour). Springer-Verlag, Berlin.

Dodge, M.C. (1977): Combined use of modelling techniques and smog chamber data to derive ozone-precursor relationships. In: *Proc. Int. Conf. on Photochemical Oxidant Pollution and Its Control*, Vol. II. EPA-600/3-770016, 881-889.

Fabian, P. (1989): *Atmosphäre und Umwelt* (Atmosphere and environment), 3rd ed. Springer-Verlag, Berlin.

Friedli, H. et al. (1986): Ice core record of the 13C/12C ratio of atmospheric CO_2in the past two centuries. *Nature*, 324, 237-239.

Neftel, A. et al. (1982): Ice core sample measurements give atmospheric CO_2 content during the past 40,000 years. *Nature*, 295, 220-223.

Schupbach, E. and Wanner, H. (ed.) (1991): *Luftschadstoffe und Lufthaushalt in der Schweiz* (Air pollutants and meteorology in Switzerland). Verlag der Fachvereine ETH Zürich, Zürich.

Heinz Wanner
Geographisches Institut
Universität Bern
Hallerstr. 12
3012 Bern
Switzerland

4.5. AGRICULTURAL LANDUSE AND LOCAL CLIMATE

MASATOSHI YOSHINO

1. Introduction

Historically, local climatology developed in response to increasing demand for agricultural production through agroclimatological knowledge (Yoshino 1987). The history of local climatology shows a close relationship to human interest in climate change in relation to landuse and land cover change in areas such as arable land, forest, grassland, and cities.

Generally, changes in land use patterns through human activities have two complementary implications in the evolution of local climates. One is the effect of land use change on climate; the other is the effect of the local climate on land use. These two impacts and their processes are dealt with in this review. In recent years, global change problems are studied within the framework of the "International Geosphere-Biosphere Programme" (IGBP), a programme whose main focus is the relationship between land use and local climate. The human dimensions of this inter-relationship are also important and occupy a major position in a global environmental change programme named "International Human Dimensions Programme" (IHDP). The present review assesses the role of local climates studies in these Programmes.

2. Land/cover as a local climate factor

2.1. *Progress of studies*

One of the earliest description of land use in a relatively small area in relation to local climatic conditions is "*Harima Fudoki* (Regional Geography for Harima)", Harima is an old name of the southwestern part of today's Hyogo Prefecture, Kinki District in Japan. The *grades of soil* in a 40 km × 60 km

M. Yoshino et al. (eds.), Climates and Societies – A Climatological Perspective, 381–400.

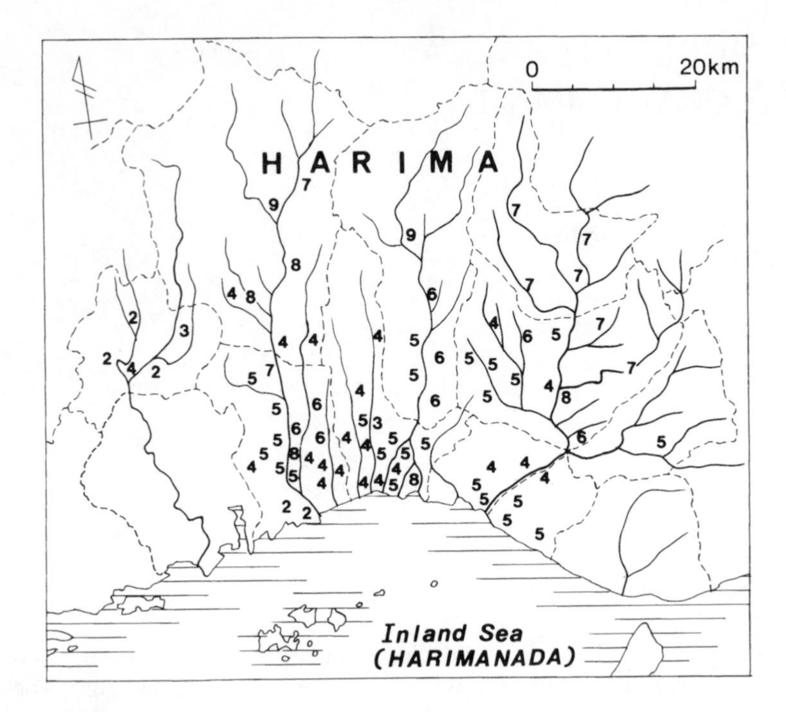

Figure 1. Distribution of the *grades of soil* (productivity by natural environments) in Harima, SW-Japan recorded in "*Harima Fudoki*" in the eighth century (Yoshino, 1975). 1: best of best, 2: medium of best, 3: worst of best, 4: best of medium, 5: medium of medium, 6: worst of medium, 7: best of worst, 8: medium of worst, and 9: worst of worst.

area are shown in Figure 1 as recorded in *Harima Fudoki*, and are seemed to represent the overall natural environment including climate suitability for agricultural production. Although judgement of productivity seems to have been based on subjective criteria, this record is noteworthy as the first treatise in which consideration was given to climatic differences in a relatively small area. There were three degrees – best, medium and worst – at the first stage, and each degree was further divided into three grades - best, medium and worst – at the second stage. Therefore, there were altogether nine grades as shown in Figure 1. This figure is important not only as a map showing conditions which prevailed 1,200 years ago, but also as a map based on observed results (Yoshino 1975).

Sir John Evelyn (1620–1706), a famous British diarist, reported an observation on the severest frost damage in the century that occurred in the winter of 1683–1684. He wrote about the "surprising" effect of cold snow cover which prevented damage, together with the facts that the shrubs and garden plants covered with straw suffered less damage from frost.

In the 18th and 19th century, there were several examples of afforestation and shelter belts in strong wind regions developed out of knowledge of the local climate. Air and ground temperature, heat balance, vertical distribution of wind velocity etc. at and above the ground surface for various land use regions were being studied by the end of the 19th century.

Effects of forests on climate were discussed also in the 19th century. Marsh (1864) was the first to write on the American experience. Aughey of Nebraska reported in 1880 that precipitation amounts increased with the expansion of farm land. Hamberg in Sweden from 1885–1896, and Voeikov in Assam and Malaya in 1887, discussed the effects of forests on precipitation and air temperature, respectively.

In the 20th century, research into local climates became more detailed and included more climatic elements. Kraus published a pioneering work "*Soil and climate in small regions*" in 1911, and Geiger summarized the phenomena in the monographs "*Climate near the ground*" (1st edition) in 1927 and "*Microclimate and plant climate*" in 1930. These developments occurred mainly in Europe with Germany and Austria as the major centres. In Japan, Misawa (1885–1937) of Nagano Prefecture, Central Japan, established a theory on "Local-climate-dependent agricultural industry (Fudo-sangyo)" during the 1930s.

Development of research methods, summerizing the previous studies (Geiger 1961; Geiger et al. 1995; Oke 1978; Yoshino 1975) could be classified chronologically into the following three periods:

1. The period when analyses were made on the effect of local conditions, such as land use pattern, at a certain climatological station. This period covered the 19th century through to the beginning of the 20th century.
2. The period when analyses were made in as much detail as possible on the horizontal distribution of climatic elements in small areas based mainly on automobile transects. The period began towards the end of the 1920s.
3. The period, when analyses are made by experiment, such as wind tunnel and computer based mathematical models. Remote sensing techniques are introduced also. These developments occurred, for the most part, after the 1970s.

2.2. *Recent development*

Recent development of the local climatology related to agricultural land use is quite striking. The items, which are considered to be important, are summarized as follows:

(i) Division of area by local climate, introducing one or multiple elements of climate observed by instruments. Because the important climatic elements and the contents, such as annual mean at the macro-scale or daily fluctuation at the local scale, are different (Hess et al. 1976), a statistical approach should be applied.

(ii) Division of area by local climate, using climatic indicator(s), such as plant phenology.

Photo 1. Windbreaks developed from knowledge of prevailing wind directions, and shade trees in between at the tea plantation located in a hilly countries of Central Sri Lanka (taken by M. Yoshino).

(iii) Establishing a rational application of local climatological knowledge. An example is seen in Photo 1 which was taken in a central mountain region (NE of Nuwara Eliya) of Sri Lanka. It shows a beautiful acacia windbreaks and shade trees standing between the windbreaks in a tea plantation on the slopes of a hilly terrain. The alignment of the windbreaks is quite suitable for reducing the prevailing strong wind and other adverse micro-topographical effects.
(iv) Statistical analysis has been developing in association with computer technology. Small differences can be detected by using suitable statistical methods. Particularly useful is the examination of small differences in two nearby regions with similar topographical features.
(v) Simulation studies using super-computer are of importance. Modeling is a theme persued in this field and one that is shared with other sciences in recent years. Many results of such studies have been published.
(vi) Map presentation and other visual methods have been developing also. Scale of maps, method of presentation, application of the results etc. are of significance. In particular, the stages and quality of crops in agroclimatological investigations should be taken into consideration for this purpose.
(vii) Remote sensing method, using satellite image data, should be utilized. Landsat and other image types are useful for problem identification, description and analyses.

3. Land use/cover change and agricultural human activities

3.1. *LUCC (land use/cover change)*

IGBP, as mentioned in the introduction of this chapter, is the international and interdisciplinary organization of the ICSU (International Council of Scientific Unions) that develops and articulates physical science programs concerned with studies on global environmental change. IGBP contains many core projects most of which are concerned with (a) conditions on Earth's ground surface, and (b) analyses of the impacts of land cover change on the planet's biogeochemistry, climate and ecology and their complex relationships. On the other hand, IHDP is the international and interdisciplinary organization of the ISSC (International Social Science Council) that develops and articulates social science programs involving studies on global environmental change. Among the core projects of IHDP, land use changes are considered to be related primarily to global environmental change driven by various forces. It was agreed that the description "land cover" is suitable for treatments involving physical science, but that "land use" is preferable from the viewpoint of social science. Because the land use/cover change (LUCC) is a point of contact for the two sciences and represents a common problem, LUCC is the most important area in global change studies. Accordingly,

Table 1. Human-induced conversions[a] in selected land covers.

Covers	Date	Area ($\times 10^6$ km^2)	Date	Area ($\times 10^6$ km^2)	% Change
Cropland[b]	1700	2.8	1980	15.0	+435
	1700	3.0	1980	14.8	+393
Irrigated Cropland	1800	0.08	1989	2.0	+2400
Closed Forest	pre-agricultural	46.3	1983	39.3	−15.1
Forest and Woodland	pre-agricultural	61.5	1983	52.4	−14.8
Grassland/Pasture	1700	68.6	1980	67.9	−1
Drained Land			1985	1.6	
Settlement					
Urban	–	–	1990	2.5[c]	
Rural	–	–	1990	2.1	

[a] The variation in dates and significant digits reflects the various sources from which the data were taken.
[b] Estimates given from two different sources. Includes some areas often classed separately as shrub and arid land.
[c] Includes substantial non built up areas. Summary based on Meyer and Turner (1992).

LUCC has been designated a core project and is the only one shared by IGBP and HDP. Research plan is now being prepared and includes foci on issues such as: situational assessment, modeling and projecting, conceptual scaling, and data development activities.

During the last three hundred years, 7,000,000 km^2 of closed forest and 9,000,000 km^2 of forest and woodland have changed (net gain) to cropland which now has a total of 12,000,000 km^2 as shown in Table 1 (Meyer and Turner 1992). Such changes as caused by human-induced conversions have affected CO_2 emission, the hydrological cycle, anomalous weather, and biodiversity. For understanding the interaction among these processes, the work of the LUCC should be and will be advanced.

Linkages between the human and physical dimensions of global environmental change have been summarized by Turner et al. (1993). A modified flow chart which emphasizes the key position held by land use and land cover in the relationship between the human and physical dimensions of the change is shown in Figure 2. The spatial dimensions of global, regional, local and micro-scales are noted on the right hand side of the figure. The driving forces (upper left hand side) are population, level of affluence, technology, political economy, political structure, attitudes and values (in particular). Environmental impact (I) is estimated roughly as

$$I = PAT$$

where P is population, A is level of affluence, and T is technology.

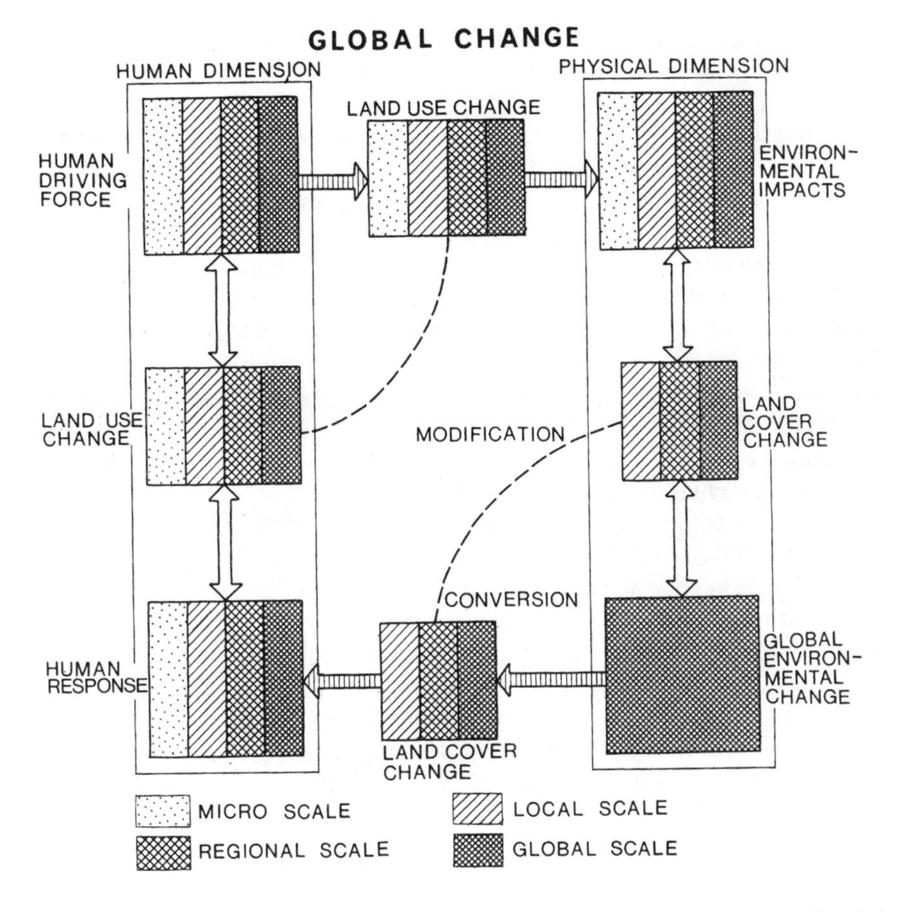

Figure 2. Flow chart showing linkages between human and physical dimensions of global change in relation to space scales.

From the standpoint of agricultural climatology, spatial scale concept is important for environmental phenomena analyses.

3.2. *Human dimensions*

In global change problems, human dimensions have been recognized as an important subject. According to a dictionary of geography (Monkhouse 1970), "environment" is the sum of the surrounding external conditions within which an organism, a community or an object exists. The geographical environment includes all factors of the environment whose relationships are considered in terms of spatial location. Consequently, from the standpoint of geography, discussions on environmental change should include its natural and human components and their spatial representation.

The spatial scale can be classified into four; micro, local, regional and global. The absolute scales are determined subjectively as discussed elsewhere (Yoshino 1975).

Agriculture and the rural area are the most fundamental space in human life. Kitamura (1993) pointed out that both are influenced by global atmospheric change. Global warming, increasing ultraviolet rays caused by the destruction of the ozone layer; acid rain; greenhouse gases such as carbon dioxide; chlorofluorocarbon; and sulfurous acid exhausted by human activities, mostly in urban and industrial areas are changing global environments. As well, crop production, forestry production, fishery and livestock are much influenced by agricultural environments and resources. The relationship between agricultural production and the causes of global environmental change should be clarified.

There are several published flow patterns which show links between human activities and environmental change. From my viewpoint, all of them have yet to take into account the space and time dimensions of the issues. Therefore, in Figure 2, referred to earlier, the spatial scales at the micro, local, regional and global levels have been added. Of particularly importance, is how to involve sociocultural concepts, such as religion, race, language, and human resources; as well as other elements including population, economy and infrastructure. Of course, local climate and agricultural land use at a local scale, which is the main topic of the present chapter, are relatively speaking, easily treatable because the two variables are fairly homogeneous at that level. But, when we treat them at a regional scale, and further at a global scale, homogeneity is out of the question, hence the difficulty experienced with attempts to discuss the topic in a general fashion.

4. Examples of local climatological observations

One of the examples related to the local climatological observation can be obtained from results of an investigation conducted in Sugadaira, Nagano Prefecture, Central Japan, where temporal observations have been made since 1944 to detect the formation of cold air lakes, nocturnal vertical structure change of temperature inversion, relationship between temperature distribution and synoptic wind conditions, nocturnal cold air drainage, etc. (Yoshino 1975, 1984; Nakamura 1989). The Sugadaira basin is shaped with a long axis of about 4 km and a short axis of about 2 km. The basin shows a very clear formation of cold air lake during anticyclonic weather in clear, calm nights. On May 8, 1959 at 05:00h, the temperature inversion between the bottom of the basin and the surrounding thermal belt was 7°C in a typical sounding. At that time, the strength of this inversion was believed to be the extreme observed case. But, in November, 1993, a more striking inversion was observed. The synoptic weather situation was an unusually prolonged anticyclone and therefore a completely cloudless, fine weather prevailed through two days before

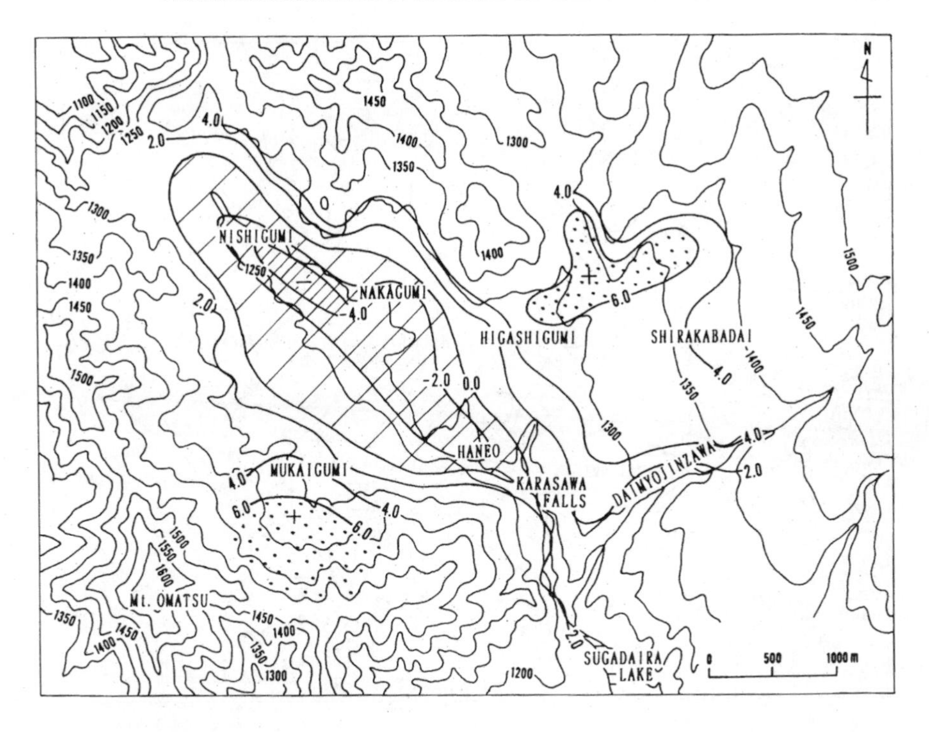

Figure 3. Air temperature (°C) distribution in the Sugadaira basin, Central Japan, on November 6, 1993, at 05:30 in the morning.

observations were made. The distribution of air temperature shown in Figure 3 was observed by Assmann thermometer at 27 points. In parallel, mobile observations by two cars equipped with thermistors were made at 47 points, but not surprisingly the distribution of temperature was quite similar.

The most striking feature revealed in Figure 3 is that the basin bottom (1,250 m a.s.l.) is below −2°C with a cold center of −4.2°C. On the other hand, the warmest part shows higher than 6°C temperature with the maximum of 6.8°C located at a height of 1,350 m a.s.l. The temperature difference was 11°C between the bottom of the basin and the thermal belt on the surrounding slopes was accordingly 11°C. This was an extraordinarily great temperature difference over a height difference of about 100 m and horizontal distance of about 1,500 m. This magnitude of inversion might be an observed extreme world record, excluding the results observed in a small doline of limestone area in Europe (Geiger 1961; Geiger et al. 1995; Yoshino 1975, 1984; Weise 1978), where human land use was normally absent. The large local difference such as shown in Figure 3 does not occur frequently, but it should be taken into consideration as a factor in land utilization, especially for agriculture. In this area, vegetables are cultivated in the bottom of the basin, excluding the marshy land along the river running along the long axis of the basin. The

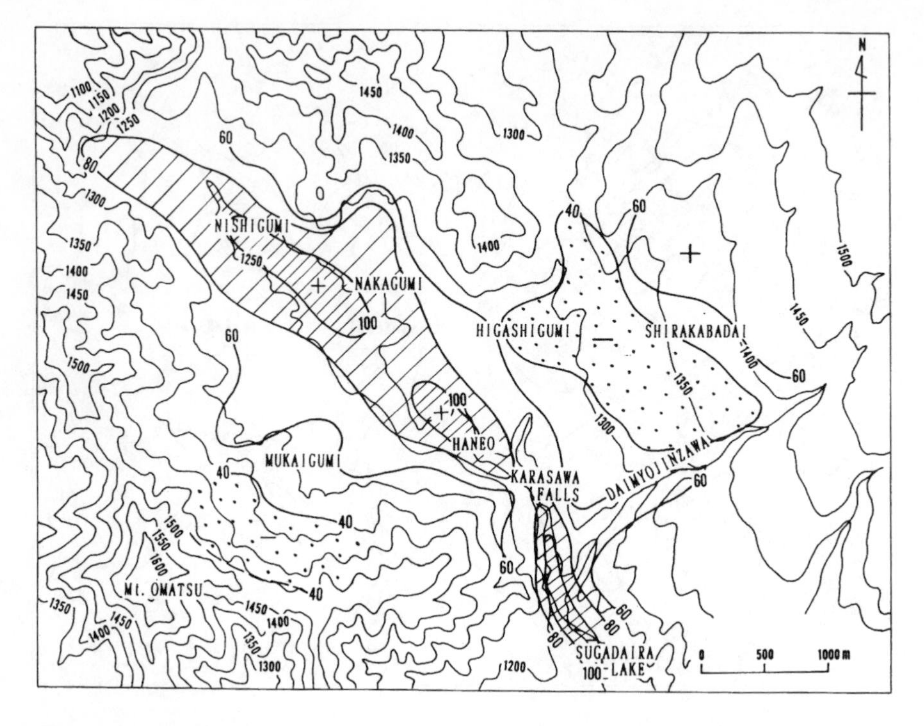

Figure 4. Relative humidity (%) distribution in the Sugadaira basin, Central Japan, on November 6, 1993, at 05:30h, which is the same observation time for Figure 3.

suitable dates for sowing in spring and harvesting in autumn are thought to be different from those used elsewhere due to such conditions.

The relative humidity (%) distribution shown in Figure 4 is of interest. The bottom of basin was wetter than 80% with 100% at the center. In contrast, there was a drier belt on the slopes where humidity was lower than 40%. This contrast is quite striking, even though the accuracy of observation should be carefully checked. The values indicate that the vapor pressure is low at the dry belt on the slope. The basin bottom where forests are developed along the river is relatively wet. In the morning of November 6, 1993, when the observation was made, fog was not formed in the basin bottom, but normally fog formation is very frequent there. These humidity characteristics of the basin are reflected in the current land use scheme by the presence of forests. In the skiing season from the end of December to the beginning of March, many cars and buses park in the basin. The polluted air from these vehicles stagnate in the cold air lake, forming a mist or smog layer. The damage to agriculture and forests is at present not serious, because the land is covered by snow in winter and the forest trees are mostly deciduous. But the air pollution poses a serious risk to the health of people who live in or near the basin bottom.

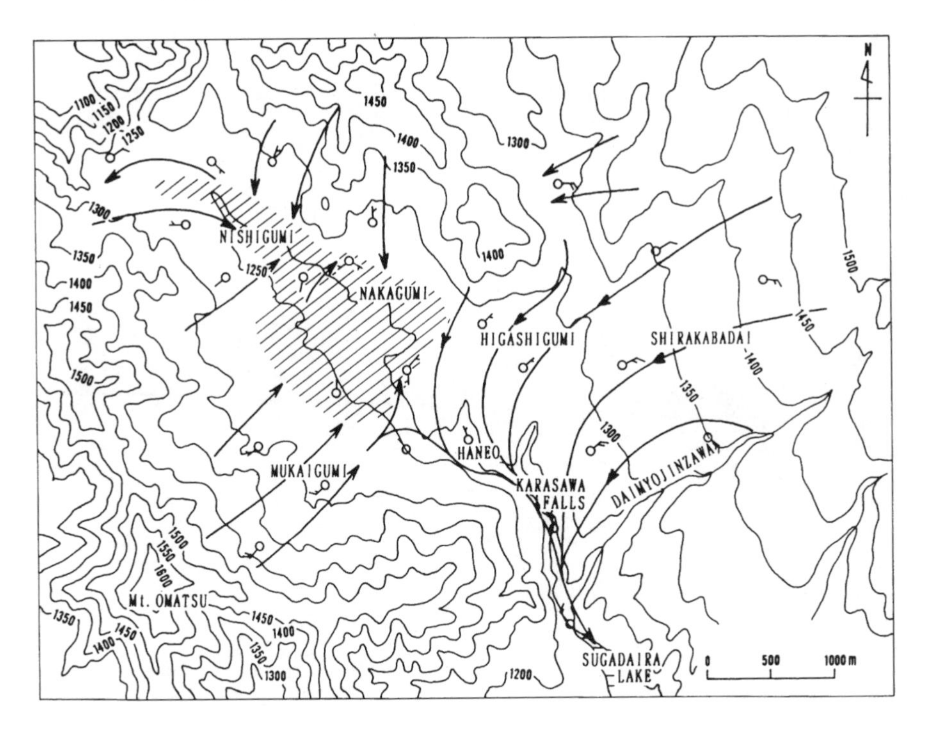

Figure 5. Distribution of wind velocity and direction, and the estimated air stream in the Sugadaira basin, Central Japan, on November 6, 1993, at 04:45–06:27h. Hatched area shows the area, where wind velocity is zero.

The wind conditions are shown in Figure 5. Local climatologically, it is very interesting to note that cold air drains from the surrounding slopes of Nishigumi, Nakagumi and Mukaigumi to the bottom of the basin where wind velocity is generally zero as shown in the hatched part of the figure. On the other hand, in the eastern part of slopes of Shirakabadai, the wind is easterly, which is the upper wind direction. This wind becomes a downslope wind at Higashigumi and the lower part of Shirakabadai, blowing down as a mountain breeze to the Sugadaira Lake in the valley. One part of the cold air drainage at Mukaigumi flows down into the basin bottom, joins the downslope wind from north at Haneo and blows down as the mountain breeze mentioned above. The cold air lake shown in Figure 3 was formed clearly under the influence of cold air drainage, but the eastern half of the area studied was under the influence of upper easterly. Therefore, it is thought that the coldest part was located approximately at the northwestern part (upper part of the river) of the basin that morning. The dry belt at Shirakabadai, which coincides with the thermal belt, can be considered as partly due to the foehn effect of the upper easterly wind coming down on the lower part of the slope. But, the existence of the dry belt and thermal belt on the slope at Mukaigumi proves that they were

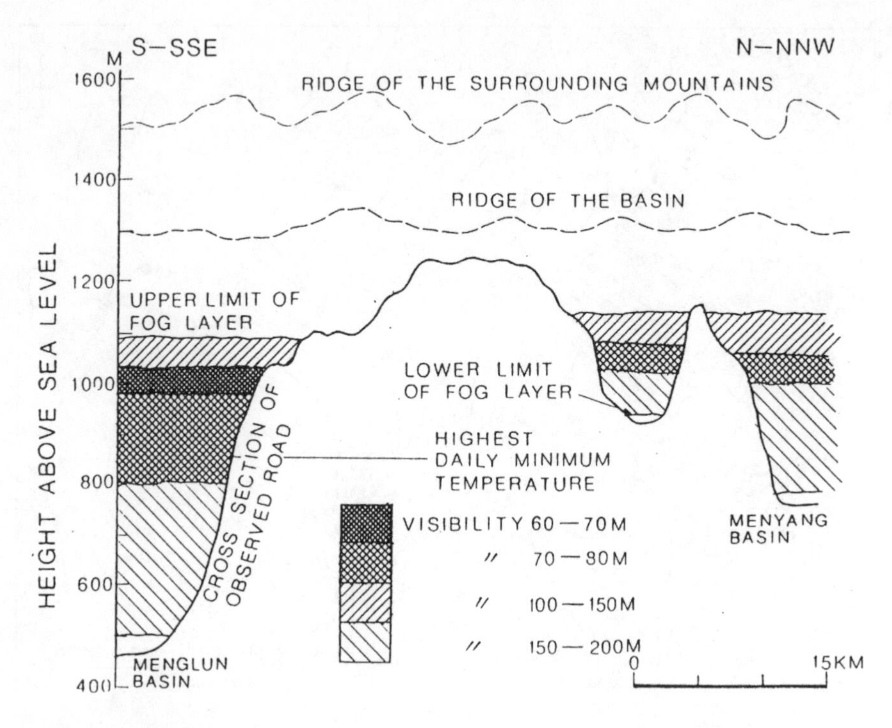

Figure 6. Fog layer structure based on visibility along the cross-section from Menglun to Mengyang, Xishuanbanna, South Yunnan, China.

caused mainly by the effect of local circulation of cold air lake formation in the basin. The results of the whole observation will be reported elsewhere.

In summary, it is concluded that the local climatological distributions should be studied in connection with local wind conditions. This is because wind fields play a basic controlling role in the distribution of air temperature, humidity, etc. at a local scale.

Another problem in basins or valleys is fog formation. In most cases, fog develops in the bottom of basins, where the cold air lake is formed. Photo 2 shows a striking fog layer in the basin under the influence of nocturnal cooling by long wave radiation. From the standpoint of land use, climate conditions are more favorable at the upper part of the slopes above the fog layer, where minority farmers cultivate, as shown in the bottom picture of Photo 2. Shifting cultivation is seen in Photo 2. An observation of a vertical distribution fog layers along a cross section from Menglun to Menyang in Xishuangbanna, South Yunnan, China, is shown in Figure 6. The upper limit of the fog layer was about 1,150 m a.s.l. Above this height, sunny fields cultivated by minority farmers were enjoying enough insolation under clear sky of the early morning. A more sophisticated observation of this phenomenon was carried out in the area by Nomoto et al. (1989).

Photo 2. (Upper) Early morning basin fog observed in Menglun, Xishuang-banna, South Yunnan, China. (Bottom) Sunny field, with minority farmers working the upper part of slopes above the fog layer in Xishuangbanna. (Both taken by M. Yoshino in December 1984.)

5. Examples of effect of local weather on agriculture and forestry

In recent years, damage to agricultural crops by the effects of unusual weather is becoming more serious. This is particularly true for rice production in Japan. The reasons are: (i) mechanization of transplanting has shortened the work period, (ii) because they are concentrating on some famous tasty varieties, and

(iii) consumers in Japan favor quality over quantity. Therefore, when unusual weather occurs, the production, yield and quality, and the associated costs, are more seriously affected than in previous years. The interannual fluctuation of rice production depends on the earlier and later periods of each stage in rice cultivation, discussed in detail elsewhere by the present writer (Yoshino 1993).

These tendencies are more striking local climatologically, because the concentration of the phenomena mentioned above is more conspicuous on a local or regional scale. One example of such damage is the effect of foehn on white head injury of paddy rice in Hokuriku District on the Japan Sea side of Central Honshu, Japan. Muramatsu (1989) studied the foehn in the area meteorologically and climatologically, and its effect on paddy rice using a physiological experiment. His findings are: (i) The white head injury of paddy rice occurs when the paddy rice encounters a foehn immediately after head sprouting. (ii) When foehn occurs at night, injury to paddy rice is more serious than when the event occurs in daytime. (iii) The reason why the night time injury is more serious is as follows: In daytime, there is an increase in the amount of water absorbed by the crop during a foehn without a corresponding increase in water stress. At night, the amount of transpiration exceeds the amount of water absorbed, hence the water stress.

Forest fires have become an important environmental issue during the last decade due to the frequent appearance of unusual weather.

The 1980 forest fire season was the worst in Canada during the period 1971–1980. More than 9,000 fires occurred across the country, burning in excess of 4.84×10^6 ha and resulting in the expenditure of more than $\$190 \times 10^6$ in fire management funds (Stocks 1983). Table 2 shows the burned area and number of fires started by lightning and by humans for the ten-year average (1970–1979) and for 1980 in westcentral Canada (Alberta, Saskatchewan, Manitoba, Ontario, and the Northwest Territories). The meteorological conditions were as follows: fairly stationary upper air high pressure system over central North America dominated in April; the duration and intensity of the high pressure system was most unusual resulting in much warmer and drier weather and three months followed with very little rain, frequent low humidities and strong winds.

The timber industry, recreational utilization, secondary industrial and commercial activities (transportation and merchandizing) are affected by fire, even though their costs are often impossible to estimate quantitatively. Similarly, wildfires became a severe ecological and financial problem in Israel during the late 1980s (Kutiel et al. 1991). A very dry, strong easterly wind brings a type of weather called "Sharav" in Israel, which occurs in most cases in the spring and autumn. Wild fire under the influence of the Sharav is of importance for the local forest climate of the Aleppo pine (*Pinus halepensis*) and kermes oak (*Quercus calliprinos*).

Table 2. The annual number of fires and area burned (in hectares) in West-Central Canada in 1980 in comparison to average figures for the 1970s. Only fires bigger than 200 hectares considered (Stocks 1983).

	1970–1979*		1980	
	Area Burned	Fires	Area Burned	Fires
Lightning				
200–1,000 ha	25,585	54.8	49,359	101
1,000–10,000 ha	215,215	60.6	373, 073	96
10,000–50,000 ha	419,125	18.4	794,297	32
> 50,000 ha	371,290	4.2	1,536,173	15
Total	1,031,215	138.0	2,752,901	244
Man-caused				
200–1,000 ha	8,283	18.1	20,294	50
1,000–10,000 ha	24,464	8.7	105,711	36
10,000–50,000 ha	33,202	1.9	260,484	9
> 50,000 ha	10,117	0.1	770,213	7
Total	76,066	28.8	1,156,702	102
Total				
200–1,000 ha	33,868	72.9	69,653	151
1,000–10,000 ha	239,679	69.3	478,784	132
10,000–50,000 ha	452,327	20.3	1,054,781	41
> 50,000 ha	381,407	4.3	2,306,385	22
Total	1,1067,281	166.8	3,909,603	346

* Figures are averaged over 10 years.

6. Application of mesh-data for local agroclimatology

Mesh-data or Geographical Information System Data are well organized in many countries throughout the world. Using these data, impacts of local climate on agricultural production can be studied. Four types of models are utilized in topoclimatologial studies, namely: (i) empirical models, (ii) empirical-statistical models, (iii) laboratory or analogue models, and (iv) numerical models (Wanner 1984).

In order to evaluate the suitable locations for barley and fruit tree cultivation in Hokuriku District, Japan, the Japan Sea side of Central Honshu, an attempt was made to detect, first, by statistical relationships between the number of days with snow cover and maximum snow depths, and third order topographical mesh-data (Yamada et al. 1987). The size of the mesh in the

Table 3. Climatological parameters or conditions defining the climatological suitability for vine growing (Hearty Burgundy); see Figure 8 (Volz 1982).

Condition 1: Cold air lakes
Vinicultures have to be situated outside of the cold air lakes (danger of forest!)
Condition 2; Temperature sums
S (T MAX. > 15°C) > 650°C (for 90% of all years)
n = days from January 1 until vintage date
S April–October > 600 kWh/m^2

third order is 45 seconds longitudinally and 30 seconds latitudinally, which translates to 1 km^2 area at the location of Hokuriku. Secondly, detailed geographical distributions of the number of days mentioned above were obtained from regression equations. Then, agroclimatic classification criteria for the planning of barley cultivation based on the number of days with snow cover were made based on the strength of the effect of snow cover on barley yield. Similarly, the results obtained from these procedures were superimposed on the mesh distribution of the number of days with snow cover and the maximum snow depth. In this way, a land use planning map was constructed.

Another example of studies designed to evaluate the climatological potential for different land use types was obtained from viniculture in the northern Swiss Jura, east of Basel (Volz, 1982). First, he observed air temperature profile along a cross-section in which a variety of micro-topography and land use categories were represented. As a second step, regression analysis relating temperature profiles to altitude for different exposures was made. Thirdly, a topoclimatological map of the northern Jura was prepared as shown in Figure 7. This map represents an important base for an initial evaluation of the impact of climate on land use planning. Although the map was made for general use in agriculture, it contains four climatological parameters/phenomena namely, summer time air temperature (a probable indicator for the vegetation duration); the average date of the last frost in spring; cold air drainage; and incoming direct-beam shortwave radiation from April to October. Finally, a map of the suitability of the climate for vine growing (Hearty Burgundy) was summarized as given in Figure 8 and Table 3.

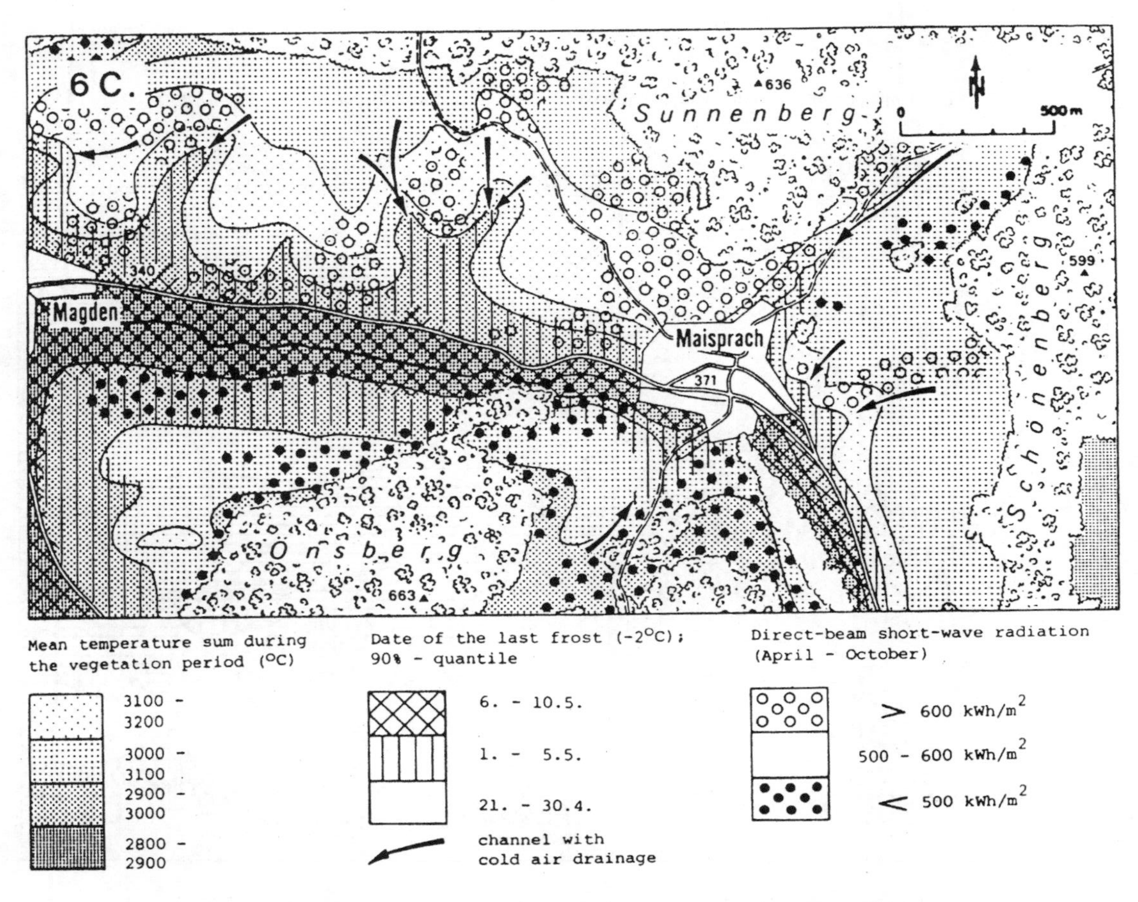

Figure 7. An abstract of topoclimatological map of the northern Jura, east of Basel (Volz 1982).

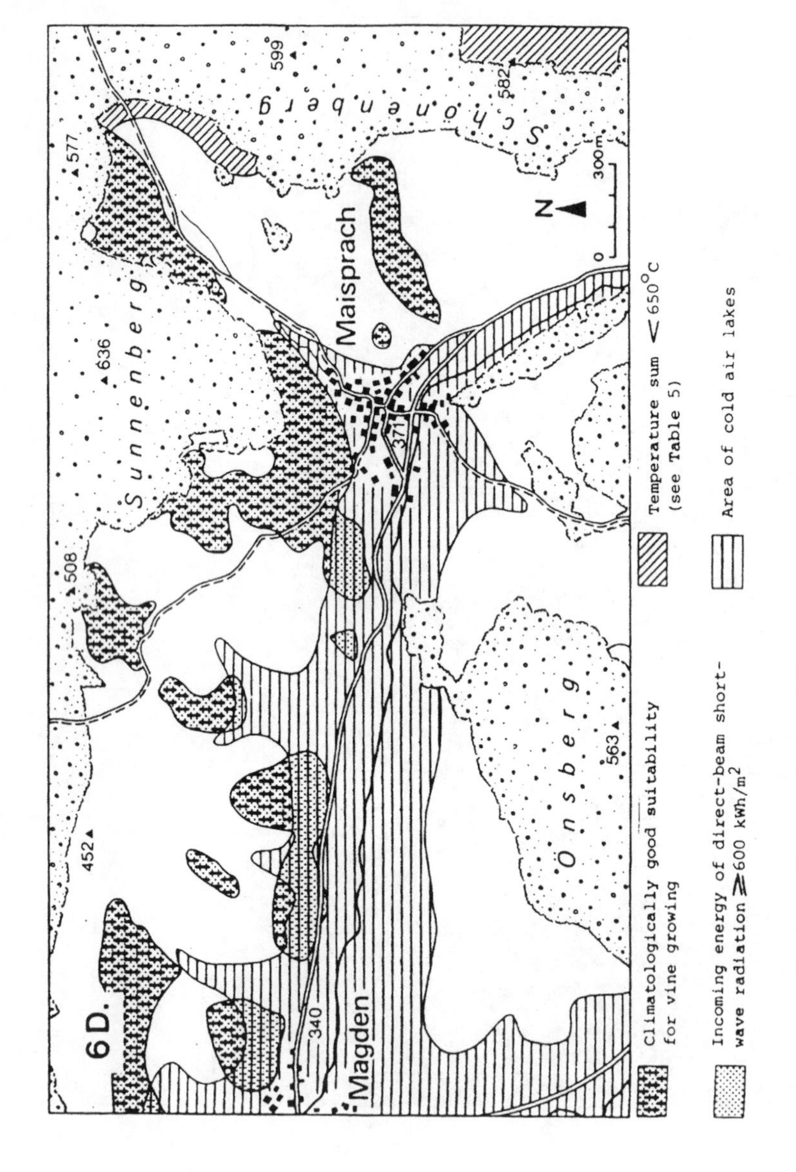

Figure 8. Vine growing suitability map for Northern Jura, east of Basel (Volz 1982).

7. Conclusion

The first part of this chapter dealt with historical developments in local agroclimatology and included a summary of methods and techniques used in the studies involved. Recent developments of local climatology related to agricultural land use are: (i) division of area according to local climates, (ii) establishment of the practical application of local climatological knowledge, (iii) statistical analysis of small local differences caused by land use/cover and micro-topographical conditions, (iv) simulation by modeling at a local scale, (v) analyses by remote sensing methods, and (vi) presentation of local scale phenomena on maps or through other visual methods.

The second part of this chapter was dedicated to dealing with problems related to the work of the IGBP and IHDP, which are on going international programmes. Human dimensions represented by land use/cover change are reviewed. Flow diagrams and other charts are utilized to show linkages between the human and physical dimensions of global change in relation to various spatial scales.

In the last part, the following examples were given: (i) local climatological observation in Sugadaira Basin involving cold air lake formation, (ii) effects of unusual local weather on agriculture and forestry such as paddy rice production and forest (wild) fire, and (iii) application of mesh-data to local agroclimatology.

Based on these descriptions, it should be stressed that research on the interactions between agricultural land use and local climates and associated problems will be taken up and conducted intensively in cooperation with the international programs listed.

References

Geiger, R. (1961): *Das Klima der bodennahen Luftschicht.* Vieweg, Braunschweig, 646 pp.

Geiger, R., Aron, R.H. and Todhuntes, P. (1995): *The climate near the ground.* Vieweg, Braunschweig, 528 pp.

Hess, M., Niedzwiedz, T. and Obrebska-Starkel, B. (1976): The method of characterizing the climate of the mountains and uplands in the macro-, meso- and micro-scale (Exemplified by Southern Poland). *Prace Geograficzne*, 43, 83–102.

Kitamura, T. (1993): The research Domain VII of IGBP Japan and its perspective. Towards the IGBP/HDP Joint Project of Land Use/Cover Change (LUCC), 20 pp.

Kutiel, H. and Kutiel, P. (1991): The distribution of autumnal easterly wind spells favoring rapid spread of forest wildfires on Mount Carmel, Israel. *GeoJournal*, 23(2), 147–152.

Meyer, W.B. and Turner, B.L. II (1992): Human population growth and global land-use/land-cover change. *Ann. Rev. of Ecol. and Systematics.*

Muramatsu, K. (1989): Development mechanism of white head injury of paddy rice caused by foehn. *Bull. Hokuriku Natl. Agric. Exp. Stn.*, 30, 131–148 (in Japanese with English abstract).

Nakamura, K. (1989): Local climatological study of the nocturnal cold air drainage on the mountain slope (Doctoral Thesis at the Inst. Geoscience, Univ. Tsukuba). *Bull. Dohto Univ., Gen. Educ.*, 8, 53–145.

Nomoto, S., Du, M. and Ueno, K. (1989): Some characteristics of cold air lakes and fog in the Jinghong and Mengyang Basins, Xishuanbanna, China. *Geog. Rev. Japan*, 62B(2), 35–46.
Oke, T.R. (1978): *Boundary layer climates*. Methuen, London, 372 pp.
Stocks, B.J. (1983): The 1980 forest fire seasons: Its impact in West-Central Canada. In: *Seventh Conf. Fire and Forest Met.*, Fort. Collins, Apr. 25–28, 1983. Amer. Met. Soc., 67–70.
Turner, B.L. II, Moss, R.H. and Skole, D.L. (eds.) (1993): *Relating land use and global land-cover change. Proposal for an IGBP–HDP Core Project*. IGBP Report, No. 24, HDP Report No. 5, 60 pp.
Volz, R. (1982): Das Gelaendeklima und seine Bedeutung fuer den landwirstschaftlichen Anbau. PhD Diss. Univ. Bern, 502 pp.
Wanner, H. (1984): Methods in applied topoclimatology. *Zuericher Geogr. Schriften*, 14, 5–17.
Weise, A. (1978): Zum Auftreten der "Warmen Hangzone" im Tiefland der D.D.R. *Zeitsch. Met.*, 28(5), 281–284.
Yamada, K., Takami, S., Iwakiri, S. and Horie, T. (1987): Climatic chart of snow cover using mesh data and its application for evaluating the best land locations in the Hokuriku District. *Bull. Hokuriku Natl. Agric. Exp. Stn.*, 29, 95–128 (in Japanese with English abstract).
Yoshino, M. (1975): *Climate in small area*. University of Tokyo Press, Tokyo, 549 pp.
Yoshino, M. (1984): Thermal belt and cold air drainage on the mountain slope and cold air lakes in the basin at quiet, clear night. *GeoJournal*, 8(3), 230–250.
Yohino, M. (1986): Some aspects of climate, geoecology and agriculture in Hainan Island and Xishuangbanna in South China. *Climat. Notes*, 35, 5–33.
Yoshino, M. (1987): Local climatology. In: *Encyclopedia of climatology*, ed. by J.E. Oliver and R.W. Fairbridge, Van Nostrand Reinhold, 551–558.
Yoshino, M. (1993): Climatic change and rice yield and production in Japan during the last 100 years. *Geogr. Rev. Japan*, 66B(1), 70–88.
Yoshino, M., Ichikawa, T. and Yamashita, S. (1984): Climate and agricultural land use in Japan. In: *Climate and agricultural land use in Monsoon Asia*, ed. by M.M. Yoshino. University of Tokyo Press, Tokyo, 317–332.

M. Yoshino
Institute of Geography
Aichi University
Toyohashi-City
441 Japan

INDEX

The GeoJournal Library

1. B. Currey and G. Hugo (eds.): *Famine as Geographical Phenomenon.* 1984 ISBN 90-277-1762-1
2. S.H.U. Bowie, F.R.S. and I. Thornton (eds.): *Environmental Geochemistry and Health.* Report of the Royal Society's British National Committee for Problems of the Environment. 1985 ISBN 90-277-1879-2
3. L.A. Kosiński and K.M. Elahi (eds.): *Population Redistribution and Development in South Asia.* 1985 ISBN 90-277-1938-1
4. Y. Gradus (ed.): *Desert Development.* Man and Technology in Sparselands. 1985 ISBN 90-277-2043-6
5. F.J. Calzonetti and B.D. Solomon (eds.): *Geographical Dimensions of Energy.* 1985 ISBN 90-277-2061-4
6. J. Lundqvist, U. Lohm and M. Falkenmark (eds.): *Strategies for River Basin Management.* Environmental Integration of Land and Water in River Basin. 1985 ISBN 90-277-2111-4
7. A. Rogers and F.J. Willekens (eds.): *Migration and Settlement.* A Multiregional Comparative Study. 1986 ISBN 90-277-2119-X
8. R. Laulajainen: *Spatial Strategies in Retailing.* 1987 ISBN 90-277-2595-0
9. T.H. Lee, H.R. Linden, D.A. Dreyfus and T. Vasko (eds.): *The Methane Age.* 1988 ISBN 90-277-2745-7
10. H.J. Walker (ed.): *Artificial Structures and Shorelines.* 1988 ISBN 90-277-2746-5
11. A. Kellerman: *Time, Space, and Society.* Geographical Societal Perspectives. 1989 ISBN 0-7923-0123-4
12. P. Fabbri (ed.): *Recreational Uses of Coastal Areas.* A Research Project of the Commission on the Coastal Environment, International Geographical Union. 1990 ISBN 0-7923-0279-6
13. L.M. Brush, M.G. Wolman and Huang Bing-Wei (eds.): *Taming the Yellow River: Silt and Floods.* Proceedings of a Bilateral Seminar on Problems in the Lower Reaches of the Yellow River, China. 1989 ISBN 0-7923-0416-0
14. J. Stillwell and H.J. Scholten (eds.): *Contemporary Research in Population Geography.* A Comparison of the United Kingdom and the Netherlands. 1990 ISBN 0-7923-0431-4
15. M.S. Kenzer (ed.): *Applied Geography.* Issues, Questions, and Concerns. 1989 ISBN 0-7923-0438-1
16. D. Nir: *Region as a Socio-environmental System.* An Introduction to a Systemic Regional Geography. 1990 ISBN 0-7923-0516-7
17. H.J. Scholten and J.C.H. Stillwell (eds.): *Geographical Information Systems for Urban and Regional Planning.* 1990 ISBN 0-7923-0793-3
18. F.M. Brouwer, A.J. Thomas and M.J. Chadwick (eds.): *Land Use Changes in Europe.* Processes of Change, Environmental Transformations and Future Patterns. 1991 ISBN 0-7923-1099-3
19. C.J. Campbell: *The Golden Century of Oil 1950–2050.* The Depletion of a Resource. 1991 ISBN 0-7923-1442-5
20. F.M. Dieleman and S. Musterd (eds.): *The Randstad: A Research and Policy Laboratory.* 1992 ISBN 0-7923-1649-5
21. V.I. Ilyichev and V.V. Anikiev (eds.): *Oceanic and Anthropogenic Controls of Life in the Pacific Ocean.* 1992 ISBN 0-7923-1854-4

The GeoJournal Library

22. A.K. Dutt and F.J. Costa (eds.): *Perspectives on Planning and Urban Development in Belgium.* 1992 ISBN 0-7923-1885-4
23. J. Portugali: *Implicate Relations.* Society and Space in the Israeli-Palestinian Conflict. 1993 ISBN 0-7923-1886-2
24. M.J.C. de Lepper, H.J. Scholten and R.M. Stern (eds.): *The Added Value of Geographical Information Systems in Public and Environmental Health.* 1995 ISBN 0-7923-1887-0
25. J.P. Dorian, P.A. Minakir and V.T. Borisovich (eds.): *CIS Energy and Minerals Development.* Prospects, Problems and Opportunities for International Cooperation. 1993 ISBN 0-7923-2323-8
26. P.P. Wong (ed.): *Tourism vs Environment: The Case for Coastal Areas.* 1993 ISBN 0-7923-2404-8
27. G.B. Benko and U. Strohmayer (eds.): *Geography, History and Social Sciences.* 1995 ISBN 0-7923-2543-5
28. A. Faludi and A. der Valk: *Rule and Order. Dutch Planning Doctrine in the Twentieth Century.* 1994 ISBN 0-7923-2619-9
29. B.C. Hewitson and R.G. Crane (eds.): *Neural Nets: Applications in Geography.* 1994 ISBN 0-7923-2746-2
30. A.K. Dutt, F.J. Costa, S. Aggarwal and A.G. Noble (eds.): *The Asian City: Processes of Development, Characteristics and Planning.* 1994 ISBN 0-7923-3135-4
31. R. Laulajainen and H.A. Stafford: *Corporate Geography.* Business Location Principles and Cases. 1995 ISBN 0-7923-3326-8
32. J. Portugali (ed.): *The Construction of Cognitive Maps.* 1996 ISBN 0-7923-3949-5
33. E. Biagini: *Northern Ireland and Beyond.* Social and Geographical Issues. 1996 ISBN 0-7923-4046-9
34. A.K. Dutt (ed.): *Southeast Asia: A Ten Nation Region.* 1996 ISBN 0-7923-4171-6
35. J. Settele, C. Margules, P. Poschlod and K. Henle (eds.): *Species Survival in Fragmented Landscapes.* 1996 ISBN 0-7923-4239-9
36. M. Yoshino, M. Domrös, A. Douguédroit, J. Paszynski and L.D. Nkemdirim (eds.): *Climates and Societies – A Climatological Perspective.* A Contribution on Global Change and Related Problems Prepared by the Commission on Climatology of the International Geographical Union. 1997 ISBN 0-7923-4324-7
37. D. Borri, A. Khakee and C. Lacirignola (eds.): *Evaluating Theory-Practice and Urban-Rural Interplay in Planning.* 1997 ISBN 0-7923-4326-3
38. J.A.A. Jones, C. Liu, M-K. Woo and H-T. Kung (eds.): *Regional Hydrological Response to Climate Change.* 1996 ISBN 0-7923-4329-8
39. R. Lloyd: *Spatial Cognition.* Geographic Environments. 1997 ISBN 0-7923-4375-1
40. I. Lyons Murphy: *The Danube: A River Basin in Transition.* 1997 ISBN 0-7923-4558-4

KLUWER ACADEMIC PUBLISHERS – DORDRECHT / BOSTON / LONDON